Lecture Notes in Physics

For information about Vols. 1–67, please contact your bookseller or Springer-Verlag.

Vol. 68: Y. V. Venkatesh, Energy Methods in Time-Varying System Stability and Instability Analyses. XII, 256 pages. 1977.

Vol. 69: K. Rohlfs, Lectures on Density Wave Theory. VI, 184 pages. 1977.

Vol. 70: Wave Propagation and Underwater Acoustics. Edited by J. Keller and J. Papadakis. VIII. 287 pages. 1977.

Vol. 71: Problems of Stellar Convection. Proceedings 1976. Edited by E. A. Spiegel and J. P. Zahn. VIII, 363 pages. 1977.

Vol. 72: Les instabilités hydrodynamiques en convection libre forcée et mixte. Edité par J. C. Legros et J. K. Platten. X, 202 pages. 1978.

Vol. 73: Invariant Wave Equations. Proceedings 1977. Edited by G. Velo and A. S. Wightman. VI, 416 pages. 1978.

Vol. 74: P. Collet and J.-P. Eckmann, A Renormalization Group Analysis of the Hierarchical Model in Statistical Mechanics. IV, 199 pages. 1978.

Vol. 75: Structure and Mechanisms of Turbulence I. Proceedings 1977. Edited by H. Fiedler. XX, 295 pages. 1978.

Vol. 76: Structure and Mechanisms of Turbulence II. Proceedings 1977. Edited by H. Fiedler. XX, 406 pages. 1978.

Vol. 77: Topics in Quantum Field Theory and Gauge Theories. Proceedings, Salamanca 1977. Edited by J. A. de Azcárraga. X, 378 pages 1978.

Vol. 78: Böhm, The Rigged Hilbert Space and Quantum Mechanics. IX, 70 pages. 1978.

Vol. 79: Group Theoretical Methods in Physics. Proceedings, 1977. Edited by P. Kramer and A. Rieckers. XVIII, 546 pages. 1978.

Vol. 80: Mathematical Problems in Theoretical Physics. Proceedings, 1977. Edited by G. Dell'Antonio, S. Doplicher and G. Jona-Lasinio. VI, 438 pages. 1978.

Vol. 81: MacGregor, The Nature of the Elementary Particle. XXII, 482 pages. 1978.

Vol. 82: Few Body Systems and Nuclear Forces I. Proceedings, 1978. Edited by H. Zingl, M. Haftel and H Zankel. XIX, 442 pages. 1978.

Vol. 83: Experimental Methods in Heavy Ion Physics. Edited by K. Bethge. V, 251 pages. 1978.

Vol. 84: Stochastic Processes in Nonequilibrium Systems, Proceedings, 1978. Edited by L. Garrido, P. Seglar and P. J. Shepherd. XI, 355 pages. 1978

Vol. 85: Applied Inverse Problems. Edited by P. C. Sabatier. V, 425 pages. 1978.

Vol. 86: Few Body Systems and Electromagnetic Interaction. Proceedings 1978. Edited by C. Ciofi degli Atti and E. De Sanctis. VI, 352 pages. 1978.

Vol. 87: Few Body Systems and Nuclear Forces II, Proceedings, 1978. Edited by H. Zingl, M. Haftel, and H. Zankel. X, 545 pages. 1978.

Vol. 88: K. Hutter and A. A. F. van de Ven, Field Matter Interactions in Thermoelastic Solids. VIII, 231 pages. 1978.

Vol. 89: Microscopic Optical Potentials, Proceedings, 1978. Edited by H. V. von Geramb. XI, 481 pages. 1979.

Vol. 90: Sixth International Conference on Numerical Methods in Fluid Dynamics. Proceedings, 1978. Edited by H. Cabannes, M. Holt and V. Rusanov. VIII, 620 pages. 1979.

Vol. 91: Computing Methods in Applied Sciences and Engineering, 1977, II. Proceedings, 1977. Edited by R. Glowinski and J. L. Lions. VI, 359 pages. 1979.

Vol. 92: Nuclear Interactions. Proceedings, 1978. Edited by B. A. Robson. XXIV, 507 pages. 1979.

Vol. 93: Stochastic Behavior in Classical and Quantum Hamiltonian Systems. Proceedings, 1977. Edited by G. Casati and J. Ford. VI, 375 pages. 1979.

Vol. 94: Group Theoretical Methods in Physics. Proceedings, 1978. Edited by W. Beiglböck, A. Böhm and E. Takasugi. XIII, 540 pages. 1979.

Vol. 95: Quasi One-Dimensional Conductors I. Proceedings, 1978. Edited by S. Barišić, A. Bjeliš, J. R. Cooper and B. Leontić. X, 371 pages. 1979.

Vol. 96: Quasi One-Dimensional Conductors II. Proceedings 1978. Edited by S. Barišić, A. Bjeliš, J. R. Cooper and B. Leontić. XII, 461 pages. 1979.

Vol. 97: Hughston, Twistors and Particles. VIII, 153 pages. 1979.

Vol. 98: Nonlinear Problems in Theoretical Physics. Proceedings, 1978. Edited by A. F. Rañada. X, 216 pages. 1979.

Vol. 99: M. Drieschner, Voraussage – Wahrscheinlichkeit – Objekt. XI, 308 Seiten. 1979.

Vol. 100: Einstein Symposion Berlin. Proceedings 1979. Edited by H. Nelkowski et al. VIII, 550 pages. 1979.

Vol. 101: A. Martin-Löf, Statistical Mechanics and the Foundations of Thermodynamics. V, 120 pages. 1979.

Vol. 102: H. Hora, Nonlinear Plasma Dynamics at Laser Irradiation. VIII, 242 pages. 1979.

Vol. 103: P. A. Martin, Modèles en Mécanique Statistique des Processus Irréversibles. IV, 134 pages. 1979.

Vol. 104: Dynamical Critical Phenomena and Related Topics. Proceedings, 1979. Edited by Ch. P. Enz. XII, 390 pages. 1979.

Vol. 105: Dynamics and Instability of Fluid Interfaces. Proceedings, 1978. Edited by T. S. Sørensen. V, 315 pages. 1979.

Vol. 106: Feynman Path Integrals, Proceedings, 1978. Edited by S. Albeverio et al. XI, 451 pages. 1979.

Vol. 107: J. Kijowski, W. M. Tulczyjew, A Symplectic Framework for Field Theories. IV, 257 pages. 1979.

Vol. 108: Nuclear Physics with Electromagnetic Interactions. Proceedings, 1979. Edited by H. Arenhövel and D. Drechsel. IX, 509 pages. 1979.

Vol. 109: Physics of the Expanding Universe. Proceedings, 1978. Edited by M. Demiański. V, 210 pages. 1979.

Vol. 110: D. A. Park, Classical Dynamics and Its Quantum Analogues. VIII, 339 pages. 1979.

Vol. 111: H.-J. Schmidt, Axiomatic Characterization of Physical Geometry. V, 163 pages. 1979.

Vol. 112: Imaging Processes and Coherence in Physics. Proceedings, 1979. Edited by M. Schlenker et al. XIX, 577 pages. 1980.

Vol. 113: Recent Advances in the Quantum Theory of Polymers. Proceedings 1979. Edited by J.-M. André et al. V, 306 pages. 1980.

Lecture Notes in Physics

Edited by J. Ehlers, München K. Hepp, Zürich
R. Kippenhahn, München H. A. Weidenmüller, Heidelberg
and J. Zittartz, Köln
Managing Editor: W. Beiglböck, Heidelberg

146

Bruce J. West

On the Simpler Aspects of Nonlinear Fluctuating

Deep Water Gravity Waves

(Weak Interaction Theory)

Springer-Verlag
Berlin Heidelberg GmbH 1981

Author

Bruce J. West
Center for Studies of Nonlinear Dynamics, La Jolla Institute
P.O.Box 1434, La Jolla, CA 92038, USA

ISBN 978-3-540-10852-8 ISBN 978-3-540-38762-6 (eBook)
DOI 10.1007/978-3-540-38762-6

Originally published by Springer-Verlag Berlin Heidelberg New York in 1981

2153/3140-543210

An Apology To Oceanographers

These lectures are based in large part on a graduate special topics course given at the Scripps Institution of Oceanography on the campus of the University of California at San Diego. Organizing these lectures provided me with an opportunity to examine the present state of our (my) understanding of the nonlinear mechanisms operative in an evolving field of wind generated water waves. In delivering the lectures it soon became clear to me that it was not possible to discuss all the physical processes, or rather all the theories of the physical processes, in a single semester graduate course on nonlinear deep water gravity waves. I therefore elected to present the simplest aspects of the water wave physics which could include the dominant physical mechanisms. Given the subjective nature of the concepts of simplicity and importance, I may or may not have included what the reader would consider as the simpler or more important physical mechanisms. In any event, the lectures are concerned with the mathematical description of those nonlinear interactions that may be classified as weak or very weak and thus may not be of immediate value to physical oceanographers interested in such strong nonlinear effects as wave breaking.

The point of view adopted in these lectures has evolved over the years through the collaborations of the author with J. A. L. Thomson and K. M. Watson on the spectral representation of deep water wave gravity waves and with K. Lindenberg and V. Seshadri on the solution of both linear and nonlinear stochastic differential equations and the determination of the importance of fluctuating parameters in the dynamic evolution of physical systems. Although much of the work discussed in these lectures is drawn from the research of the author in collaboration with the above mentioned scientists, a conscious effort has been made to point out the limitations of particular studies in their application to ocean physics ànd highlight where additonal contributions to the understanding of particular physical processes can be made. These problem areas are summarized in the last lecture and may hopefully provide a stimulus for their study and resolution in the not too distant future.

This work was supported in part by the Office of Naval Research

TABLE OF CONTENTS

1. Introduction to Water Waves

In this series of lectures we describe the physical and mathematical properties of the movement of the ocean surface when the air flow at the air-sea interface is turbulent. The viewpoint we adopt is that the ocean surface is a weakly interacting field of nonlinear waves whose equations of motion are determined by a Hamiltonian. A Hamiltonian formulation of the surface dynamics is preferred over the more familiar use of Bernoulli's equation and the kinematic boundary condition[1-6] because a Hamiltonain description is more readily generalized to discussions of non-deterministic flows. Observations of wind generated water waves in wave tanks, as well as on the open ocean, indicate that the measured vertical displacement of the free surface and the surface fluid velocity fluctuate in both space and time.[7] The measurements suggest that the wave generation process cannot be described as deterministic. The surface responds to the turbulent fluctuations in the air flow in such a way that the surface properties of the fluid are also stochastic.* Kitiagorodskii and others[4-12] discuss the experimental basis for our understanding of the wave generation process and review the mechanisms for the transfer of energy from a fluctuating wind field to the water surface. The consensus of opinion is that the properties of the water waves are most consistently represented by a *stochastic* rather than a deterministic *nonlinear wave field.*[8-10] It is, therefore, necessary to develop a formalism capable of treating both nonlinear and stochastic interactions and this we attempt to do in these lectures.

---

*We shall use the terms stochastic, random, fluctuating and noisy interchangeably in these lectures.

We disclaim the direct application of the theory developed herein to physical oceanography except under restricted conditions. Our purpose is to isolate and construct models for the physical mechanisms that have been identified in the air-sea interaction and in the evolution of the wind generated water wave field. By studying each of these mechanisms from a single point of view a coherent picture of the state of our limited knowledge will emerge. It is worth stressing at the outset that the problems which remain are fundamental and their resolution will not be immediate. Such questions that persist relate to the convergence of perturbation expansions, the validity of the application statistical mechanical concepts in geophysical flows, the validity of closure hypotheses and so on. These questions will be pointed out and discussed as they arise.

Given our disclaimer about physical oceanography it is still useful to have in mind the variety of wave structures to be found on the ocean surface. For the sake of clarity of presentation, ocean waves are separated into four regions of interest. These regions are determined by wavelength (or frequency) and have quite different characteristics. The longest waves are the tides and tsunamis ("tidal waves") which have a wave period of from 10^3 to 10^4 seconds and have phase velocities on the order of 500 mph, see Figure (1.1). The extremely long wavelengths make these waves uninteresting in the present discussion in which we are concerned with rather localized wave interactions. The longest waves of interest to us are called swells and have periods on the order of 10 seconds and wavelengths between 60 and 400 meters. These waves are sinusoidal and can travel, without loss of energy, for thousands of miles in deep water. It is by means of these long wavelength gravity waves that the effect of storms are found at great distances from storm centers.[13]

The surface gravity waves of intermediate size ride on these much longer waves and can interact with them. These latter waves are seen to be sharply peaked and in large number. They are locally generated by the wind and can be seen to grow, increase in slope and eventually break under the wind's influence. The generation of surface waves by wind is an incompletely understood phenomenon,[9] as is the mechanism for the wave breaking. One element in the wave breaking process involves the gravitational stability of the breaking wave. Stokes[14] determined that a gravity wave becomes unstable when its steepness, i.e., the ratio of its height to its length, exceeds one to seven. Therefore a wave which is seven meters long will break when its height exceeds one meter. The quantity H/λ, where H is the wave height and λ the wave length, measures how close a given wave is to breaking. A related quantity we refer to frequently is the wave slope, i.e., the horizontal spatial gradient of the vertical surface displacement.

The smallest waves present on the ocean surface are ripples or capillaries, which are fractions of a centimeter long. These waves can interact strongly with the surface gravity waves in the neighboring wavenumber regime. The waves in this spectral region are strongly damped by molecular viscosity and can be generated both by the wind and the longer gravity waves.[15] The dispersion relation for these waves, i.e., the relation between frequency and wavenumber, involves surface tension as well as the acceleration of gravity.[1-6]

To round out our present classification of the ocean surface effects, we must introduce the notion of a local variable surface current. These currents can arise from large scale ocean current systems, tides, river outlets, wind-driven currents, and internal waves. Internal waves are unique in that they give rise to travelling current patterns. The effect of an internal wave at the ocean's surface is to increase the flow of water above its crest and decrease the flow above its trough, thereby giving rise to a position dependent or translating surface current pattern.[16] Low amplitude, small wavelength surface gravity waves can interact quite strongly with such current patterns.[17,18] This is manifest on the sea surface by the formation of banded regions of surface calm known as "slicks". Whether the formation of these regions is produced by the internal waves concentrating organic surface material in such regions thereby enhancing the surface viscosity in a position dependent way[16] or whether it is due to a direct mechanical interaction of the surface waves with internal waves[19] is still an unresolved question. It seems in fact that both mechanisms may be operative simultaneously.[20,5]

To capture the sequence in which the preceding wave types enter a region of the sea, let us consider a glassy ocean surface in which there are no currents or swell from distant regions. If a uniform wind is slowly turned on, one observes the generation of ripples on the water. In principle, it is the shearing stress between the wind and water which produces the initial growth of these waves. In a fairly short time interval, the ripples grow in amplitude and reach a steady state with the dissipative effects of surface viscosity. On a longer time scale intermediate size gravity waves are generated. This time interval is dependent on the velocity of the wind and the magnitude of the fluctuations in the air flow. Waves of many sizes are generated; those with small wavelength are, however, quickly damped by viscosity. Those with longer wavelength, as they grow, become more efficient in extracting energy from the wind since the surface elevation acts as a sail on which the wind can push. The waves which are the most efficient in extracting energy from the wind are those which have a phase velocity comparable to the wind speed, i.e., a given phase point on the wave moves with the mean wind speed. These waves are located in wavenumber space at the peak of the ocean wave spectrum.[7,9]

The continued growth of these primary gravity waves is of course limited by the stability criteria mentioned earlier, i.e., $H/\lambda \approx 1/7$. Thus the energy being supplied by the wind to a given wavenumber must be redistributed to other wavenumbers. The interaction between stable gravity waves is weak, however, so energy can be fed in by the wind more rapidly than it can be redistributed by the nonlinear interactions among the waves. This excess energy is dissipated in two ways: (i) the formation of *white caps* in which the gravity wave becomes unstable and breaks and (ii) the generation of *parasitic capillaries*, i.e., short waves which are radiated from the crests of sharply-peaked gravity waves and which are subsequently viscously damped.[15,21]

If the wind speed is insufficient to cause the sharp cresting of the gravity waves which lead to (i) and (ii) above, these effects may still be induced by means of a number of different mechanisms. The capping or breaking of waves can be caused by two fully-developed waves running together so that their superposition forms a sharply-crested peak which is unstable. Breaking can also occur by relatively short gravity waves passing over the crest of a swell so that, again, the crest of the combined waves is unstable. It should be pointed out that breaking and the formation of parasitic capillaries are complementary mechanisms and where one is found so generally is the other. It is probably the case, however, that the latter mechanism is operative when one exceeds the stability limit by only a small amount. The separation of air flow[21,22] and the formation of stagnation points near the leading edge of gravity waves[33] may also be important in the generation and breaking processes, but these mechanisms will not be discussed here.

Another effect of the wind is to induce a wind drift current at the ocean surface produced by the continuity of stress across the air-sea interface.

The magnitude of the wind drift current is approximately 3% of the wind speed. This is the dominant local current source for a newly aroused spectrum. The nonlinear interaction among the gravity waves, however, generates a second type of surface current called the Stokes drift. This current increases in proportion to the mean square surface slope, so that as the spectrum approaches its steady state, this current dominates the wind drift current. In Figure 1.2 (taken from Wu[25]) the local surface current is shown as a function of fetch, i.e., distance from the wind source. The net surface current is seen to be a constant at reasonable fetches, but the relative magnitude of the wind and nonlinear contributions are seen to change as the fetch increases.[25]

As one increases the wind speed, longer and longer wavelength waves are generated. However, in the normal sea, a significant fraction of the wave spectrum is usually assumed to be saturated, i.e., to be in a dynamic steady state with the wind. In this region the continued energy influx from the wind is balanced by a rapid nonlinear energy transfer eventually resulting in dissipation. These strong nonlinear effects are the breaking waves and are not describable by any model based on perturbation theory (at least in the normally applied sense).

(a) Experimental Support of Theory

To describe the processes of wave generation, evolution and the subsequent development of wave instabilities we find it convenient to express the observables at the ocean surface in series expansions of the eigenfunctions of the linearized system. The expansion coefficients in such series are constants in the linearized system, but are variable in the nonlinear system. Because the linear water wave system is harmonic, the eigenfunctions are simple sines and cosines and the series expansions are just Fourier series. The expansion coefficients are referred to as mode amplitudes and are interpreted as the independent waves in a linear wave field. Correspondingly, the nonlinear system is referred to as a nonlinear wave field and the nonlinearities are interpreted as couplings or scatterings of the once linear waves.[11] The Hamiltonian for this system is a series in which the nonlinear terms appear as products of the mode amplitudes. These nonlinear interactions induce variations in both the amplitudes and phases of the linear waves in the equations of motion. For a weakly nonlinear system such as water waves this induced variation is much slower than the harmonic variation of the linearized system.[26-28]

The gravity wave field is a conservative Hamiltonian system so that Hamilton's equations of motion provide a deterministic description of the evolution of the wave field.[29-32] If we assume that this field is well represented by N degrees of freedom, where N may be large but finite, the system can be represented by N coupled, deterministic, nonlinear rate equations for the mode amplitudes. Moser[33] gives a general mathematical discussion of the separation of the interactions in such Hamiltonian systems into resonant and non-resonant groups. The non-resonant interactions provide for a stable evolution in the phase space of the system, whereas the resonant interactions lead to instabilities. One of the properties of the gravity wave field to which we will devote a great deal of attention is the existence of such resonances.

In a qualitative way a resonance is a matching between both the space and time scales of the wave of interest and the scales of the nonlinear interactions among the other waves. The existence of such resonances in water wave fields was explicitly pointed out by Phillips[26]. He showed that just as for resonances in a linear system, the resonant nonlinear interactions among gravity waves produces an initial secular growth of new waves. Benney[27] extended these arguments to show how the nonlinear interaction also leads to an eventual quenching of this apparent instability. He also discussed the conservation laws for such a wave system in the absence of viscosity. Chirikov[34], in his discussion of the *general properties* of nonlinear systems, points out that the oscillations induced by such nonlinear resonances are *always* bounded as distinct from linear resonances which are unbounded in general. The dependence of frequency on the energy, i.e., the nonlinear dispersion relation, is the cause of the nonlinear resonant motion being bounded. The nonlinearity in the system therefore acts to stabilize the system motion and inhibit instabilities. The identification of this nonlinear stabilization has been made in a number of systems, e.g., in the stability of charged particle motion in accelerators. The phenomenon of nonlinear stabilization of resonances is well known and has attracted the attention of a number of investigators.[34-37]

The definitive experimental verification of the existence of nonlinear resonant interactions among deep water gravity waves was made in the mid-sixties.[38,39] The experiment was made at the suggestion of Longuet-Higgins[28] and will be described in detail in a later lecture. In essence the purpose of the experiment was to measure the generation and growth of a resonantly excited wave due to the nonlinear interaction of four gravity waves. This newly created mode, the fourth member of the interaction, was observed to have the properties predicted by nonlinear theory. For small amplitude waves the frequency of the wave is independent of the amplitude. As the amplitude increases the frequency becomes amplitude (energy) dependent and the evolution properties of the wave are modified. In particular the stability properties of the finite amplitude wave appear to differ from those of the infinitesimal (small) amplitude wave.

In apparent contradiction to the stabilization provided by nonlinear interactions noted above,[34] Benjamin and Feir[40] observed what was apparently a disintegration of a mechanically generated finite amplitude deep water wave. Benjamin[41] interpreted this breakup as being due to a sideband instability of the finite amplitude wave. The mechanism he postulated, based on a linear stability analysis, was a resonant coupling between the primary wave mode and modes with frequencies located at sidebands of the primary wave frequency. The resonant (exponential) growth of these sideband waves was observed to be just as predicted from linear theory at early times. The general conclusion was that finite amplitude water waves are unstable. Additional theoretical support for this viewpoint was provided by Benney[42], who stressed, as he had done earlier, that this type of growth does not imply a persistence of the instability, but should rather be interpreted as a developing incoherence in the nonlinear wave. Numerical calculations to simulate the experiment of Benjamin and Feir was made by Chu and Mei[43] and others[44,45] indicate a quenching of the growth of these sidebands as formally predicted by Benney.

The Benjamin-Feir experiment was quite dramatic and the implications of the results were explored by various people for a decade. An even more dramatic experiment has recently been conducted by Lake, Yuen, Rungaldier and Ferguson[46], however, in which the sideband instability in the evolution of a nonlinear wave train is found *not* to lead to either a disintegration of the wave train as *observed* by Benjamin and Feir nor to a loss of coherence as proposed by Benney. Instead there appears to be a balancing between the dispersive properties of the fluid and the coherence properties of the nonlinear interaction.[44] The energy is transferred from the primary mode to the sidebands for a determinable length of time and then is recollected back into the primary mode. The primary mode is, therefore, modulated strongly when the sidebands are at their maximum amplitude and weakly modulated when the initial state has been reconstituted. The wave train exhibits a kind of behavior first observed in the calculation of an anharmonic lattice by Fermi, Pasta and Ulam[47] and is referred to as the FPU recurrence phenomenon of which more will be said later. It is not possible to adequately review the experimental background on which our theoretical understanding of the generation and evolution of water waves is based and hope to also provide a presentation of the theory itself. The author has, therefore, elected to mention only those experiments which have had (in his mind) a direct bearing on the particular aspect of the theory being discussed. It is acknowledged that this is unjust and my only hope is that such excellent treatments of the experimental frontier and its impact on theory, such as given by Phillips[5], will continue to be updated.

(b) Perspectives on Mathematical Techniques

The purpose of any scientific investigation is to reduce the description of a complicated process to the smallest number of fundamental or dominating principles. Applying this desideratum to the study of water waves we find that the principle of least action is sufficient to determine the equations of motion for the surface of a simple fluid. The principle of least action[48] was Hamilton's reduction of analytic dynamics to a single principle by building on the analytic foundations provided by Lagrange.[49] Herein we assume an inviscid (usually), irrotational fluid described by a potential function and employ the principle of least action to construct the nonlinear equations of motion. We recall that the wave field of interest is being driven by the turbulent wind so that a few words on the stochastic aspect of the wave dynamics are in order to reconcile the concepts of fluctuations and potentials.

The evolution of a wind generated field of water waves is often discussed as if two distinct types of motion were occurring simultaneously, i.e., deterministic and stochastic. There is, of course, only a single motion present which has associated with it widely separated space and time scales on which it is evolving. The macroscopic equations of motion for a fluid, as developed by Lagrange[49] using potential flow, did not take into account this wide scale separation.* The classical concept of a potential has built into it the idea of an averaging over finite physical regions of the medium to determine the potential at a "point". The infinite self-energies that arise in the classical field of a point charge in electricity and magnetism or the point vortices in two dimensional fluid flow arise from the inconsistency of the classical concept of a potential with a point value measurement.[50] It is, therefore, necessary to distinguish between fluctuations which are microscopic and those which are mesoscopic.

Statistical physics provides a fundamental basis for the concept of a potential by examining the *measurable consequences* of the microscopic equations of motion.[50] In the present context microscopic refers to the classical forces acting on the individual water molecules and the macroscopic equations of evolution are obtained by averaging over these inter-particle forces. This averaging is implicit in the velocity potential as discussed for example by Green[51], and only recently has explicit averaging over these fluctuations been carried out using classical statistical mechanics to obtain the equations of motion for a fluid, see e.g., Oppenheim and Keyes[52] or the review by Fox[53]. The inter-particle forces in this picture give rise to fluctuations on the macroscopic space-time scale of the potential. Averaging over these fluctuations produces a finite dissipation of energy in the macroscopic equations of motion, i.e., the observed viscosity.[50,54,55] Instead of explicitly averaging, this effect is usually modeled by a single parameter, the coefficient of viscosity, and any further effects of microscopic fluctuations are ignored.

---

*Except, of course, parametrically through the introduction of Lagrange multipliers providing the transport coefficients through equations of constraint.

The heuristic coefficient of viscosity distinguishes between the conservative fluid flow equations of Euler[56] and the dissipative equations of Navier and Stokes[57] The viscosity, therefore, models the transfer of energy from the macroscopic degrees of freedom in the system to the microscopic degrees of freedom which have been eliminated. This irreversible effect in fluids was not considered by Lagrange, and will be discussed more fully in a later lecture.

We do not pursue the fundamental approach to fluid dynamics in these lectures, but instead we use a combination of theoretical physics and heuristic modeling. In this latter approach the inter-particle force in the above discussion is replaced by a stochastic function whose statistical properties are assumed to be known. The evolution of some average property of the fluid, such as a velocity potential, is then studied by means of a Langevin equation. Herein we refer to any system of rate equations driven by an additive, state-independent, fluctuating flux as a Langevin equation after the physical scientist who first introduced such descriptions into physics.[58]

An equation of the Langevin form was suggested by Landau and Lifshitz[3] to describe the phenomenon of turbulent fluid flow. Their suggested modification of the Navier-Stokes equations has been strongly criticized because of the presumed direct effect of microscopic particle motion on the macroscopic fluid equations. A more acceptable treatment of this idea has been recently put forward by Mashizama and Mori[59] in which a formal averaging of the microscopic degrees of freedom (coarse graining) is done explicitly. The resulting equations have the *form* proposed by Landau and Lifshitz, but the source of the fluctuations is quite different, i.e., the fluctuations are due to the process of coarse graining resulting in the dynamic description with mesoscopic fluctuations. Although we do not discuss this formalism here, its existence lends credence to some of the more phenomenological statements we will later make about the form of the equations of motion.

The dynamics of the fluid flow used in these lectures is described by a *nonlinear stochastic rate equation* whose analytic solution for an arbitrary but specific set of initial conditions is *not* known. The fluctuations in the Langevin equation model of the gravity wave field are not microscopic, but are rather mesoscopic, i.e., their space-time scales are very much longer than the characteristic times for molecular dynamics, but still very much shorter than the time scales for the variations in the macroscopic variables. The macroscopic space-time scales are those of the linear wave frequencies and wavelengths and the energy transfer times due to nonlinear resonant interactions. This model of the wave dynamics simplifies the analysis from that for a problem which is unsolvable, to one which is on the present boundaries of sciences ability to understand. The mathematical problem is one of solving a stochastic nonlinear differential equation with many degrees of freedom. The physics problem is to construct a self-consistent description of the processes known to be operative at the air-sea interface within the context of the mathematical model.

Our approach divides the equations of evolution into two categories; that which describes the deterministic evolution of the system and that which describes the effect of fluctuations. For a given set of initial conditions the solution to the deterministic equation of motion determines the motion of the wave field for all time at all points in space. In the solution of the stochastic equations of motion this is not true. The wave field fluctuates from point to point in space and from time increment to time increment. The physically meaningful properties of this latter solution are the *mean* surface displacement, *mean* velocity, power spectral density, etc. In short, the moments of the solution to a stochastic differential equation are the physically interesting quantities. These latter equations have all the complexity of the nonlinear equations of the deterministic system with the additional complication of having fluctuations. Therefore, we study the former system first. The deterministic system describes the average evolution of the system and the weakly nonlinear interactions generate fluctuations about this average motion.

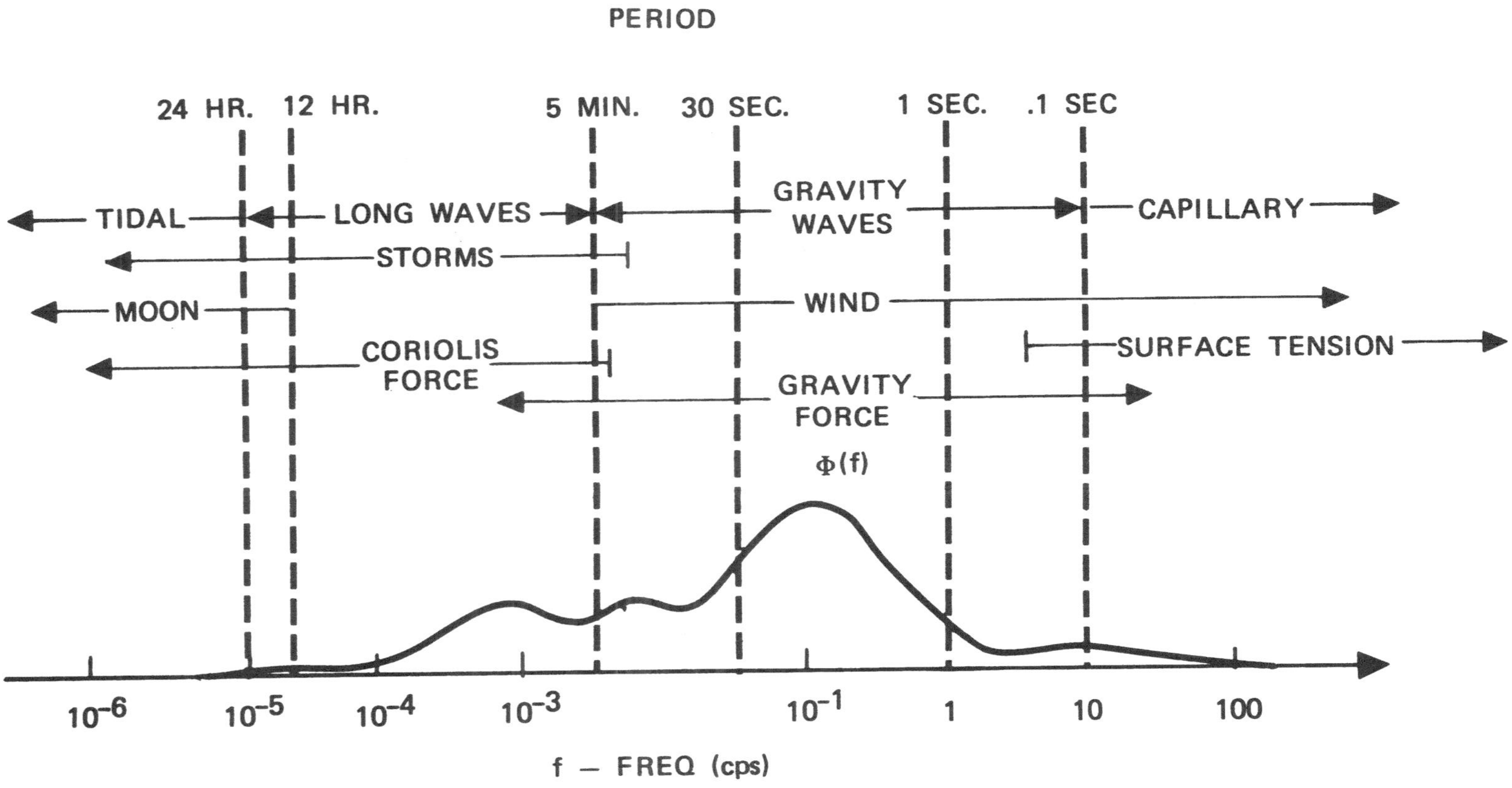

Figure 1.1: Schematic of energy contained in surface waves of the oceans. The function $\Phi(f)$ is the energy spectrum at frequency f.

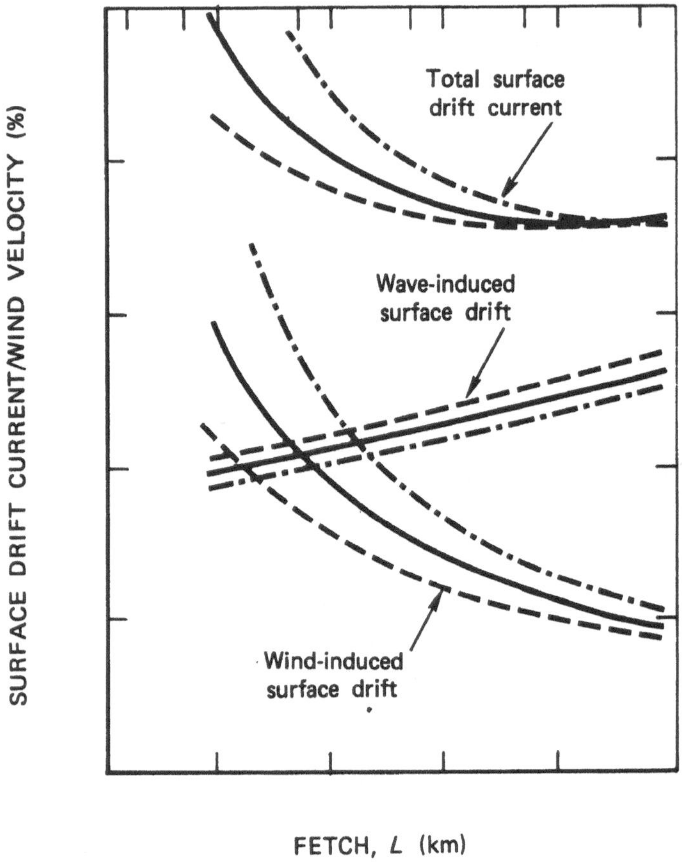

Figure 1.2: Variation of wind-induced and wave-induced surface drift currents with fetch L. The wind speeds in m/sec are indicated by: _ _ _, 5; ___, 10; and __ _ __, 20.

2. More Comments on Mathematical Techniques

Physicists have traditionally separated the analysis of the evolution of a deterministic field of waves into two apparently distinct categories. The first category is that of ray* theory and is concerned with a determination of the path along which a wave packet propagates. Such theories are used extensively in quantum mechanics to solve the Schrödinger equation in configuration space[60]. They were originally based on analogies with the classical equations for a linear wave field, see e.g., Eckart[61]. The second category is that of mode coupled theory and is concerned with a determination of the time rate of change of the amplitudes of a superposition of linear eigenfunctions of the equations of motion. The latter theories arise in the Heisenberg representation of the equations of motion in quantum mechanics[62]. The original application of this method to nonlinear water waves by Hasselmann[11] was in fact suggested by the *form* of such nonlinear equations developed in solid state physics by Pierels[63].

In ray theory one has position and time dependent wave vectors and frequencies and the ocean surface is considered to be a superposition of a number of spatially localized wave packets[64,65]. Each of these wave packets is distinguished by a characteristic wave vector $\underline{k}$ and frequency ω which are related by means of a dispersion relation $\omega = \Omega(\underline{k})$. These wave packets move along trajectories defined by Hamilton's equations

$$\frac{d\underline{x}}{dt} = \nabla_{\underline{k}}\Omega \ ; \qquad \frac{d\underline{k}}{dt} = -\nabla_{\underline{x}}\Omega \tag{1}$$

---
*We use the terms ray, characteristic curve, ray path, trajectory, etc., interchangeably in our discussions.

and are the paths along which energy naturally propagates. Here $d\underline{x}/dt$ is the group velocity of the wave packet and $d\underline{k}/dt$ is the refraction rate. In nonlinear systems the dispersion relation $\Omega(\underline{k})$ depends on the energy (or amplitude) of the wave packets as well as the wave vector. Although the ray equations (1) are still valid for some nonlinear systems, it was not until Whitham[66] developed his average Lagrangian technique that a *systematic* procedure for developing such ray equations in general nonlinear dispersive systems became available. Unfortunately we do not have time to review Whitham's technique here and refer the student to his excellent monograph.[67]

Whether or not Whitham's average Lagrangian technique is an appropriate description of certain of the nonlinear interaction of deep water waves on the ocean surface is still subject to controversy. The *dominant wave* theories of Lake and Yuen[68] and of Plate[66] presuppose the validity of a narrow band theory such as Whitham's and indeed there does appear to be some experimental support for this viewpoint. The *theoretical basis* for such a position *has not* been established, however, in fact analysis favors a broad band spectral process such as presumed in the mode coupled theories. It should also be pointed out that the experimental support for the mode coupled point of view is somewhat more direct.[7,12,38-40]

The mode oriented models describe the ocean surface as a superposition of wave trains. In a linear ocean this would be a superposition of sines and cosines. Such models focus their attention on the transfer of energy among the different modes and the principle of superposition gives the spatial structure. If all the modes in such a description are contained in a narrow spectral region about a central wave number $\underline{k}_0$, then this and the ray theories become equivalent.

All mode coupled theories have a formal similarity, making their study of interest as a branch of applied mathematics independently of any particular application. A great deal of effort has been expended in their study by both mathematicians and physicists. The result is a body of literature that is available for application in the investigation of nonlinear water wave fields from which we freely draw, see e.g., Refs. 70 and 71 and references therein.

Before we become immersed in the mathematical details of describing the nonlinear water wave field, it is useful to point out a number of recurrent concepts which are basic to the analysis. Consider a column vector with N components

$$\underline{Z}(\underline{x},t) \equiv (Z_1, Z_2, \ldots, Z_N) \tag{2}$$

which satisfies the system of differential equations

$$\left(i \frac{\partial}{\partial t} + \underline{\underline{\Omega}}\right) \underline{Z} = \underline{F}(\underline{Z},t) \ , \tag{3}$$

where $\underline{\underline{\Omega}}$ is a linear operator and $\underline{F}$ is a nonlinear vector function of $\underline{Z}$. For deep water gravity waves $\underline{Z}$ is a two component vector[29-32,45] whose elements are the surface displacement and velocity potential at the free surface. In a more general context, since one is also interested in coupled wave systems, there can be more than these two components in $\underline{Z}$. For example, in the study of the coupling of surface and internal waves $\underline{Z}$ has two components to describe the fluid surface and two to describe the fluid interior.[72,73] In a similar way if one is interested in the interaction of distinct long and short waves at the surface, then a model in which each of the scales contributes two components to $\underline{Z}$ can be constructed, see e.g., Benney[74] and Newell.[75]

The matrix $\underline{F}(\underline{Z},t)$ is a nonlinear function of the vector $\underline{Z}$ and describes the nonlinear interaction among the modes in the gravity wave field.

The quantity Ω in (3) is a linear, Hermitian operator. We suppose that a complete set of eigenvectors $\underline{\chi}_{\underline{k}}(\underline{x})$ of $\underline{\underline{\Omega}}$ exist with eigenvalues ω_k, such that for the gravity wave field in isolation

$$\underline{\underline{\Omega}}\underline{\chi}_{\underline{k}}(\underline{x}) = \omega_k \underline{\chi}_{\underline{k}}(\underline{x}) \; . \tag{4}$$

The completeness property of the eigenfunctions yields the relation

$$\sum_{\underline{k}} \underline{\chi}_{\underline{k}}(\underline{x})\underline{\chi}_{\underline{k}}^{t}(\underline{x}') = \delta(\underline{x} - \underline{x}')\,\underline{\underline{I}}; \quad \underline{\underline{I}} = \begin{pmatrix} 10 \\ 01 \end{pmatrix} \tag{5}$$

where $\underline{\chi}_{\underline{k}}^{+}$ is the complex conjugate adjoint of $\underline{\chi}_{\underline{k}}$ and $\delta(\underline{x})$ is the Dirac delta function. Using the Hermitian property of $\underline{\underline{\Omega}}$ it is a simple matter to show that the eigenvalues ω_k are real and that the eigenfunctions $\underline{\chi}_{\underline{k}}$ are orthonormal when the eigenvalues are nondegenerate, i.e., $\omega_k \neq \omega_{k'}$.[76] Thus we may expand the vector $\underline{Z}$ in terms of the complete orthonormal set of eigenvectors $\underline{\chi}_{\underline{k}}$ and the real non-degenerate eigenvalues ω_k as

$$\underline{Z}(\underline{x},t) = \sum_{\underline{k}} A_{\underline{k}}(t)\underline{\chi}_{\underline{k}}(\underline{x})e^{-i\omega_k t} \tag{6}$$

where the $A_{\underline{k}}(t)$ are the complex mode amplitudes corresponding to the linear eigenmodes of (4) and $\chi_{\underline{k}}(\underline{x}) = \chi_{-\underline{k}}^{+}(\underline{x})$.

The original equation of evolution (3) may be transformed using this eigenfunction expansion into a set of mode rate equations. Forming the scalar product of both sides of (3) with $\chi_{\underline{k}}^{+}$, i.e., multiplying (3) by $\chi_{\underline{k}}^{+}(x)$ and integrating over $\underline{x}$, leads to the set of coupled first order equations for the mode amplitudes $A_{\underline{k}}(t)$, i.e.,

$$i\dot{A}_{\underline{k}}(t) = Q_{\underline{k}}(A_{\underline{k}_1}, A_{\underline{k}_2}, \ldots, A_{\underline{k}_N}; t) \; ; \quad \underline{k} = \underline{k}_1, \underline{k}_2, \ldots, \underline{k}_N \tag{7}$$

where $Q_{\underline{k}}$ is the projection of $\underline{F}(\underline{z},t)$ onto the basis set $\chi_{\underline{k}}(\underline{x})$. If the system were linear, then $Q_{\underline{k}} = 0$ and the mode amplitudes $\{A_{\underline{k}}\}$ define a set of constants. The nonlinearity $\underline{F}(\underline{Z},t)$ in (3), however, induces a time variation in the mode amplitudes specified by $Q_{\underline{k}}$ in (7). Here we have assumed that the equations of motion for N discrete modes provide an adequate description of the evolution of the fluid surface. Therefore, (7) represents N coupled first order equations to represent the dynamics of the physical system. This representation is analogous to the harmonic oscillator description of a solid[63] or electromagnetic field often used in physics, see e.g., Ref. 77 or 78. We in fact are able to draw heavily from the formalism of solid state physics to trace the evolution of deep water gravity waves.[11,45]

The function $Q_{\underline{k}}$ in (7) consists of restricted sums over products of $\underline{A}$'s and has an explicit time dependence which appears in exponential factors such as $e^{\pm i\omega_k t}$. The product form of the $Q_{\underline{k}}$ represents the nonlinearity present in physical space as a coupling between the linear eigenmodes of the system. The mode coupling is considered as "weak" when

$$\left|\frac{Q_{\underline{k}}}{\omega_k}\right| \ll 1 \; , \qquad \underline{k} = \underline{k}_1, \underline{k}_2, \ldots \underline{k}_N \quad . \tag{8}$$

When (8) is satisfied, we expect the mode amplitudes to change with time more slowly than the exponential factors in (7). This implies that the linear response of the system is much stronger than the nonlinear response.

It was noted that $Q_{\underline{k}}$ in (7) contains the time explicitly in products of exponential factors such as

$$\exp\,(i\omega_k t)\ \prod_{\ell} \exp\,(-is_\ell \omega_\ell t) \quad ; \qquad s_\ell = +1 \text{ or } -1 \ . \tag{9}$$

When the phase in (9) satisfies the condition

$$\Delta\omega \equiv \left| \omega_k - \sum_{\ell} s_\ell \omega_\ell \right| \simeq 0 \tag{10}$$

a "resonance" is said to occur. The condition of "smallness" in (8) usually requires that $\Delta\omega$ be comparable to, or smaller than, $|Q_{\underline{k}}|$. It is clear that if $\Delta\omega$ is large, then since it is the phase of $Q_{\underline{k}}$, as one integrates (7) in time the phase changes very rapidly. This rapid phase change results in many changes of sign of this term in a small time interval. Summing these alternating terms, one would expect the net value of such a sum to be small. On the other hand, for a term with a small $\Delta\omega$, i.e., a *resonant* term, the phase change is very slow and the net contribution is a coherent sum which *can* be considerable. Indeed, theorems concerning mode coupling, as well as available examples such as the anharmonic oscillator imply that for weak coupling, significant mode coupling only occurs through resonant terms in (7).[33,34,79]

A problem simpler than that of water waves, but one which is characteristic of nonlinear phenomena described by mode coupling was studied some time ago by Fermi, Pasta and Ulam[47] as mentioned before. The physical process they were studying was the diffusion of heat in a solid body. They reasoned that the solid could be represented by an assembly of harmonic oscillators, a model introduced 43 years earlier by Debye[80]. A weak nonlinearity (anharmonicity) in this coupled system should eventually lead to a state in which each oscillator (molecule) has the same amount of energy. This process of energy redistribution is called equi-partitioning.

Since even the one-dimensional model of a system of weakly, nonlinearly coupled harmonic oscillators could not (and cannot) be solved analytically, a computer simulation of the above system was made. The computer code was constructed at Los Alamos and appropriately named MANIC-I. There have since been a number of generations of simulation codes, the most recent being MANIC-IV. Surprisingly, the expected energy equi-partitioning or ergodic mixing[81] of the energy among the linear eigenmodes failed to occur, there bing a periodic return of the system to the initial state. This not only failed to describe the physical process of heat diffusion, but also stimulated considerable interest in a number of related problems. It was not until 1975 that it was shown by Muira[82] that the particular discrete nonlinear lattice that Fermi, Pasta and Ulam had chosen to study was a discrete version of the Korteveg de Vries equation. The recurrence of the initial distribution was evidence of the soliton-like structure within the system they were attempting to integrate numerically. It required the development of the inverse scattering technique of Gardner, Greene, Kruskal and Muira[83] to resolve the paradox by showing analytically the existence of the soliton solutions.

The nonlinearities in the surface wave field are manifest in the products of mode amplitudes in the modal representation. These products couple the different modes together and induce energy transfers at rates determined by the coupling coefficients. This type of interaction is often referred to as mode mixing when the energy transfer process results in a *net* redistribution of the energy. For this energy transfer to occur the interaction must satisfy the resonance condition (10). We find that two types of resonances occur for water waves so that one does expect mode mixing. A resonance can occur in the interaction of three gravity-capillary waves and a separate resonance can occur in the interaction of four gravity waves[84].

The equations of motion for the rays (1) and the individual modes (7) are quite different, although in the limit of small wave amplitudes where the index $\underline{k}$ corresponds to a wavevector and all N wavevectors lie near a central value $\underline{k}_0$, the two descriptions become equivalent. A complication not included in either of these two equations is the effect of fluctuations on the physical quantities of interest, i.e., on the fluid velocity and vertical surface deflection. If the source of the fluctuations is external, such as the turbulent behavior of the air flow driving the water waves, then perturbation concepts must be abandoned in describing the full dynamics of the surface since instabilities in the water wave growth are possible. The linear air-sea coupling model of Phillips[85] and Miles[86,87] between the air flow and the surface waves contains a resonant energy flux from the wind to the wave field. The air-sea coupling parameter is constant in the Miles-Phillips model[87] and predicts an initial growth rate for gravity waves on the deep ocean which are between a factor of five to eight below that observed in field data[88]. A phenomenological model of the air-sea coupling which includes the fluctuations in the air flow by *postulating* a stochastic coupling parameter has been developed by West and Seshadri[89] and is discussed subsequently. This latter model predicts a wave number dependent growth rate consistent with a limited data base of field measurements[90].

If the source of the fluctuations is internal to the wave field, they can often be described by a set of coordinates $\underline{\alpha}$ whose time scale is very much shorter than that of the mode amplitudes $\underline{A}$ in (7). The $\underline{\alpha}$-waves oscillate through many periods before there is appreciable change in the $\underline{A}$-waves, therefore it is possible to solve the equations of motion for the $\underline{\alpha}$'s assuming the $\underline{A}$'s are constant. These solutions can then be used to eliminate the $\underline{\alpha}$'s altogether from the equation of motion for the $\underline{A}$'s. This procedure is known as adiabatic elimination and has the dynamic effect of introducing fluctuations in the $\underline{A}$-wave equations produced by the eliminated degrees of freedom in the dynamic description.[91] Lax[92] shows that under quite general conditions one obtains from this argument a linear equation of the form

$$\frac{d\underline{A}(t)}{dt} + \underline{\underline{\Delta}}\underline{A}(t) = \underline{f}(t) \tag{8}$$

where $\underline{f}(t)$ is a stochastic vector related to the "dissipation matrix" $\underline{\underline{\Delta}}$ by the fluctuation-dissipation relation

$$\langle \underline{f}(t)\underline{f}^{\dagger}(t')\rangle = 2\underline{\underline{D}}\delta(t - t') \ , \tag{9}$$

where the dagger indicates a complex conjugate transpose matrix. Here for the sake of discussion the fluctuations are assumed to be given by a zero-centered, delta correlated Gaussian process and the diffusion matrix $\underline{\underline{D}}$ in (9) is given by the generalized Einstein relation[92]

$$2\underline{\underline{D}} \equiv \underline{\underline{\Delta}}\langle \underline{A}\underline{A}\rangle + \langle \underline{A}\underline{A}\rangle \, \underline{\underline{\Delta}}^{\dagger} \ . \tag{10}$$

The brackets in (9) and (10) indicate a long time average or, invoking the ergodic hypothesis, an average over an ensemble of realizations of the fluctuating force $\underline{f}(t)$.

Equation (8) is referred to as the Langevin equation and with a diagonal matrix $\underline{\underline{\Delta}}$ has the *form* of the Miles-Phillips model for the wind generated growth of a field of water waves. Combining the models summarized in (7) and (8) we obtain the *nonlinear stochastic rate equation* for $\Lambda_{\underline{k}\underline{k}'} = \lambda_{\underline{k}}\delta_{\underline{k}-\underline{k}'}$

$$\dot{A}_{\underline{k}}(t) + (\lambda_{\underline{k}} + i\omega_{k})A_{\underline{k}}(t) = f_{\underline{k}}(t) + Q_{\underline{k}}(\underline{A};t) \quad . \tag{11}$$

Seshadri and West[93] argue that (11) with $Q_{\underline{k}} = 0$ describes the linearized interaction between long wavelength gravity waves and short wavelength gravity-capillary waves, the latter being driven by a fluctuating wind field. The fluctuating flux $f_{\underline{k}}(t)$ and mean flux parameter $\lambda_{\underline{k}}$ arise from the elimination of the short waves from the hydrodynamic equations of evolution for the long waves. The techniques for directly integrating a *nonlinear stochastic differential equation* such as (11) even when they exist are still very much the province of the applied mathematician rather than the physical scientist. Techniques based on the phase space evolution of the probability density $P(\underline{a},t|\underline{a}_0)$ (where $P(\underline{a},t|\underline{a}_0)d\underline{a}$ is the probability that the dynamic variable $\underline{A}(t)$ has a value in the interval $(\underline{a},\underline{a}+d\underline{a})$ at time t given an initial value $\underline{a}_0$) rather than on the dynamic equations are more familiar in the physical sciences. We will review these methods in later lectures and draw on the review aritcles of Lax[91,92] and van Kampen[94].

3. The Hamiltonian for an Isolated Gravity Wave Field

One of the fundamental principles which guides the analysis of any dynamic physical process is that certain quantities will remain conserved during any allowable interaction. In physical systems as complicated as a wave field, where interactions occur everywhere and at all times, it is only the existence of such conservation laws that allow one to describe the evolution of the wave field. In particle mechanics two of the fundamental conservation laws are those of energy and momentum. In continuum mechanics there is in addition to these two conservation laws the conservation of mass.[1,3] Each of these laws can be summarized under a more general property of physical systems contained in a theorem due to Noether[95] and which now bears her name. In essence, Noether's theorem states that for each symmetry of a field there is a corresponding conserved quantity. This theorem has been applied extensively in both relativity and quantum field theory, but it applies with equal validity to the classical wave fields of continuum mechanics. Lanczos[96] points out that Noether's theory can be viewed as an application of the theory of ignorable variables, i.e., to each symmetry there is associated an ignorable coordinate.

We have introduced the term *field* to describe water wave motion on the surface of a fluid. A field is formally defined in classical mechanics as a system with an infinite number of degrees of freedom. A function which associates a scalar ϕ to each point in space-time $(\underline{r},t)$ is called a scalar field, e.g., the velocity potential. A function which associates a vector $\underline{v}$ to each point in space-time $(\underline{r},t)$ is called a vector field, e.g., the velocity field which is the spatial gradient of the velocity potential field. It is not always necessary to represent the water wave system by an infinite number of degrees of freedom. In practice a large but finite number of discrete modes may be sufficient to determine the evolution of the physical system to a given degree of accuracy. For example, imposing specific boundary conditions on the field variables over some large area of ocean Σ_0 reduces the number of degrees of freedom in the water wave field. If the boundary conditions on the field functions are assumed to be periodic, then the set of modes is in fact discrete rather than continuous. For periodic boundary conditions over a surface area Σ_0 the continuum representation of the wave field is given by the limit as the area Σ_0 becomes infinite and the modes in wave vector space become everywhere dense. We assume that such a limit is well defined in the following lectures on the water wave field.

In this and subsequent lectures, we consider the flow of an incompressible fluid in three dimensional space $\underline{r} = (x,y,z)$ described by the velocity field $\underline{v}(\underline{r},t)$. The velocity field is determined in complete generality by a velocity potential $\phi(\underline{r},t)$ and a vorticity function $\underline{\omega}(\underline{r},\underline{t})$,

$$\underline{v}(\underline{r},t) = \nabla\phi(\underline{r},t) + \underline{\omega}(\underline{r},t) \tag{1}$$

where $\nabla \equiv (\partial_x,\partial_y,\partial_z)$. The velocity field is a sum of an irrotational part described by the potential function $\phi(\underline{r},t)$ and a rotational part described by the vorticity field $\underline{\omega}(\underline{r},t)$ generated by the viscosity in the fluid. In two dimensions the rotational component of the fluid velocity defines the scalar vorticity intensity function $\omega(\underline{x},t)$ by

$$\omega(\underline{x},t)\ \underline{\hat{e}}_z \equiv \nabla_{\underline{x}} \times \underline{v}(\underline{x},t) \tag{2}$$

where $\underline{x} = (x,y)$ and $\underline{\hat{e}}_z$ is a unit vector pointing along the positive z-axis. In two dimensions, we can introduce a stream function $\psi(\underline{x},t)$ such that

$$\underline{v}(\underline{x},t) = \nabla_{\underline{x}}\phi(\underline{x},t) + \nabla_{\underline{x}} \times [\underline{\hat{e}}_z\psi(\underline{x},t)]$$

which using (2) results in the relation

$$\nabla_{\underline{x}}^2\psi(\underline{x},t) = -\omega(\underline{x},t) \ . \tag{3}$$

In the absence of vorticity ($\omega = 0$) the stream function satisfies Laplace's equation, i.e., $\nabla^2\psi = 0$, the solution of which generates a family of equipotential surfaces, i.e., surface of constant ψ in the fluid. In particular ψ is constant along the free surface. The velocity field is therefore irrotational on these surfaces of constant ψ and we assume

$$\underline{v}(\underline{r},t) = \nabla\phi(\underline{r},t) \tag{4}$$

throughout this work, i.e., the water wave field is assumed to be irrotational in both two and three dimensions.

We note that hydrodynamic turbulence is related to vorticity[97,98] and since we are restricting our analysis to irrotational flow, we will not discuss turbulence here. The fluctuations in velocity we discuss later are related to the wave field only and do not depend on the existence of classical turbulent flow. The dispersion relation between the frequency and wavelength of water waves ties together the spatial and temporal motion in the wave field. Such a relation does not exist in a turbulent flow field. The two types of fluctuating behavior, that for the wave field and that for turbulence, can therefore be quite different.[99]

Here we describe the motion of the ocean surface using an idealized fluid. The ideal fluid of classical physics consists of an infinitely divisible continuous substance described by a mass density distribution function $\rho(\underline{r},t)$. For a single type of fluid $\rho(\underline{r},t)$ is continuous in space $\underline{r} = (\underline{x},z)$ and time t, but can be discontinuous at the interface of two types of fluids. We define the density of a fluid in an infinitesimal volume element dV by

$$dm = \rho(\underline{r},t)\, dV \tag{5}$$

where dm is the quantity of mass in the volume element dV. The total mass in the volume V is given by the definite integral

$$m = \int_V \rho(\underline{r},t)\, dV \ . \tag{6}$$

In (6) the density $\rho(\underline{r},t)$ is a function of time. For the mass m to be constant the volume V must also be a function of time, i.e., to move with the fluid. If the volume is fixed with respect to a coordinate system referenced external to the fluid, then the total mass will vary in time. Some mass will flow in and some will flow out at the boundaries of the volume V. If S is the surface bounding the volume V; $\hat{\underline{n}}$ a unit vector normal to that surface pointing outward and $\underline{v}$ the fluid velocity; then $\rho\ \underline{v}\cdot\hat{\underline{n}}\ dS$ is the mass flux through the differential surface area dS of the boundary per unit time. The total change in mass per unit time is, therefore

$$\frac{dm}{dt} + \int_S \rho\ \underline{v}\cdot\hat{\underline{n}}\ dS = 0 \ . \tag{7}$$

The surface integral in (7) can be transformed to a volume integral using Gauss' Integral Theorem to replace the surface integral by a volume integral,

$$\int_S \underline{A}\cdot\hat{\underline{n}}\ dS = \int_V \nabla\cdot\underline{A}\ dV \ . \tag{8}$$

Using (8) and (6) we can write the integral (7) as the law of conservation of mass

$$\int_V \left[\frac{\partial\rho}{\partial t} + \nabla \cdot (\rho \underline{v})\right] dV = 0 \; . \tag{9}$$

Equation (9) must hold for an arbitrarily chosen volume V which is only possible if the integrand vanishes. In this way we obtain the equation of continuity

$$\frac{\partial\rho}{\partial t} + \nabla \cdot (\rho \underline{v}) = 0 \; . \tag{10}$$

The conservation of mass, therefore, imposes the continuity condition (10) on the fluid density.[96]

In practical considerations of the fluid surface motion, one may consider the density of water to be a constant. The effects of density variations become important in the discussion of internal waves, but this topic will not concern us here.[5,6,100] Therefore, we assume that the density in (9) is constant in space and time and remove ρ from the gradient to obtain

$$\nabla \cdot \underline{v} = 0 \tag{11}$$

as the form of the conservation of mass which will be the most useful in our discussion. In all subsequent analysis we will assume (11) and therefore have an *incompressible* ideal fluid.

Our model ocean, therefore, consists of a large basin of incompressible irrotational fluid described by a velocity potential $\phi(\underline{x},z,t)$ and surface deflection $z = \zeta(\underline{x},t)$. The basin is assumed to be large in lateral extent with horizontal coordinates $\underline{x}$ and very much deeper than the longest characteristic scale of the surface motion. These two assumptions allow us to ignore the effect of the fixed boundaries of our basin on the motion of the surface and indeed to completely separate the motion of the surface from the interior fluid motion.

We here employ a Hamiltonian formulation of the dynamics because this formalism is compatible with both a deterministic and stochastic description of the surface motion. The compatibility is a consequence of the property that Hamilton's equations define a set of canonical variables for the physical system and that a "volume" element of the phase space is invariant under a canonical transformation.[81] The invariance property provides a natural measure for a probability density with which to describe the evolution of the system, an association that will be used in later lectures. For the moment we are concerned with constructing Hamilton's equations of motion for the water surface.

The surface deflection $\zeta(\underline{x},t)$ and velocity potential at the free surface of the fluid $\phi_s(\underline{x},t)$ constitute a set of canonical field variables.[30-32,72] Zakharov[101] showed that the total energy is a constant for free waves with generalized coordinates $\zeta(\underline{x},t)$ and generalized momentum $\phi_s(\underline{x},t)$ in the Hamiltonian H_g. The total Hamiltonian is

$$H_g = \int_{\Sigma_0} d^2x \left\{ \int_{-B}^{\zeta(\underline{x},t)} \frac{1}{2} \nabla\phi(\underline{x},z,t) \cdot \nabla\phi(\underline{x},z,t)\, dz + \frac{1}{2} g\zeta^2(\underline{x},t) \right\} \qquad (12)$$

where the first term in the brackets is the total kinetic energy in a column of water extending from the ocean bottom at $z = -B$ to the free surface $z = \zeta(\underline{x},t)$ and the second term is the potential energy of the free surface. The equations of motion are given by the variational equations

$$\frac{\partial\phi_s(\underline{x},t)}{\partial t} = -\frac{\delta H_g}{\delta\zeta(\underline{x},t)} ; \qquad \frac{\partial\zeta(\underline{x},t)}{\partial t} = \frac{\delta H_g}{\delta\phi_s(\underline{x},t)} . \qquad (13)$$

However to make use of (13) we must express (12) in terms of the canonical field variables $\phi_s(\underline{x},t)$ and $\zeta(\underline{x},t)$.

Consider the vector identity for the kinetic energy term in (12)

$$\nabla_{\underline{x}} \cdot \int_{-B}^{\zeta} \phi \nabla_{\underline{x}} \phi dz = \phi \nabla_{\underline{x}} \phi \cdot \nabla_{\underline{x}} \zeta \Big|_{z=\zeta} + \int_{-B}^{\zeta} [\nabla_{\underline{x}} \phi \cdot \nabla_{\underline{x}} \phi + \phi \nabla_{\underline{x}}^2 \phi] \, dz \, . \qquad (14)$$

We have used the boundary condition $\nabla_{\underline{x}} B = 0$ and defined $\nabla_{\underline{x}} \equiv (\partial_x, \partial_y)$ to obtain the R.H.S. of (14) and replacing the potential ϕ by ϕ_s at the free surface we rewrite (14) as

$$\int_{-B}^{\zeta} \nabla_{\underline{x}} \phi \cdot \nabla_{\underline{x}} \phi \, dz = -\phi_s \nabla_{\underline{x}} \phi_s \cdot \nabla_{\underline{x}} \zeta \Big|_{z=\zeta} + \nabla_{\underline{x}} \cdot \int_{-B}^{\zeta} \phi \nabla_{\underline{x}} \phi \, dz - \int_{-B}^{\zeta} \phi \nabla_{\underline{x}}^2 \phi \, dz \, . \qquad (14')$$

Substituting (14') into (12) and noting that

$$\int_{\Sigma_0} d^2x \, \nabla_{\underline{x}} \cdot \int_{-B}^{\zeta} \phi \nabla_{\underline{x}} \phi \, dz = 0$$

we obtain

$$H_g = \int_{\Sigma_0} \frac{1}{2} d^2x \left\{ -\phi \nabla_{\underline{x}} \phi \cdot \nabla_{\underline{x}} \zeta \Big|_{z=\zeta} + \int_{-B}^{\zeta} \left[\left(\frac{\partial \phi}{\partial z} \right)^2 - \phi \nabla_{\underline{x}}^2 \phi \right] dz + g\zeta^2 \right\} . \qquad (15)$$

Integrating the z-derivative term by parts and recalling that $\nabla^2 \phi = 0$ throughout the fluid volume for potential flow we obtain

$$H_g = \int_{\Sigma_0} \frac{1}{2} d^2x \left\{ \phi_s(\underline{x},t) \left[\frac{\partial \phi}{\partial z} - \nabla_{\underline{x}} \phi \cdot \nabla_{\underline{x}} \zeta \right]_{z=\zeta(\underline{x},t)} + g\zeta^2(\underline{x},t) \right\} . \qquad (16)$$

Alternatively we can use the kinematic boundary condition at the surface

$$\frac{\partial \zeta(\underline{x},t)}{\partial t} + \nabla_{\underline{x}} \phi(\underline{x},t) \cdot \nabla_x \zeta(\underline{x},t) = \frac{\partial \phi}{\partial z} \, ; \qquad z = \zeta(\underline{x},t) \qquad (17)$$

to write

$$H_g = \frac{1}{2}\int_{\Sigma_0} d^2x \left\{\phi_s(x,t)\,\frac{\partial\zeta(\underline{x},t)}{\partial t} + g\zeta^2(\underline{x},t)\right\} \tag{18}$$

as mentioned by Monin, Kamenkovich and Kort.[102]

Note that we have neglected the high frequency effects of viscosity and surface tension and the coupling to the external wind field. The Hamiltonian (18) is based on the fluid motion being irrotational since all forces acting on the fluid considered here are conservative and therefore reversible. An irreversible force such as molecular viscosity would generate a dissipative current in the fluid and would be rotational, i.e., the dissipative current would generate vorticity at the fluid surface. Although viscosity is outside a conservative Hamiltonian formalism it can be incorporated into the equations of motion in an a posteriori way. In the main we are more concerned with gravity waves than with gravity-capillary waves so that we will not dwell on this last point, but we do return to it later.

The time variation of the free surface in (18) can be replaced by expressing the kinematic boundary condition (17) as the operator equation

$$\frac{\partial\zeta(\underline{x},t)}{\partial t} = T(\kappa,\zeta)\phi_s(\underline{x},t) \ . \tag{17'}$$

To obtain $T(\kappa,\zeta)$ Watson and West[103] employ a special application of potential theory with Dirichlet boundary conditions to the derivative of the velocity potential $\phi(\underline{x},z,t)$. To apply the method we introduce the velocity potential on the reference plane z = 0 and define

$$\phi_0(\underline{x},t) \equiv \phi(\underline{x},z,t)\Big|_{z=0} \ . \tag{19}$$

Since specifying the potential on a closed surface defines a unique potential problem we assert that the velocity potential at the free surface can be written as the Taylor series expansion about the z = 0 plane

$$\phi_s(\underset{\sim}{x},t) = \sum_{n=0}^{\infty} \frac{\zeta^n}{n!} \left.\frac{\partial^n \phi}{\partial z^n}\right|_{z=0} . \tag{20}$$

The vertical derivative in (20) can be replaced by the operator κ using the fact that $\phi(x,z,t)$ satisfies Laplace's equation in the fluid interior, i.e.,

$$\kappa^2 \phi \equiv -\nabla_{\underset{\sim}{x}}^2 \phi . \tag{21}$$

Formally, the operator κ replaces $\partial/\partial z$ in Laplace's equation and has the form

$$\kappa \equiv \sqrt{-\nabla_{\underset{\sim}{x}}^2} . \tag{22}$$

This operator is of interest because it enables us to replace the vertical derivative operation with a horizontal derivative operation and delete reference to the z coordinate altogether. The quantity κ is defined to operate only on Fourier series such that if an arbitrary function $f(\underset{\sim}{x})$ has a Fourier series expansion

$$f(\underset{\sim}{x}) = \sum_{\underset{\sim}{k}} f_{\underset{\sim}{k}} \, e^{i\underset{\sim}{k}\cdot\underset{\sim}{x}} \tag{23}$$

then operating with κ yields

$$\kappa f(\underset{\sim}{x}) = \sum_{\underset{\sim}{k}} k f_{\underset{\sim}{k}} \, e^{i\underset{\sim}{k}\cdot\underset{\sim}{x}} . \tag{24}$$

Miles[31] also found the operator κ useful in his analysis of the equations of motion for shallow water waves.

The velocity potential at the free surface can be expressed in terms of the operator $O(\underline{x},t)$ defined by the series expansion

$$\phi_s(\underline{x},t) = \sum_{n=0}^{\infty} O_n\phi_0(\underline{x},t) = O(\underline{x},t)\phi_0(\underline{x},t) \tag{25}$$

where

$$O_n \equiv \frac{1}{n!}\,\zeta^n(\underline{x},t)\kappa^n \tag{26}$$

and is obtained by replacing $\partial/\partial z$ by κ in (16). In a similar manner we can write the vertical fluid velocity at the fluid surface as a Taylor series with an additional vertical derivative

$$\left.\frac{\partial\phi}{\partial z}\right|_{z=\zeta(\underline{x},t)} = \sum_{n=0}^{\infty} Q_n\phi_0(\underline{x},t) = Q(\underline{x},t)\phi_0(\underline{x},t) \tag{27}$$

where

$$Q_n \equiv \frac{1}{n!}\,\zeta^n(\underline{x},t)\kappa^{n+1} \ . \tag{28}$$

Now to express $\partial\phi/\partial z|_{z=\zeta}$ in terms of ϕ_s we need only express the intermediary velocity potential ϕ_0 in terms of ϕ_s by inverting the operator $O(\underline{x},t)$ in (25), i.e., assuming the inverse of $O(\underline{x},t)$ exists we have

$$\phi_0(\underline{x},t) = O^{-1}(\underline{x},t)\phi_s(\underline{x},t) \tag{29}$$

and write (27) as

$$\left.\frac{\partial\phi}{\partial z}\right|_{z=\zeta(\underline{x},t)} = Q(\underline{x},t)O^{-1}(\underline{x},t)\phi_s(\underline{x},t) \ . \tag{30}$$

The operator $O(\underline{x},t)$ represents the projection of the velocity potential defined on the z = 0 reference plane onto the free surface. The inverse operator $O^{-1}(\underline{x},t)$ projects the velocity potential from the free surface back onto the reference plane. For moderate surface slopes the inverse operator O^{-1} can be expressed as a perturbation series in the surface displacement, i.e.,

$$O^{-1}(\underline{x},t) = 1 - O_1(\underline{x},t) - O_2(\underline{x},t) + O_1^2(\underline{x},t) + O_2^2(\underline{x},t) + O_1(\underline{x},t)O_2(\underline{x},t) + O_2(\underline{x},t)O_1(\underline{x},t) + \ldots \tag{31}$$

The vertical velocity (30) can then be written

$$\begin{aligned}\left.\frac{\partial\phi}{\partial z}\right|_{z=\zeta(\underline{x},t)} &= Q_0\phi_s + (Q_1 - Q_0O_1)\phi_s + (Q_2 - Q_1O_1 - Q_0O_1^2)\phi_s \\ &\quad - (Q_2O_1 + Q_1O_2 - Q_1O_1^2 - Q_0O_1O_2 - Q_0O_2O_1)\phi_s + \ldots \\ &= \kappa\left[\phi_s - \zeta\kappa\phi_s + \zeta\kappa(\zeta\kappa\phi_s) - \frac{1}{2}\zeta^2\kappa^2\phi_s\right] + \zeta\kappa^2\phi_s - \zeta\kappa^2(\zeta\kappa\phi_s) \\ &\quad + \frac{1}{2}\zeta^2\kappa^3\phi_s + O(\zeta^3\phi_s)\end{aligned} \tag{32}$$

to third order in the vertical surface displacement. A general series expression in terms of O_n and Q_n can be written for the vertical velocity, but this is not of interest to us here. We can now express the operator in the surface boundary condition (17') as

$$\begin{aligned}T(\kappa,\zeta) &= -\nabla_{\underline{x}}\zeta\cdot\nabla_{\underline{x}} + [1 + \nabla_{\underline{x}}\zeta\cdot\nabla_{\underline{x}}\zeta]\Big[\kappa + (\zeta\kappa^2 - \kappa\zeta\kappa) \\ &\quad + \left(\kappa\zeta\kappa\zeta\kappa - \zeta\kappa^2\zeta\kappa - \frac{1}{2}\kappa\zeta^2\kappa^2 + \frac{1}{2}\zeta^2\kappa^3\right) + O(\zeta^3)\Big]\end{aligned} \tag{33}$$

where the factor $(1 + \nabla_{\underline{x}}\zeta\cdot\nabla_{\underline{x}}\zeta)$ arises from the differential relation

$$\left.\nabla_{\underline{x}}\phi\right|_{z=\zeta} = \nabla_{\underline{x}}\phi_s - \nabla_{\underline{x}}\zeta\left.\frac{\partial\phi}{\partial z}\right|_{z=\zeta}$$

substituted into the kinematic boundary condition.

Inserting (17') into the Hamiltonian for the wave field we obtain the perturbation series

$$H_g = H_2 + H_3 + H_4 + \ldots \tag{34}$$

where the subscripts indicate the order of products of the field variables in the integrals. The Hamiltonian representation (34) clearly specifies the wave dynamics as an ordering of the *nonlinear interactions* in powers of the surface slope. Each term in (34) is independent of the vertical coordinate z and the explicit expressions for H_2, H_3 and H_4 are discussed in the next lecture. Also this perturbation theory is contrasted with the more familiar perturbation series expansions of the velocity potential such as given in Stoker[2] or by Hasselmann[11]. One of the questions which still haunts investigators is whether (25), and therefore (34), is convergent or not.

An expansion of the surface wave field given by Stokes[104] was shown to be convergent by Levi-Civita[105]. This expansion was made in terms of the harmonics of a fundamental mode for the field, i.e., a dominant wave. The higher harmonics of this fundamental are *not* free waves in this expansion, but have the same phase velocity as the fundamental and are bound to it. The nonlinear wave represented by this series is isolated and is in a steady state. The convergence properties for the transients of this wave are not known. However, there has been some recent extensions of Stokes original work by Weber and Barrick[106,107] to a spectrum of nonlinear waves in which the second order correction to the wave height and the first non-zero correction to the dispersion relation are obtained. The dynamics of the energy transfer process of interest to us here was not considered in their work so we postpone further discussion of their analysis until later. It should also be mentioned that Luke[108] has given an alternate derivation of the equations of motion in configuration space including the boundary conditions from a single stationary pressure condition on the free surface. We present a modified version of his derivation in a later lecture.

4. Deep Water Eigenmodes

As outlined in the second lecture, the mode coupled description of the surface wave dynamics relies on the intuition that scientists have developed about the evolution of conservative systems. This intuition is in large part based on the linear analysis of dynamic systems and has resulted in the viewpoint that the world is essentially linear and that linear wave fields are an adequate description of most phenomena. Further, that the nonlinear interactions in fields, when present, are weak and may be treated as low order products of linear mode amplitudes. We note that some *known* strong nonlinear interactions will be excluded from discussion when this viewpoint is adopted. For example, it seems unlikely that wave breaking can be described by the type of mode coupled theory contemplated here. Therefore, although this philosophy excludes the study of some phenomena, we will show that many of the physical mechanisms operative at the ocean surface can be described by this method. Precisely how one applies this philosophy to the nonlinear water waves described by the Hamiltonian (3.34) is discussed in this and later lectures.

Until recently the most complete dynamic theory using the method of coupled modes was that due to Hasselman[11,9]. He introduced the Fourier mode expansion for the surface displacement and velocity potential into the dynamic equations for the ocean surface, i.e., Bernoulli's equation and the kinematic boundary condition. In this method each surface property is expanded in a perturbation series using some system quantity as a smallness parameter. For water waves the slope of the sea surface is such a smallness parameter. The nonlinear interactions are then given at each order in perturbation theory by the product of an appropriate number of first order terms. The region of validity of such a perturbation expansion is limited, however. To emphasize this limitation consider the solution to Laplace's equation in the fluid interior, i.e.,

$$\phi(\underline{x},z,t) = \frac{1}{2} \sum_{\underline{k}} \phi_{\underline{k}}(t)\, \chi_{\underline{k}}(\underline{x},z,t) + cc \tag{1}$$

where the $\chi_{\underline{k}}(\underline{x},z,t)$ are the linear eigenfunctions of $\nabla^2 \chi_{\underline{k}} = 0$; $\phi_{\underline{k}}(t)$ is the amplitude of the $\underline{k}$-th eigenfunction and cc indicates the complex conjugate of the preceding series. We impose the boundary condition $\phi = 0$, at $z = -\infty$ and periodic boundary conditions over some large horizontal area Σ_0 yielding the linear eigenfunctions $\chi_{\underline{k}}(\underline{x},z,t) = \exp\,[kz + i\underline{k}\cdot\underline{x}]$, i.e., (1) becomes a Fourier expansion. In Hasselman's analysis[29] the potential (1) is expanded in a Taylor series about the reference surface $z = 0$, so that if $\zeta(\underline{x},t)$ is the deviation of the surface from that reference plane, then

$$\phi(\underline{x},z = \zeta,t) = \frac{1}{2} \sum_{\underline{k}} \phi_{\underline{k}}(t)\, e^{i\underline{k}\cdot\underline{x}} \sum_{n=0}^{\infty} \frac{(k\zeta)^n}{n!} + cc \tag{2}$$

which is valid for $k\zeta < 1$. The surface displacement $\zeta(\underline{x},t)$ itself consists of many wave vectors, however, having itself the Fourier series representation

$$\zeta(\underline{x},t) = \frac{i}{2} \sum_{\underline{k}} \zeta_{\underline{k}}(t)\, e^{i\underline{k}\cdot\underline{x}} + cc \tag{3}$$

where $\zeta_{\underline{k}}(t)$ is the Fourier mode amplitude for the surface displacement. Thus one cannot guarantee the convergence of (2) in general. This is shown in detail for the trivial case of two surface waves.

Consider the superposition of two linear one-dimensional waves on which to test the convergence of the expansion (2);

$$\zeta(x,t) = a_1 \cos (k_1 x - \omega_1 t) + a_2 \cos (k_2 x - \omega_2 t) \tag{4}$$

with the judiciously chosen values of amplitude and wave numbers $a_1 = 10$ m, $k_1 = .01\ m^{-1}$ and $a_2 = .1$ m, $k_2 = 1\ m^{-1}$. Notice that we have selected both waves to have the same slope, i.e., the product of wavenumber and amplitude is .1 for each. The maximum amplitude of the surface displacement is essentially that of the longer wave, i.e., $\zeta_{max} \approx 10$ m. In the sum on wavenumber in (2) we have two values, k_1 and k_2, so that the velocity potential at the extreme of the surface displacement is

$$\phi(x,\zeta_{max},t) \cong \phi_{k_1}(t) \cos k_1 x \sum_{n=0}^{\infty} \frac{(k_1\zeta_{max})^n}{n!} + \phi_{k_2}(t) \cos k_2 x \sum_{n=0}^{\infty} \frac{(k_2\zeta_{max})^n}{n!} . \tag{5}$$

In the first term we have $k_1\zeta_{max} = .1$ so the series expansion converges rapidly. However $k_2\zeta_{max} = 10$ and the second series converges very slowly. The expansion of the shorter wave about the $z = 0$ surface, therefore, leads to a divergent result unless all the terms in the series are kept. This is not entirely unexpected since the shorter wave physically rides atop the longer wave and is displaced quite far from the $z = 0$ plane. An expansion in the vicinity of the reference plane for a wave whose wavelength is *shorter* than the displacement from the plane would naturally lead to nonphysical results.

The above example stresses the fact that the truncated expansion of the velocity potential about the $z = 0$ reference surface is valid for some water wavelengths but not for others. The surface wave field on the ocean contains a broad range of wavelengths and estimates of how much of this range can be reasonably included in an expansion of the form (2) is central in determining the utility of such a theory. Such estimates have been made, e.g., by Holliday[109] in his discussion of the evolution of gravity-capillary waves.

We note that by using the canonical variables $\phi_s(\underline{x},t)$ and $\zeta(\underline{x},t)$ that the explicit expansion of the velocity potential about the z = 0 reference plane is avoided in the Hamiltonian. It was necessary to use this expansion as an intermediate step in obtaining the expansion for the vertical velocity, but the final expression for the Hamiltonian (3.34) is an expansion about the free surface *not* the reference surface and is correct to third order in ϕ_s and ζ. To obtain explicit expressions for the contributions to the Hamiltonian we represent the surface field by the complex amplitude $Z(\underline{x},t)$ defined on a rectangular ocean of area Σ_0 (with periodic boundary conditions) and write

$$Z(\underline{x},t) \equiv \sum_{\underline{k}} \sqrt{2/V_k}\, B_{\underline{k}}(t)\, e^{i\underline{k}\cdot\underline{x}} \tag{6}$$

where

$$\phi_s(\underline{x},t) = \frac{1}{2} V_\kappa \, [Z(\underline{x},t) + Z^*(\underline{x},t)] \tag{7a}$$

$$\zeta(\underline{x},t) = \frac{i}{2} [Z(\underline{x},t) - Z^*(\underline{x},t)] \, . \tag{7b}$$

The operator V_κ weights each mode by the phase speed of a small amplitude gravit wave in deep water ($V_k \equiv \omega_k/k$) and ω_κ weights each mode by the corresponding gravity wave frequency ($\omega_k = \sqrt{gk}$). The factor $\sqrt{2/V_k}$ appears in the expansion for the complex amplitude $Z(\underline{x},t)$ to ensure that the $B_{\underline{k}}$'s constitute a set of canonical variables. This normalization of the mode amplitudes is often overlooked in such Fourier expansions.

The lowest order contribution to the Hamiltonian (3.34) is (with ρ_0=1)

$$H_2 \equiv \frac{1}{2} \int d^2x \, \{\phi_s(\underline{x},t)\kappa\phi_s(\underline{x},t) + g\zeta^2(\underline{x},t)\} \tag{8}$$

where we have used the constant operator part of (3.33) in (3.18). Inserting the series expansion for ϕ_s and ζ given by (7a) and (7b) into (8) and using Parsavel's theorem we obtain

$$H_2 = \sum_{\underline{\ell}} \omega_{\ell} B_{\underline{\ell}} B^*_{\underline{\ell}} \tag{9}$$

which is the energy for a linear wave field, familiar from its use in quantum mechanics.[76,77] The quantization (discretization of wavelengths) arises from the imposition of periodic boundary conditions over the surface area.

The next order term in the Hamiltonian (3.34) comes about from the terms in (3.33) that are linear in the surface deflection yielding the third order expression

$$H_3 \equiv \frac{1}{2} \int d^2x \, \{-\phi_s(\underline{x},t)\nabla_{\underline{x}}\zeta(\underline{x},t)\cdot\nabla_{\underline{x}}\phi_s(\underline{x},t) + \phi_s(\underline{x},t)\zeta(\underline{x},t)\kappa^2\phi_s(\underline{x},t)$$

$$- \phi_s(\underline{x},t)\kappa\zeta(\underline{x},t)\kappa\phi_s(\underline{x},t)\} . \tag{10}$$

In our rectangular two-dimensional space the exponentials in the Fourier expansion (6) satisfy the relations

$$\frac{1}{\Sigma_0} \int d^2x \, e^{i(\underline{k}-\underline{k}')\cdot\underline{x}} = \delta_{\underline{k}-\underline{k}'} \; ; \qquad \delta_{\underline{k}-\underline{k}'} = \begin{cases} 1 & \underline{k} = \underline{k}' \\ 0 & \underline{k} \neq \underline{k}' \end{cases}$$

$$\frac{1}{\Sigma_0} \sum_{\underline{k}} e^{i\underline{k}\cdot(\underline{x}-\underline{x}')} = \delta(\underline{x} - \underline{x}') \tag{11}$$

where $\delta_{\underline{k}-\underline{k}'}$ is the Kronecker and $\delta(\underline{x} - \underline{x}')$ the Dirac delta function. Using (11) we again apply Parsavel's theorem and obtain from (10)

$$H_3 = \sum_{\underline{\ell},\underline{m},\underline{p}} \{\delta_{\underline{\ell}+\underline{m}+\underline{p}} V_{\underline{\ell}\underline{m}\underline{p}} B_{\underline{\ell}} B_{\underline{m}} B_{\underline{p}} + \delta_{\underline{\ell}+\underline{m}-\underline{p}} V^{\underline{p}}_{\underline{\ell}\underline{m}} B_{\underline{\ell}} B_{\underline{m}} B^*_{\underline{p}}$$

$$+ \delta_{\underline{\ell}-\underline{m}-\underline{p}} V^{\underline{m}\underline{p}}_{\underline{\ell}} B_{\underline{\ell}} B^*_{\underline{m}} B^*_{\underline{p}} + \delta_{\underline{\ell}+\underline{m}+\underline{p}} V^{\underline{\ell}\underline{m}\underline{p}} B^*_{\underline{\ell}} B^*_{\underline{m}} B^*_{\underline{p}}\} \tag{12}$$

where we note the restriction on wave vectors in each of the three-wave interactions.

Finally from the quadratic terms in the boundary condition operator (3.33) we obtain for the four wave interaction term in the Hamiltonian (3.34)

$$H_4 \equiv \frac{1}{2}\int d^2x\ \{\phi_s(\underline{x},t)\kappa\zeta(\underline{x},t)\kappa\zeta(\underline{x},t)\kappa\phi_s(\underline{x},t) - \phi_s(\underline{x},t)\zeta(\underline{x},t)\kappa^2\zeta(\underline{x},t)\kappa\phi_s(\underline{x},t)$$

$$- \frac{1}{2}\phi_s(\underline{x},t)\zeta^2(\underline{x},t)\kappa^3\phi_s(\underline{x},t) + \phi_s(\underline{x},t)[\kappa\phi_s(\underline{x},t)]\nabla_{\underline{x}}\zeta(\underline{x},t)\cdot\nabla_{\underline{x}}\phi_s(\underline{x},t) \quad . \tag{13}$$

Again applying Parsavel's theorem to (13) we obtain

$$H_4 = \sum_{\underline{\ell},\underline{m},\underline{p},\underline{g}} \{\delta_{\underline{\ell}+\underline{m}+\underline{p}+\underline{g}} V_{\underline{\ell}\underline{m}\underline{p}\underline{g}} B_{\underline{\ell}} B_{\underline{m}} B_{\underline{p}} B_{\underline{g}} + \delta_{\underline{\ell}+\underline{m}+\underline{p}-\underline{g}} V^{\underline{g}}_{\underline{\ell}\underline{m}\underline{p}} B_{\underline{\ell}} B_{\underline{m}} B_{\underline{p}} B^*_{\underline{g}}$$

$$+ \delta_{\underline{\ell}+\underline{m}-\underline{p}-\underline{g}} V^{\underline{p}\underline{g}}_{\underline{\ell}\underline{m}} B_{\underline{\ell}} B_{\underline{m}} B^*_{\underline{p}} B^*_{\underline{g}} + \delta_{\underline{\ell}-\underline{m}-\underline{p}-\underline{g}} V^{\underline{m}\underline{p}\underline{g}}_{\underline{\ell}} B_{\underline{\ell}} B^*_{\underline{m}} B^*_{\underline{p}} B^*_{\underline{g}}$$

$$+ \delta_{\underline{\ell}+\underline{m}+\underline{p}+\underline{g}} V^{\underline{\ell}\underline{m}\underline{p}\underline{g}} B^*_{\underline{\ell}} B^*_{\underline{m}} B^*_{\underline{p}} B^*_{\underline{g}}\} \tag{14}$$

in which the restriction on wavevectors during the four-wave interactions are explicitly indicated.

The evolution of the isolated gravity wave field is determined by Hamilton's equations of motion in the modal representation as

$$\dot{B}_{\underline{k}} = -i\,\frac{\partial H_g}{\partial B^*_{\underline{k}}} \ ; \qquad \dot{B}^*_{\underline{k}} = i\,\frac{\partial H_g}{\partial B_{\underline{k}}} \ . \tag{15}$$

Using the series expansion (3.34) for H_g and the expressions (9), (12) and (14) for the individual terms in the Hamiltonian we obtain the mode rate equations

$$\dot{B}_{\underline{k}}(t) + i\omega_k B_{\underline{k}}(t) = T^{(2)}_{\underline{k}}(t) + T^{(3)}_{\underline{k}}(t) \ . \tag{16}$$

The functions $T_{\underline{k}}^{(2)}$ and $T_{\underline{k}}^{(3)}$ represent the quadratic and cubic interactions among the gravity waves, respectively. The similarity in structure of these equations to those in quantum field theory motivated Hasselmann[11,29] to consider such an interpretation for the water wave field. The function $T_{\underline{k}}^{(2)}$ is given by

$$T_{\underline{k}}^{(2)}(t) \equiv \sum_{\underline{\ell},\underline{p}} \delta_{\underline{k}-\underline{\ell}-\underline{p}} \{\Gamma_{\underline{\ell}\underline{p}}^{\underline{k}} B_{\underline{\ell}} B_{\underline{p}} + \Gamma_{\underline{\ell}}^{\underline{k},-\underline{p}} B_{\underline{\ell}} B_{\underline{p}}^{*} + \Gamma^{\underline{k},-\underline{p},-\underline{\ell}} B_{-\underline{\ell}}^{*} B_{-\underline{p}}^{*}\} \tag{17}$$

with the real coefficients $\Gamma_{\underline{\ell}\underline{p}}^{\underline{k}} = -iV_{\underline{\ell}\underline{p}}^{\underline{k}}$, $\Gamma_{\underline{\ell}}^{\underline{k},-\underline{p}} = -2iV_{\underline{\ell}}^{\underline{k},-\underline{p}}$ and $\Gamma^{\underline{k},-\underline{\ell},-\underline{p}} = -3iV^{\underline{k},-\underline{\ell},-\underline{p}}$ and describes the interaction of two surface waves $\underline{\ell}$ and $\underline{m}$ with the wave of interest $\underline{k}$. The function $T_{\underline{k}}^{(3)}$ is given by

$$\begin{aligned} T_{\underline{k}}^{(3)}(t) \equiv i \sum_{\underline{\ell},\underline{p},\underline{n}} \delta_{\underline{k}+\underline{n}-\underline{\ell}-\underline{p}} &\{\Gamma_{\underline{\ell},\underline{p},-\underline{n}}^{\underline{k}} B_{\underline{\ell}} B_{\underline{p}} B_{-\underline{n}} + \Gamma_{\underline{\ell}\underline{p}}^{\underline{k}\underline{n}} B_{\underline{\ell}} B_{\underline{p}} B_{\underline{n}}^{*} \\ &+ \Gamma_{\underline{\ell}}^{\underline{k},-\underline{p},\underline{n}} B_{\underline{\ell}} B_{-\underline{p}}^{*} B_{\underline{n}}^{*} + \Gamma^{\underline{k},-\underline{p},-\underline{\ell},\underline{n}} B_{-\underline{\ell}}^{*} B_{-\underline{p}}^{*} B_{\underline{n}}^{*}\} \end{aligned} \tag{18}$$

with the real coefficient $\Gamma_{\underline{\ell}\underline{p}}^{\underline{k}\underline{n}} = -2V_{\underline{\ell}\underline{p}}^{\underline{k}\underline{n}}$ and the other four wave interactions are non-resonant in frequency as discussed below and, therefore, not of interest to us here. Equation (18) requires the interaction of three surface waves $\underline{\ell}$, $\underline{p}$ and $\underline{n}$ to stimulate $\underline{k}$. Most of these interactions involve the periodic interchange of energy among the waves. A *net* energy transfer from one region of wavenumber space to another is only accomplished by means of an interaction in which the three frequencies ω_ℓ, ω_n and ω_p sum to give the frequency of the fourth water wave ω_k. Based on an analogy with harmonically driven linear systems Phillips[26] referred to such a frequency matching condition as a resonance and indeed such interactions yield a secular growth of the amplitude of the $\underline{k}$-mode at early times. Phillips also showed that for gravity waves only the single term $B_{\underline{\ell}} B_{\underline{p}} B^*_{\underline{n}}$ can satisfy the frequency resonance condition necessary for this net energy exchange. This resonance phenomenon in nonlinear systems will be discussed in detail in subsequent lectures.

The reality of the Hamiltonian implies certain symmetry properties in the interaction coefficients. These are

$$V^{\underline{k}}_{\underline{p}\underline{\ell}} = V^{\underline{k}}_{\underline{\ell}\underline{p}} = (V^{\underline{\ell}\underline{p}}_{\underline{k}})^* \;; \qquad V^{\underline{k}\underline{n}}_{\underline{\ell}\underline{p}} = (V^{\underline{k}\underline{n}}_{\underline{\ell}\underline{p}})^* \;; \qquad V^{\underline{p}\underline{k}}_{\underline{\ell}} = V^{\underline{k}\underline{p}}_{\underline{\ell}} = (V^{\underline{\ell}}_{\underline{k}\underline{p}})^* \;;$$

$$V^{\underline{k}\underline{n}}_{\underline{\ell}\underline{p}} = V^{\underline{\ell}\underline{p}}_{\underline{k}\underline{n}} = V^{\underline{n}\underline{k}}_{\underline{\ell}\underline{p}} = V^{\underline{n}\underline{k}}_{\underline{p}\underline{\ell}} \;; \qquad V^{\underline{k}\underline{p}\underline{\ell}} = V^{\underline{k}\underline{\ell}\underline{p}} = (V_{\underline{k}\underline{\ell}\underline{p}})^* \;. \tag{19}$$

The detailed forms of these coefficients obtained from (8), (10) and (13) for the three wave interactions are

$$V_{\underline{\ell}\underline{m}\underline{p}} = \frac{-i}{24}\left(\frac{2}{V_\ell V_p V_m}\right)^{1/2} [V_\ell V_m(m\ell + \underline{m}\cdot\underline{\ell}) + V_\ell V_p(\ell p + \underline{\ell}\cdot\underline{p}) + V_m V_p(mp + \underline{m}\cdot\underline{p})] \tag{20a}$$

$$V^{\underline{p}}_{\underline{\ell}\underline{m}} = \frac{i}{8}\left(\frac{2}{V_\ell V_p V_m}\right)^{1/2} [V_\ell V_m(m\ell + \underline{m}\cdot\underline{\ell}) - V_\ell V_p(\ell p - \underline{\ell}\cdot\underline{p}) - V_m V_p(mp - \underline{m}\cdot\underline{p})] \tag{20b}$$

$$V^{\underline{m}\underline{p}}_{\underline{\ell}} = \frac{i}{8}\left(\frac{2}{V_\ell V_p V_m}\right)^{1/2} [V_\ell V_m(m\ell - \underline{m}\cdot\underline{\ell}) + V_\ell V_p(\ell p - \underline{\ell}\cdot\underline{p}) - V_m V_p(mp + \underline{m}\cdot\underline{p})] \tag{20c}$$

and for the possibly resonant four wave interactions,

$$\begin{aligned} V^{\underline{p}\underline{n}}_{\underline{\ell}\underline{m}} = \frac{-1}{2^3\,[V_p V_n V_\ell V_m]^{1/2}} \Big\{ & V_\ell V_n \ell n(\ell + n - |\underline{\ell}+\underline{m}| - |\underline{m}-\underline{n}|) \\ & + V_m V_n mn(m+n-|\underline{\ell}+\underline{m}| - |\underline{\ell}-\underline{n}|) + V_m V_p mp(m+p-|\underline{\ell}+\underline{m}| - |\underline{\ell}-\underline{p}|) \\ & + V_\ell V_p \ell p(\ell+p-|\underline{\ell}+\underline{m}| - |\underline{m}-\underline{p}|) - V_\ell V_m \ell m(\ell+m-|\underline{\ell}-\underline{n}| - |\underline{m}-\underline{n}|) \\ & - V_p V_n pn(p+n-|\underline{p}-\underline{m}| - |\underline{m}-\underline{n}|) \Big\} \end{aligned} \tag{21}$$

5. Resonant Interactions and Dynamic Equations

In the preceding lecture the dynamics of the nonlinear wave field is represented in terms of the time derivative of the amplitude of a discrete set of eigenmodes of the inviscid linear wave field. The nonlinearities appear as products of mode amplitudes in this representation and for the discrete set of modes can be interpreted as the elastic scattering of water waves. The quadratic terms in the equations of motion represent the scattering of three waves, the cubic terms the scattering of four waves, etc. The interactions are elastic because of the finite number of degrees of freedom in this representation. Energy initially concentrated in a few degrees of freedom will be transferred by the nonlinear interaction further and further from its initial mode. Eventually, however, the finite boundaries of the system will be experienced and the energy will be reflected back towards the initial state. In a system with an infinite number of degrees of freedom this energy transfer can remain uni-directional in $\underline{k}$-space. The energy transferred to very small spatial scales will eventually be damped by viscosity, so that in a continuum model of this process viscosity must be included. In the discrete model considered in these lectures we ignore this effect although we will return to this question again later.

In his formulation of the mode coupled equations describing the interaction of water waves, Hasslemann[11] implicitly assumed the existence of canonical field variables. This was made explicit in the later development of the interaction Hamiltonian to describe the interaction of the wave field with other fields such as the turbulent air flow at the fluid surface.[29] Later investigators in the main restricted their analysis to configuration space. It is more natural for our purposes to examine the evolution of a Hamiltonian system consisting of a finite number of degrees of freedom in phase space.

We write the Hamiltonian H_g as a linear part H_2 and nonlinear perturbation V, i.e.

$$H_g(\underline{q},\underline{p}) = H_2(\underline{q},\underline{p}) + \lambda V(\underline{q},\underline{p}) \tag{1}$$

where λ is a parameter that measures the *strength* of the perturbation and $(\underline{p},\underline{q})$ is a set of canonical variables. The Hamiltonian (1) in the simplest case, i.e. $\lambda = 0$, can be used to describe the motion of an element of fluid near a free surface. If $q_{\underline{k}}$ is interpreted as the coordinate of a parcel of fluid displaced an infinitesimal distance from its equilibrium position and released, gravity pulls the fluid parcel back to equilibrium. The surface following the fluid parcel undergoes simple harmonic motion with wavelength $2\pi/|\underline{k}|$. The frequency of oscillation (ω_k) that this process initiates, is dependent on the wavelength of the wave generated and the accleration of gravity (g),[106]

$$\omega_k = \sqrt{gk} \quad . \tag{2}$$

Both the surface displacement and fluid velocity are harmonic with frequency given by (2) so that the fluid parcel has closed circular orbits in physical space. For a finite amplitude wave where nonlinear effects become important [$\lambda \neq 0$ in (1)] these frequencies, i.e. those of displacement and velocity, are not equal and the orbits do not close. This results in a drifting of fluid near the surface, i.e. a surface current induced by the nonlinear interaction.[104]

In the normal mode representation discussed earlier the linear Hamiltonian has the form given by (4.9) and Hamilton's equations of motion are given by (4.16). The homogeneous solutions to (4.16) are harmonic in time with frequency ω_k, i.e. $B_{\underline{k}}(t) = B_{\underline{k}}(0)\, e^{-i\omega_k t}$ when $\lambda = 0$. The nonlinearity therefore modifies this solution such that the new variable

$$b_{\underline{k}}(t) = e^{i\omega_k t} B_{\underline{k}}(t) \quad , \tag{3}$$

which is constant when $\lambda = 0$, varies slowly in time when $\lambda \neq 0$. Therefore defining the function

$$Q_{\underline{k}}(\underline{b}) = e^{i\omega_k t} \frac{\partial V}{\partial B_{\underline{k}}^*} \tag{4}$$

(4.16) can be written as done in the second lecture

$$i\,\dot{b}_{\underline{k}}(t) = \lambda Q_{\underline{k}}(\underline{b}) \quad . \tag{5}$$

Now that we know that the function $Q_{\underline{k}}$ consists of the nonlinear interactions in $T_{\underline{k}}^{(2)}$ and $T_{\underline{k}}^{(3)}$ given by (4.17) and (4.18), respectively, we recall the comments made in lecture 2.

In particular we recall that the condition that the nonlinear interactions be "weak" is

$$\frac{\lambda|Q_{\underline{k}}|}{\omega_k} = \frac{\left|e^{i\omega_k t}[T_{\underline{k}}^{(2)} + T_{\underline{k}}^{(3)}]\right|}{\omega_k} \ll 1 \qquad \text{all modes} \tag{6}$$

so that when (6) is satisfied, we expect the mode amplitude $b_{\underline{k}}(t)$ to change more slowly than the exponential factors in $Q_{\underline{k}}$.[45] For example, the four waves with wave vectors $\underline{k}, \underline{\ell}$, $\underline{m}$ and $\underline{n}$ could result in an exponential factor of the form $\exp i[\omega_k + \omega_n - \omega_\ell - \omega_m]t$. When the phase in (6) satisfies the condition†

$$\Delta\omega \equiv \omega_k + \omega_n - \omega_\ell - \omega_m \simeq 0 \tag{7}$$

†We note that $\omega_k + \omega_n \pm \omega_\ell \neq 0$ for gravity waves when $\underline{k} + \underline{n} \pm \underline{\ell} = 0$ and so three wave resonances are not present here.[26]

a *resonance* is said to occur. The condition of *smallness* in (6) usually requires that $\Delta\omega$ be comparable to , or smaller than the nonlinear term $\lambda|Q_{\underline{k}}|$. It is clear that if $\Delta\omega$ is large, since it is the phase of $Q_{\underline{k}}$, then as one integrates (5) in time the phase changes very rapidly. This rapid variation results in many changes of sign of the term in a small time interval. Adding these alternating terms, one would expect the net value of such a sum to be small. On the other hand, for a term with a small $\Delta\omega$, i.e., a *resonant* term, the phase change is very slow and the net contribution is a coherent sum which can be substantial.

Returning to the system of mode rate equations generated by the Hamiltonian (1) we note that classical dynamics prescribes how to solve such a set of equations. Hamilton-Jacobi theory and canonical perturbation theory were developed to integrate such equations of motion. However, it is not uncommon for Hamilton-Jacobi theory to yield divergent perturbation expansions when applied to nonlinear mechanical systems.[34,35,37] It is useful for us to understand the fundamental limitations of applying classical techniques to the analysis of such nonlinear physical phenomena. A review of some features of Hamilton-Jacobi theory provides a qualitative picture of much of the dynamics of the water wave field.

THe Hamiltonian H_2 given by (4.9) is the energy of an assembly of uncoupled oscillators, e.g. the surface water wave field modeled as a linear superposition of sines and cosines with infinitesimal amplitudes. The normal modes for different physical systems are distinguished by the frequencies $\omega_{\underline{k}}$, so that the following discussion is not dependent on the particular dispersion relation of the oscillator field. That there are N degrees of freedom and 2N equations of motion for the linear system determined by H_2 implies there are N first integrals of the motion. To see this explicitly we introduce the variables $(\underline{J},\underline{\Theta})$ by

$$B_{\underline{k}}(t) = \sqrt{J_{\underline{k}}(t)} \exp [i\, \Theta_{\underline{k}}(t)] \tag{8}$$

which is the polar coordinate representation of the oscillator motion.

One advantage of the polar representation is that the Hamiltonian for the linear system is cyclic in the angle variable $\underline{\Theta}$, i.e., it is independent of $\underline{\Theta}$ which is then an ignorable coordinate in Noether's theorem. Using the Hamiltonian (4.9) and substituting in the new variables (8) yields

$$H_2(J) = \sum_{\underline{k}} \omega_k J_{\underline{k}} = \underline{\omega}_0 \cdot \underline{J} \tag{9}$$

where $\underline{\omega}_0 = (\omega_{k_1}, \omega_{k_2}, \ldots \omega_{k_N})$ is a constant vector. The transformation (8) is canonical, implying that the new variables $(\underline{J},\underline{\Theta})$ obey the Poissen bracket relations (defined below) and are also canonical, eg. see Corben and Stehle.[110] The resulting equations of motion for the unperturbed system in terms of the new canonical variables $(\underline{J},\underline{\Theta})$ are

$$\dot{\underline{J}} = [\underline{J},H_2] \equiv \nabla_{\underline{\Theta}}\underline{J}\cdot\nabla_{\underline{J}}H_2 - \nabla_{\underline{\Theta}}H_2\cdot\nabla_{\underline{J}}\underline{J} = 0$$

$$\dot{\underline{\Theta}} = [\underline{\Theta},H_2] \equiv \nabla_{\underline{\Theta}}\underline{\Theta}\cdot\nabla_{\underline{J}}H_2 - \nabla_{\underline{\Theta}}H_2\cdot\nabla_{\underline{J}}\underline{\Theta} = \nabla_{\underline{J}}H_2 \quad . \tag{10}$$

For linear systems the action in each of the N degrees of freedom is a constant of the motion, i.e., N first integrals resulting from $\dot{\underline{J}} = 0$. The angle variables increase linearly with time since $\underline{\omega}_0$ is a constant vector, i.e., $\underline{\Theta}(t) = \underline{\omega}_0 t + \underline{\Theta}(0)$, and the physical state of the system is unchanged under a displacement of 2π in each of the angle variables, i.e., $\Theta_k \to \Theta_k + 2\pi$. For a system with one degree of freedom the action can be viewed as a radius vector and the angle as a rotation angle about a closed orbit. This orbit is in the two-dimensional phase space for the system and does not correspond to the orbital motion of a fluid parcel in a physical water wave. For a system with two degrees of freedom the motion in the four-dimensional phase space is on the surface of a doughnut (torus), the action variables are the major and minor radii and the rotations of these radii the corresponding angle variables. The action-angle description in general displays the dynamics of a mechanical system with N degrees of freedom as a N-dimensional torus embedded in a 2N dimensional phase space.[33]

The above discussion is predicated on the equations of motion for a linear system. However, the interpretation is equally valid for nonlinear systems which are integrable.[35,37] A Hamiltonian system is said to be integrable if there exists N independent analytic single-valued first integrals, i.e., N functions that are constant along each trajectory in the 2N dimensional phase space for the system. In this case a canonical transformation from a set of variables $(\underline{q},\underline{p})$ to the new coordinates $(\underline{Q},\underline{P})$ can be constructed such that the system is cyclic in $\underline{Q}$ and the Hamiltonian $\bar{H}$ in the new variables is independent of $\underline{Q}$ so that

$$\dot{\underline{Q}} = \nabla_{\underline{P}}\bar{H} \quad ; \quad \dot{\underline{P}} = -\nabla_{\underline{Q}}\bar{H} = 0 \quad . \tag{11}$$

Thus the new momenta $\underline{P}$ are the above first integrals and are constant along each trajectory of the system, i.e., $\underline{P} = \underline{c}$ a constant vector. Since $\underline{P}$ is constant along a trajectory, the system has the additional integrals,

$$\underline{Q}(t) = t\nabla_{\underline{c}}\bar{H}(\underline{c}) + \underline{\delta} \tag{12}$$

where $\underline{c}$ and $\underline{\delta}$ determine the 2N constants of integration needed to define a solution of (11) [See Berry[111] or Tabor[37] for a more complete discussion].

The dynamics of the system are determined if we can express the new coordinates ($\underline{Q},\underline{P}$) in terms of the old. As Moser[33] points out, the Hamilton-Jacobi theory was in fact designed for this purpose. The generating function for the canonical transformation to the new representation of the integrable system is

$$S(\underline{q},\underline{P}) = \int_{\underline{q}_0}^{\underline{q}} \underline{p}(\underline{q},\underline{P})\cdot d\underline{q} \tag{13}$$

and replaces the coordinate transformation (8) for the linear oscillators. The generating function (13) is easily shown to satisfy the equations [see Corben and Stehle, Sect. 61, for a complete discussion]

$$\underline{Q} = \nabla_{\underline{P}} S(\underline{q},\underline{P}) \quad ; \quad \underline{p} = \nabla_{\underline{q}} S(\underline{q},\underline{P}) \tag{14}$$

so that any function satisfying (14) gives rise to a canonical transformation, provided one can express ($\underline{q},\underline{p}$) in terms of ($\underline{Q},\underline{P}$).

The time independent Hamiltonian $H(\underline{q},\underline{p})$ is expressed using (13) and (14) a $H(\underline{q},\nabla_{\underline{q}} S)$ and is related to the Hamiltonian in the ($\underline{Q},\underline{P}$) system by

$$H(\underline{q},\nabla_{\underline{q}} S) = \bar{H}(\underline{P}) \quad . \tag{15}$$

The condition (15) is satisfied for any canonical transformation cyclic in $\underline{Q}$, however, the generating function for the transformation (13) is generally multi-valued. The action-angle representation $(\underline{Q},\underline{P}) \equiv (\underline{\Theta},\underline{J})$ is a particular choice of variables defined such that

$$J_{\underline{k}} \equiv \frac{1}{2} \int_{\gamma_{\underline{k}}} \underline{p}(\underline{q},\underline{c})\cdot d\underline{q} \tag{16}$$

where $\gamma_{\underline{k}}$ is a complete circuit of the torus, i.e., $q_{\underline{k}} \to q_{\underline{k}}(0)$ in (13), [see, e.g., Berry[111] or Corben and Stehle[110], Chapter 11]. This attractive viewpoint of a mechanical system is reinforced since it is well-known that a solution to the Hamilton-Jacobi system of equations exists locally. Moser stresses that the existence of action-angle variables is a global, not a local question however. This point is the fatal flaw in Hamilton-Jacobi theory. An interesting example of a harmonic system which is locally stable, but globally unstable is given in Tabor[110] and also in Lindenberg, Seshadri and West[112].

We now apply Hamilton-Jacobi theory to the Hamiltonian given by (1). H_2 is again the Hamiltonian for the linear water wave field and V the nonlinear interaction among the gravity waves. For an integrable system the unperturbed Hamiltonian can be expressed as a function of the action variables only, however, for (1) we have

$$H(\underline{J},\underline{\Theta}) = H_2(\underline{J}) + \lambda V(\underline{J},\underline{\Theta}) \quad . \tag{17}$$

In this representation $\underline{J}$ and $\underline{\Theta}$ are still canonical variables, but they are no longer action-angle variables since the total Hamiltonian is *not* independent of $\underline{\Theta}$.

We assume the $(\underline{q},\underline{p})$ coordinates in (1) can be expressed in terms of the $(\underline{J},\underline{\Theta})$ coordinates of (17) by the Fourier expansions

$$\underline{q} = \sum_{\underline{m}} \underline{q}_{\underline{m}}(\underline{J}) e^{i\underline{m}\cdot\underline{\Theta}} \quad ; \quad \underline{p} = \sum_{\underline{m}} \underline{p}_{\underline{m}}(\underline{J}) e^{i\underline{m}\cdot\underline{\Theta}} \tag{18}$$

where $\underline{m}$ is an integer vector with positive and negative entries. We note that for the present system to be integrable there must exist a canonical transformation to action-angle variables $(\overline{\underline{J}},\overline{\underline{\Theta}})$ such that

$$H(\underline{J},\underline{\Theta}) = \overline{H}(\overline{\underline{J}}) \quad . \tag{19}$$

Hamilton-Jacobi theory prescribes this canonical transformation by means of the generating function $S(\underline{\theta},\underline{\bar{J}})$ from (13) to be

$$\underline{\bar{\theta}} = \nabla_{\underline{\bar{J}}} S(\underline{\theta},\underline{\bar{J}}) \quad ; \quad \underline{J} = \nabla_{\underline{\theta}} S(\underline{\theta},\underline{\bar{J}}) \tag{20}$$

which must therefore be solved in terms of the perturbation parameter λ in (17).

The lowest order solution to the Hamilton-Jacobi equations must satisfy the system with $\lambda = 0$, so that $\underline{\theta} = \underline{\bar{\theta}}$ and $\underline{J} = \underline{\bar{J}}$. Therefore, we write the generating function as the series expansion assumed to be analytic at $\lambda = 0$,

$$S(\underline{\theta},\underline{\bar{J}}) = \underline{\theta}\cdot\underline{\bar{J}} + \lambda S_1(\underline{\theta},\underline{\bar{J}}) + \lambda^2 S_2(\underline{\theta},\underline{\bar{J}}) + \ldots \tag{21}$$

and which reduces to the identity transformation when $\lambda = 0$. S_1 is the generator of the first order canonical transformation, S_2 the second order canonical transformation, etc. To use this expansion in the expression for the Hamiltonian (17) we use the Hamilton-Jacobi equations (20) to write

$$\bar{H}(\underline{\bar{J}}) = H_2(\underline{\bar{J}} + \lambda\nabla_{\underline{\theta}} S_1 + \lambda^2\nabla_{\underline{\theta}} S_2 + \ldots) + \lambda V(\underline{\bar{J}} + \lambda\nabla_{\underline{\theta}} S_1 + \lambda^2\nabla_{\underline{\theta}} S_2 + \ldots) \quad . \tag{22}$$

Expanding (22) about the action $\underline{\bar{J}}$ we obtain

$$\bar{H}(\underline{\bar{J}}) = H_2(\underline{\bar{J}}) + \lambda[\nabla_{\underline{\theta}} S_1\cdot\nabla_{\underline{\bar{J}}} H_2 + V(\underline{\bar{J}},\underline{\theta})] + \mathcal{O}(\lambda^2) \quad . \tag{23}$$

For the $\lambda = 0$ system we know that the $\underline{J}$ gradient of H_2 yields the constant frequency vector $\underline{\omega}_0$. In the new representation we have the new frequency column vector

$$\nabla_{\underline{\bar{J}}} H_2 = \underline{\Omega}(\underline{\bar{J}}) \tag{24}$$

independent of $\bar{\underline{\theta}}$. The $\underline{\theta}$ gradient in (23) is removed by Fourier expanding the generating function S_1 and the perturbation Hamiltonian V as

$$S_1(\underline{\theta},\underline{\bar{J}}) = \sum_{\underline{m}\neq\underline{0}} S_{1\underline{m}}(\underline{\bar{J}})e^{i\underline{m}\cdot\underline{\theta}} \quad ; \quad V(\underline{\theta},\underline{\bar{J}}) = \sum_{\underline{m}} V_{\underline{m}}(\underline{\bar{J}})e^{i\underline{m}\cdot\underline{\theta}} \quad . \tag{25}$$

The $\underline{m} = \underline{0}$ (constant term) of the generating function has arbitrarily been set equal to zero.

The Hamiltonian (23) can now be written

$$\bar{H}(\underline{\bar{J}}) = H_2(\underline{\bar{J}}) + \lambda[i \sum_{m\neq 0} \underline{m}\cdot\underline{\Omega}(\underline{\bar{J}})S_{1\underline{m}}(\underline{\bar{J}})e^{i\underline{m}\cdot\underline{\theta}} + \sum_{\underline{m}} V_{\underline{m}}(\underline{\bar{J}})e^{i\underline{m}\cdot\underline{\theta}}] + O(\lambda^2) \quad . \tag{26}$$

Equating terms with the same phase variation yields

$$\bar{H}(\underline{\bar{J}}) = H_2(\underline{\bar{J}}) + \lambda V_0(\underline{\bar{J}}) + O(\lambda^2) \quad ; \quad \underline{m} = 0 \tag{27}$$

$$S_{1\underline{m}}(\underline{\bar{J}}) = i \frac{V_{\underline{m}}(\underline{\bar{J}})}{\underline{m}\cdot\underline{\Omega}(\underline{\bar{J}})} \quad ; \quad \underline{m} \neq 0 \quad . \tag{28}$$

The generating function for the new action-angle variables to first order in the interaction parameter is therefore

$$S(\underline{\theta},\underline{\bar{J}}) = \underline{\theta}\cdot\underline{\bar{J}} + i\lambda \sum_{\underline{m}\neq\underline{0}} \frac{V_{\underline{m}}(\underline{\bar{J}})}{\underline{m}\cdot\underline{\Omega}(\underline{\bar{J}})} e^{i\underline{m}\cdot\underline{\theta}} + O(\lambda^2) \quad . \tag{29}$$

We have the Hamiltonian (27) solely in terms of the action $\underline{\bar{J}}$ and the generator of the transforamtion (29) in terms of the perturbation Fourier components. The generating function can in principle be obtained to any order in λ by repeating the above procedure. The variables $(\bar{\underline{\theta}},\bar{\underline{J}})$ are the action-angle variables to $O(\lambda^2)$.

The problem is not solved however. The denominator of the correction term in (29) can vanish

$$\underline{m}\cdot\underline{\Omega} = 0 \tag{30}$$

giving rise to an infinite term, i.e., a divergence, in the series (21). If the linear eigenfrequencies are commensurable, i.e., *all* the frequencies are a multiple of some fundamental frequency, then terms of the form (30) exist. Lower order resonances of (30) can also occur. For example, if N-1 of the frequencies are commensurable but incommensurable with the N*th* say; then for $\underline{m} = (m_1, m_2, \ldots, m_{N-1}, 0)$ a resonance still occurs, etc. Such terms are explicitly excluded from analyses in most standard texts on classical dynamics, e.g. Goldstein[113] and Corben and Stehle[110]. For completely incommensurable frequencies the situation does not improve, since it is always possible to represent an irrational number to any desired degree of accuracy by the ratio of two integers (a rational number), i.e., there are values of $\underline{m}$ which make the denominator in (29) arbitrarily small.

The above divergence was pointed out by Brillouin[114] as one of the outstanding problems in physics. An example due to him is given in Figure 5.1 for a system with the two unperturbed frequencies ω_1 and ω_2. In the (ω_1, ω_2) plane the diagonal element corresponds to the condition (30) for continuous $\underline{m}$. A singularity occurs when this diagonal crosses a lattice site with integer values of m_1 and m_2. To avoid a divergence, the Fourier coefficient of the perturbation (V_{m_1, m_2}) must vanish when this happens. This condition obtains when the variables of the problem are separable, so that the system is integrable and only the matrix elements of the perturbation along the horizontal ($m_1 = 0$) and vertical ($m_2 = 0$) lines are non-zero and the divergence does not occur.

As we increase the range of values of $\underline{m}$ in (29) we encompass ever smaller terms $\underline{m}\cdot\underline{\Omega}$ in the sum. This places in doubt the whole question of the convergence of (29) in either the parameter λ or the individual sums on $\underline{m}$. This is in essence the *problem of small divisors* that fills the celestial mechanics literature. This problem does not arise in systems with one degree of freedom.

A theorem due to Kolmogorov,[115] and Arnold[116] and Moser[117] (the KAM theorem) establish rigorously that for linearly independent frequencies, i.e., $\underline{m}\cdot\underline{\Omega} \neq 0$ in (30) the dynamic system is integrable. In the interaction Hamiltonian V there are terms which satisfy this condition so they would not be expected to modify the dynamics of the system significantly. On the other hand, these are also terms for which $\underline{m}\cdot\underline{\Omega} = 0$, i.e., resonant terms, and these are expected to destroy the integrability conditions of the KAM theorem and thereby lead to much different dynamical behavior. It is by means of these resonant terms that energy is transferred from one degree of freedom to another in a complicated system. It will therefore be assumed throughout this work that in deterministic calculations the resonant interactions in the water wave field are dominant. The non-resonant interactions on the other hand can always be removed by means of an appropriate canonical transformation. Hamilton-Jacobi theory, which is valid for these non-resonant terms is just one of a number of techniques we will use in subsequent discussions to treat the non-resonant interactions of deep water waves.

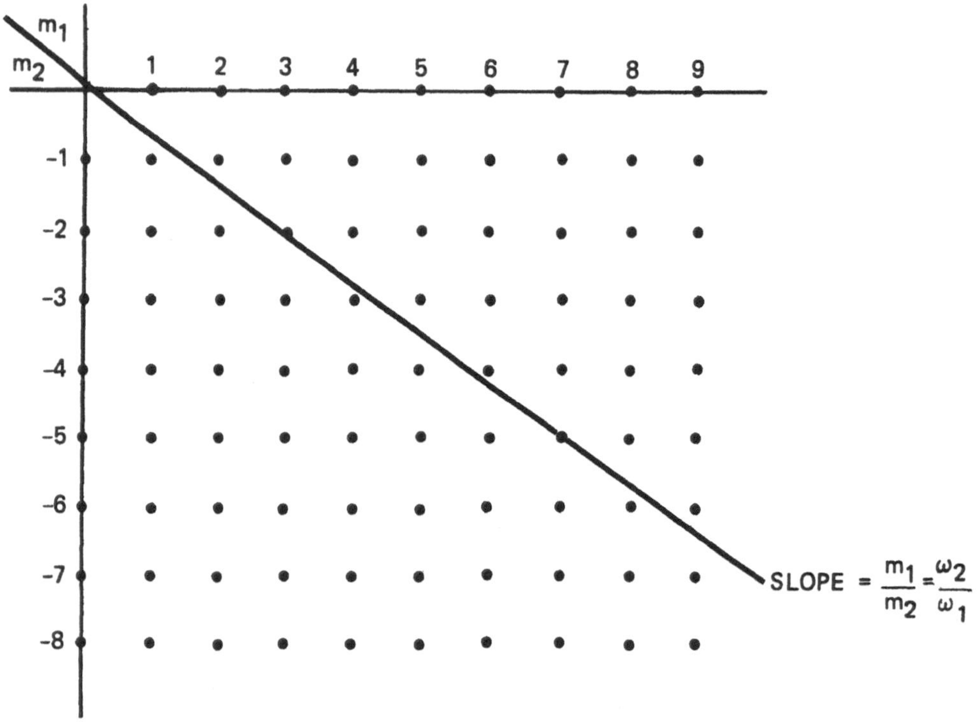

Figure 5.1 A two dimensional lattice with indices m_1 and m_2 is indicated. The straight line corresponds to continuous frequency values. The intersection of this line with a grid point indicates the ratio of two frequencies given by the ratio of two integers.

6. Resonances Among Gravity Waves

In the preceding lecture we concluded that nonlinear resonances in general mechanical systems describable by a Hamiltonian markedly alter the dynamic behavior of the system. In a field of nonlinear water waves the dispersion relation requires a matching in both wave vectors and frequencies for such a resonance to occur.[26-28] Such resonant interactions are presumed to dominate the dynamics of the wave field as was first demonstrated in the wave tank experiments of Longuet-Higgins and Smith[38] and also of McGoldrick *et al.*[39] In the analysis of Phillips[26] and Longuet-Higgins[28] it was shown that a mode not initially present in the system could be generated by a four wave resonance, i.e., the $T^{(3)}$ term in (4.16). They also showed that the dispersion relation for gravity waves prohibits resonances involving three waves, so that the effect of the $T^{(2)}$ term in (4.16) is vitiated. The newly excited mode was shown to grow linearly with time, i.e., to be proportional to t. Although this is true of the initial growth, as it must be for a resonance phenomenon, it was Benney[27] who showed how this growth is quenched by the nonlinear interactions. If the growth of this excited mode were not quenched, the amplitude would diverge in time and become an infinite source of energy. The growth of the new mode modifies the amplitudes of the existing modes and this change in turn modifies the growth of the new mode, forming a feedback system. The stabilizing property of nonlinear terms is a generic property of weak nonlinear interactions.[34]

Experimental observations of these resonantly generated modes were made by Longuet-Higgins and Smith and the research group at Johns Hopkins University as described by Phillips.[118] The experiment was conducted with three mechanically generated waves selected so as to resonantly generate a fourth wave.

To first order the $T^{(2)}$ terms in (4.16) do not contribute to the dynamics and with two of the wave vectors set equal, say $\underline{k}_1 = \underline{k}_2$, we have the resonant quartet of waves

$$2\underline{k}_1 = \underline{k}_3 + \underline{k}_4 \quad ,$$

$$2\omega_1 = \omega_3 + \omega_4 \quad . \tag{1}$$

The experiments were configured with the $\underline{k}_1$ and $\underline{k}_3$ waves being mechanically generated along adjacent walls of a square tank, so that $\underline{k}_1$ and $\underline{k}_3$ were perpendicular to each other. The third member of this set, $\underline{k}_4$, was then generated by the resonance mechanism and its linear growth in time was observed as a linear growth in distance of propagation along the tank. We will briefly describe the results of these experiments after we have discussed some further properties of the wave field.

In Figure (6.1) the dispersion relation for gravity waves is used to construct the dispersion curve in (k,ω) space. In two horizontal spatial dimensions the dispersion relation is

$$\omega_k = \sqrt{gk}\ ; \qquad k = \sqrt{k_x^2 + k_y^2} \tag{2}$$

and maps out a surface in the three dimensional $(\underline{k},\omega)$ space. This latter surface is defined by rotating the dispersion curve in Figure (6.1) about the frequency axis and appears as a funnel or trumpet shaped figure. Weber and Barrick[106,107] show by means of expanding the frequency in a perturbation series in addition to the perturbation expansions of the free surface and velocity potential that "forced" waves exist off the linear dispersion surface. They obtain corrections to the frequency which are quadratic in the first order surface displacements thereby yielding waves throughout the $(\underline{k},\omega)$ space.

These latter waves are not free waves, however, but are forced by the non-linear interactions among the free waves and cannot exist independently of the free waves. These forced waves cannot transfer energy off the linear dispersion curve. Their existence has been noted by a number of investigators[119-121] all of whom were concerned with the determination of this effect from field observations of wind generated gravity waves.

A standard technique for determining this effect is to measure the phase velocity of a given wavelength wave as a function of frequency for a given wind speed. The phase velocity should change as the ambient wave amplitudes increase with fetch thereby modifying the frequency and consequently the phase speed of the wave. This effect has been observed in tank experiments on wind generated deep water waves by Ramamonjiarisoa and Coantic[122] and others[123,124] in which the fetch-limited phase speed of wind generated waves becomes independent of the wave frequency, i.e., it approaches a constant value dependent only on the dominant frequency ω_0 as determined by the wind speed, but not the frequency of the measured wave. Phillips[5] has suggested that this lack of dependence of the phase velocity on frequency (non-dispersive behavior) indicates that free waves are not a significant contributor to the measured waves in this region. The non-dispersive behavior is, therefore, associated with the forced waves above and was first observed in the classical Stokes wave in which the harmonics of the fundamental wave all travel at the phase speed of the fundamental. Komen's[125] quantitative estimates indicate that this is a proper interpretation of the observed secondary spectral peak in wind wave data at the second harmonic of the spectral peak $\omega_k \cong 2\omega_0$. An alternative explanation has been given by Longuet-Higgins[126] in terms of two free waves traveling symmetrically with respect to the wind direction and where the angle they make with this direction is frequency dependent. In any event there exists this controversy on the interpretation of the modification in the phase velocity for which Komen[127] proposes an experimental resolution determined by a measurement of the cross-correlation function using two wave probes.

All free waves in the water field are tied to the dispersion surface. A resonance among four free gravity waves is achieved by considering a constant vector $\underline{K}$ and scalar Ω, such that the difference between two points on the dispersion surface $(\underline{k}_1,\omega_1)$ and $(\underline{k}_3,\omega_3)$ is given by

$$(\underline{k}_1,\omega_1) - (\underline{k}_3,\omega_3) = (\underline{K},\Omega) \quad . \tag{3}$$

The two additional vectors $(\underline{k}_2,\omega_2)$ and $(\underline{k}_4,\omega_4)$ form a resonant quartet with these vectors if they trace out curves with the terminal points of the vector $(\underline{K},\Omega)$ being their origin.

In Figure (6.1) the terminal points of the vector $(\underline{k}_1,\omega_1)$ and $(\underline{k}_3,\omega_3)$ define the points A and B, respectively. The vector $(\underline{k}_2,\omega_2)$ eminates from A and $(\underline{k}_4,\omega_4)$ from B since they must satisfy the dispersion relation, i.e.,

$$(\underline{k}_4,\omega_4) - (\underline{k}_2,\omega_2) = (\underline{K},\Omega) \quad . \tag{4}$$

Since both $\underline{k}_2$ and $\underline{k}_4$ are constrained to lie on the dispersion surface their lengths are not arbitrary. Phillips[26] finds that using the points A and B as focii the two vectors $\underline{k}_2$ and $\underline{k}_4$ trace out the figure-eight curves indicated in Figure (6.2).

The wavevector resonance condition is an exact result from the restriction on the spatial Fourier transforms used to construct $T^{(3)}$ in (4.16). The frequency resonance condition is less rigorous but the physical arguments are compelling. If the frequencies are mismatched by an amount, $\Delta\omega$ say, there is an oscillatory exchange of energy among the modes as discussed in previous lectures. By integrating the equations of motion (4.16) over a time long compared to the mismatch period $(2\pi/\Delta\omega)$ the net transfer of energy by such terms will vanish. It is only when $\Delta\omega \approx 0$ that the possibility of resonance and energy transfer among the modes exists. Exactly how near zero $\Delta\omega$ must be for this to occur is an open question. Studies, such as those made by Chirikov[34] attempt to determine the *width* of such resonances in terms of a measure of the strength of the nonlinearity. The concept of a resonance width $(\Delta\omega)$ will be used in discussing the elimination of nonresonant terms in the equation of evolution and in the numerical integration of their equations.

Following Phillips[118] we consider a set of four interacting wave trains which satisfy the resonance conditions;

$$\underline{k}_1 + \underline{k}_2 = \underline{k}_3 + \underline{k}_4$$

$$\omega_1 + \omega_2 = \omega_3 + \omega_4 \quad . \tag{5}$$

The equations of motion (4.16) are expressed in simple form by using the mode amplitude with the rapid time dependence removed, i.e.,

$$b_{\underline{k}}(t) = e^{-i\omega_k t} B_{\underline{k}}(t) \tag{6}$$

as

$$\dot{b}_1 = i(\Gamma_{11}|b_1|^2 + \Gamma_{12}|b_2|^2 + \Gamma_{13}|b_3|^2 + \Gamma_{14}|b_4|^2)\, b_1 + i\omega_1\Gamma b_2^* b_3 b_4$$

$$\dot{b}_2 = i(\Gamma_{21}|b_1|^2 + \Gamma_{22}|b_2|^2 + \Gamma_{23}|b_3|^2 + \Gamma_{24}|b_4|^2)\, b_2 + i\omega_2\Gamma b_1^* b_3 b_4$$

$$\dot{b}_3 = i(\Gamma_{31}|b_1|^2 + \Gamma_{32}|b_2|^2 + \Gamma_{33}|b_3|^2 + \Gamma_{34}|b_4|^2)\, b_3 + i\omega_3\Gamma b_1 b_2 b_4^*$$

$$\dot{b}_4 = i(\Gamma_{41}|b_1|^2 + \Gamma_{42}|b_2|^2 + \Gamma_{43}|b_3|^2 + \Gamma_{44}|b_4|^2)\, b_4 + i\omega_4\Gamma b_1 b_2 b_3^* \qquad (7)$$

where

$$b_j \equiv b_{\underline{k}_j}(t)\,\sqrt{\omega_j} \quad ; \qquad \omega_j\Gamma_{ij} \equiv \frac{2}{1+\delta_{ij}}\,\Gamma^{\underline{k}_i\underline{k}_j}_{\underline{k}_i\underline{k}_j} \; ; \; i,j = 1,\, 2,\, 3,\, 4 \qquad (8)$$

and

$$\gamma\Gamma = \Gamma^{\underline{k}_1\underline{k}_2}_{\underline{k}_3\underline{k}_4} = \Gamma^{\underline{k}_2\underline{k}_1}_{\underline{k}_3\underline{k}_4} = \Gamma^{\underline{k}_3\underline{k}_4}_{\underline{k}_1\underline{k}_2} = \Gamma^{\underline{k}_4\underline{k}_3}_{\underline{k}_1\underline{k}_2} \quad ; \; \gamma = [\omega_1\omega_2\omega_3\omega_4]^{1/2} \qquad (9)$$

The coupling coefficients Γ are real constants and depend on the configuration of the four vectors as is indicated by (4.18). These four waves constitute a closed interacting system; energy is exchanged among the members of the set, but none is coupled to waves outside the set.

The reality of Γ implies that the leading effect of the nonlinear interaction is a phase shift. To demonstrate this we rewrite (7), neglecting the term coupling the three different modes, as

$$\dot{b}_\ell = i\,\Gamma_\ell|b_\ell|^2\, b_\ell + i\sum_{m(\neq\ell)} \Gamma_{m\ell}|b_m|^2 b_\ell \quad ; \quad \ell,m = 1,2,3,4 \qquad (10)$$

here

$$\Gamma_\ell \equiv \Gamma^{\ell\,\ell}_{\underline{\ell}\,\underline{\ell}} = -\ \ell^3/\omega_\ell\, V_\ell \quad . \tag{11}$$

An integration of (10), assuming that the quadratic term in the sum is constant in the time interval t, is* $A_\ell = b_\ell{}^{(t)}\,[2/V_\ell]^{1/2}$

$$A_\ell(t) = A_\ell(o)\ \exp\left[i(\Gamma_\ell|A_\ell(o)|^2 + \sum_{m(\neq\ell)} \Gamma_{m\ell}|A_m(o)|^2)t\ \omega_\ell\ V_\ell^2/2\right] \tag{12}$$

where $b_\ell(o)$ is the constant initial value of the *ℓ-th* mode amplitude. Writing the mode amplitude in terms of a modulus $R_\ell(t)$ and phase $\Theta_\ell(t)$,

$$A_\ell(t) = R_\ell(t)\ e^{i\Theta_\ell(t)} \tag{13}$$

and comparing with (7), we see that the modulus is constant to first order and the phase has a linear time dependence. This, of course, modifies the linear wave dispersion relation and therefore, the phase and group velocities of the *ℓth* mode.

The dispersion relation (Ω_ℓ) obtained by neglecting the $m\neq\ell$ terms in (12) and using (11) is

$$\Omega_\ell = \omega_\ell\left(1 + \frac{1}{2}\,\ell^2|A_\ell(o)|^2\right) \tag{14}$$

which is just the dispersion relation for a Stokes wave to second order. The first nonlinear correction is therefore, the self interaction of a given mode and results in the phase speed $V_\ell \equiv \omega_\ell/\ell$ being modified by the nonlinear interaction

$$\dot{\Theta}_\ell \cong V_\ell\left(1 + \frac{1}{2}\,\ell^2|A_\ell(o)|^2\right) \quad . \tag{15}$$

*This argument can be made rigorous by introducing an expansion parameter and doing a two time scale calculation, but the same lowest order result will obtain.

The off-diagonal interactions are those which couple a given mode ℓ to the other modes in the wave field and also modifies the phase velocity, i.e.,

$$\dot{\Theta}_\ell \cong V_\ell \left[1 + \frac{1}{2} \ell^2 \, |A_\ell(o)|^2 + \sum_{m(\neq\ell)} \Gamma'_{\ell m} |A_m(o)|^2 \right] . \tag{16}$$

Note that the modification of the phase velocity is a quadratic function of the modulus of the mode amplitudes; the correction is therefore proportional to the energy. In the continuum limit Watson and West[103] obtain such a correction for a test wave in the presence of an ambient spectrum.

Let us now return to the experiments mentioned earlier[38,39] in which interaction among these four waves can be described, following Phillips,[118] using (4.16) restricted to this four wave situation. We choose an amplitude a_1 which is the sum of b_1 and b_2 with the latter two waves differing in phase by π, i.e., $b_2 = i\, b_1$. In the experiment essentially all the energy was initailly concentrated in the a_1 and b_3 waves, so that from (7) we have

$$\frac{db_4}{dt} \cong \frac{i}{2} \omega_4 \Gamma \, b_3^* \, a_1^2 \tag{17}$$

since $a_1 = (1 + i)\, b_1 => b_1 b_2 = \frac{i}{2} a_1^2$. The time required for the exchange of energy due to the nonlinear interaction is fairly long on the scale of the wave frequencies, so that for most of the experiment the amplitudes b_3 and a_1 are constant and (17) can be integrated to yield the expression linear in time

$$b_4 \cong \frac{i}{2} \Gamma \, \omega_4 b_3^* a_1^2 t \quad . \tag{18}$$

Replacing the time by the ratio of the distance traveled by the $\underline{k}_4$ wave (x) and its group velocity ($\omega_4/2k_4$), (18) becomes

$$b_4 \cong i\, k_4 \Gamma\, b_3^* a_1^2\, x \quad . \tag{19}$$

Both the experiments of Longuet-Higgins and Smith[38] and McGoldrick et al[39] confirm the prediction of the generation of the $\underline{k}_4$-mode as indicated in Figure (6.3) where the linear growth of this new mode with distance is indicated. See Phillips[5,118] for an overview of these experiments. Now Longuet-Higgins'[28] calculations show that with the waves $\underline{k}_1$ and $\underline{k}_3$ at right angles with a frequency ratio of ω_1/ω_3 = 1.736 the coupling coefficient is $i\, k_4\Gamma/k_3 k_1^{\,2}$ = 0.442 and the experimentally measured values are within 20% of this value.

It is clear that the diagonal terms in the set of equations (7) contribute to the phase shift, the remaining term, that involving the three other waves in the system determines the actual interchange of energy among the modes. This interaction provides the resonant growth mechanism given in (7). Benney[27] showed that this resonance is only "apparent" in that although it provides the initial growth rate of a newly generated mode, it gives no indication of the long time effect of the interaction on for example, the amplitudes of the other modes or the level to which the new mode will grow. He showed that the closed system indicated by (7) in fact admits to a number of simple integrals. We can show how the growth of each mode is bounded by multiplying the members of (7) by the appropriate complex conjugate amplitude and adding each such member to its complex conjugate to obtain the rate equations

$$\frac{dE_\ell}{dt} = i(P-P^*)\,\omega_\ell + i\,(Q_\ell - Q_\ell^*) \tag{20}$$

Here E_ℓ is the "energy density" in the ℓ-*th* mode defined by

$$E_\ell = \frac{1}{2} |A_\ell|^2 \quad ; \tag{21}$$

Q_ℓ is defined by

$$Q_\ell \equiv E_\ell \sum_{m=1}^{4} \Gamma_{\ell m} E_m \tag{22}$$

and

$$P \equiv \tfrac{1}{4}\Gamma \, (V_1 \; V_2 \; V_3 \; V_4)^{\frac{1}{2}} A_1{}^* A_2{}^* A_3 A_4 \tag{23}$$

Now Q_ℓ is a real quantity since the coupling coefficient $\Gamma_{\ell m}$ is real, so that (20) reduces to

$$\frac{dE_1}{dt} = i \, (P-P^*)\omega_1 \tag{24}$$

It is only the imaginary part of the P term which changes the energy in the first mode. We obtain a similar equation for the rate of change in the energy of mode $\ell = 2$, so that

$$\frac{d}{dt}\left(\frac{E_1}{\omega_1} - \frac{E_2}{\omega_2}\right) = 0 \tag{25}$$

which integrates to

$$\frac{|A_1(t)|^2}{\omega_1} - \frac{|A_2(t)|^2}{\omega_2} = \text{constant} \quad . \tag{26a}$$

In a similar way we have

$$\frac{dE_3}{dt} = - \quad i\ (P-P^*)\omega_3 \tag{27}$$

resulting in the further integrals

$$\frac{|A_1(t)|^2}{\omega_1} + \frac{|A_3(t)|^2}{\omega_3} = \text{constant} \quad , \tag{26b}$$

$$\frac{|A_1(t)|^2}{\omega_1} + \frac{|A_4(t)|^2}{\omega_4} = \text{constant} \quad . \tag{26c}$$

Note that the ratio of the modulus of the mode amplitude squared divided by its frequency is the *action* of that mode.

The equations in the set (26) specify the sharing of the total action among the four modes in this closed system. The growth of action in a particular mode must be compensated for by the depletion in the action of another mode, such that the conservation of action in the system (26) is maintained. The nonlinear interaction does not merely modify the phase velocity of a finite amplitude wave, but also induces changes in the mode amplitudes. One pair of amplitudes losing energy at rates determined by their frequencies and the other pair gaining energy at rates determined by their frequencies.

If we now add the four terms of the form

$$\frac{dE_j}{dt} = \quad i\ (P-P^*)\omega_j\ s_j \quad ; \qquad j = 1,2,3,4 \tag{28}$$

where $s_j = \pm 1$, we obtain

$$\frac{d}{dt}\left(\sum_{j=1}^{4} E_j\right) = \left(\omega_1 + \omega_3 - \omega_3 - \omega_4\right) i(P-P^*) \tag{29}$$

which vanishes because of the resonance restriction on the frequencies. The total energy in the system is therefore conserved in addition to the total action.

Finally by introducing the wave *momentum density* by

$$\underline{M}_j = \hat{\underline{k}}_j E_j/V(k_j) \quad ; \qquad j = 1,2,3,4 \tag{30}$$

which by taking a time derivative, identifying $V(k_j)$ as the phase velocity of the mode $\underline{k}_j$, and using (26) becomes

$$\frac{d\underline{M}_j}{dt} = i\,(P-P^*)\underline{k}_j\, s_j \quad . \tag{31}$$

The total rate of change in momentum is therefore

$$\frac{d}{dt}\left(\sum_{j=1}^{4} \underline{M}_j\right) = \left(\underline{k}_1+\underline{k}_2-\underline{k}_3-\underline{k}_4\right) \; i(P-P^*) \tag{32}$$

which vanishes because of the resonance restriction on the wavevectors. The total momentum is therefore also conserved. Bretherton[128] has obtained an analytic solution for this conservative system in terms of elliptic functions.

Note that by assuming these are four wavetrains we have been able to freely commute the frequencies and wavevectors with the time derivatives. If instead we had been discussing wavepackets, the frequencies and wavevectors would not be constant, but rather functions of time and position and much more care would be required in the analysis. Ray path methods, especially Whitham's average Lagrangian method,[67] are more convenient to use in this latter case.

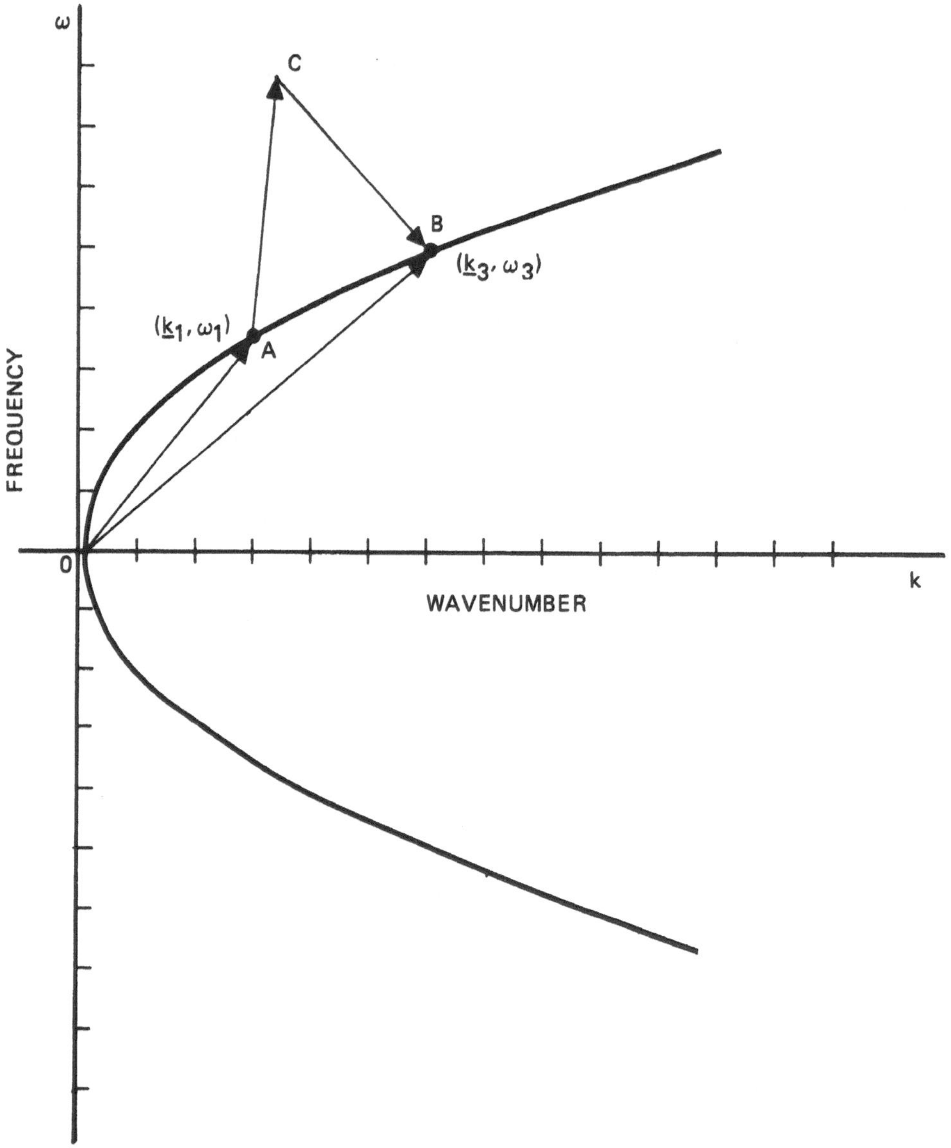

Figure 6.1: The dispersion relation $\omega = \sqrt{gk}$ is indicated by the solid curve. The scale is arbitrary. The two vectors emanating from the origin 0 define $\overline{OA} = (\underset{\sim}{k}_1, \omega_1)$ and $\overline{OB} = (\underset{\sim}{k}_3, \omega_3)$ and couple to a second pair of wave vectors $\overline{AC}$ and $\overline{CB}$ to form a resonant quartet.

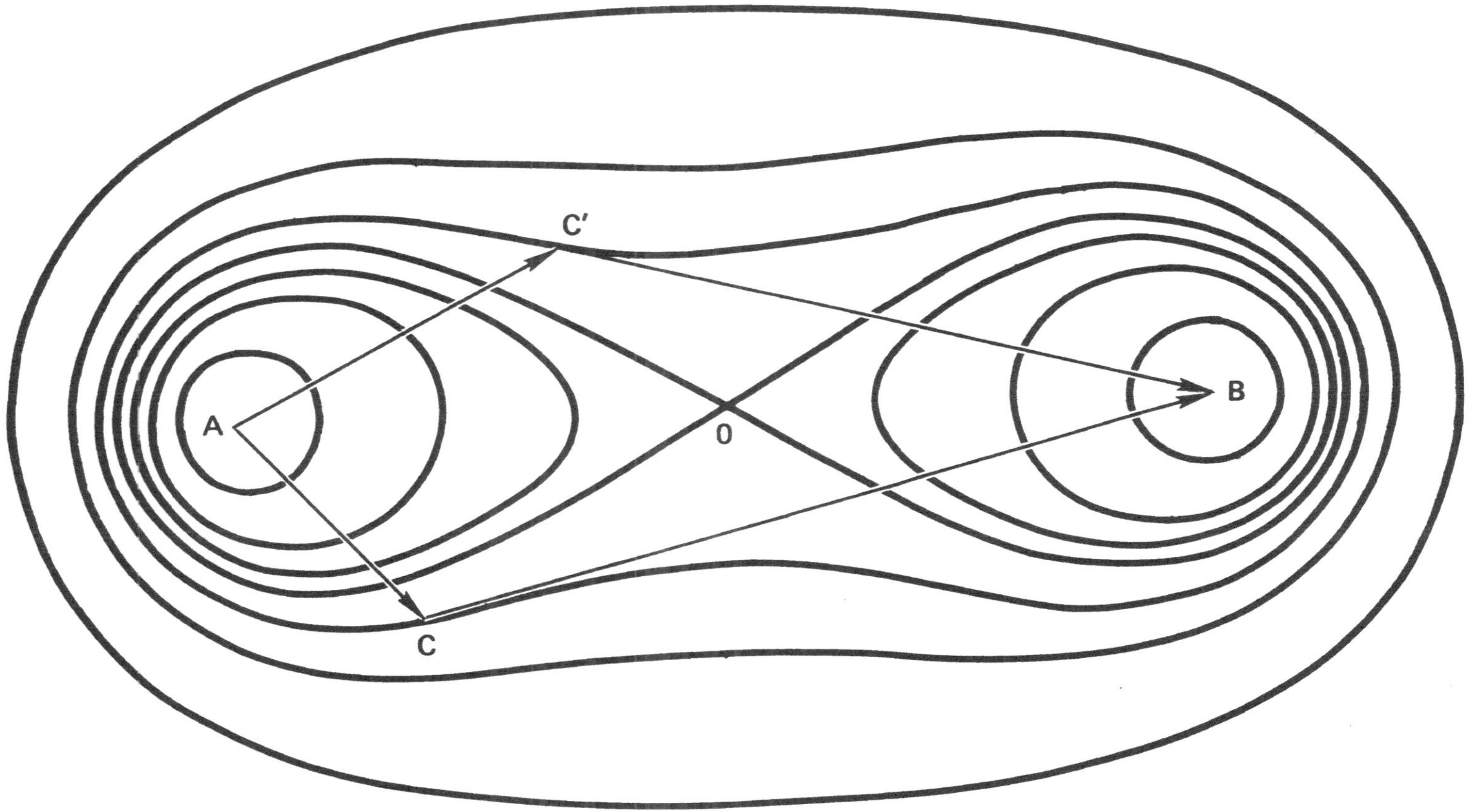

Figure 6.2 The family of curves in the wave number plane are defined by the resonance condition (5.). The scale is arbitrary. If C ,C' and any pair of points on one member of the family, the vectors $\overline{AC}$, $\overline{CB}$, $\overline{AC'}$ anc $\overline{C'B}$ form a resonant quartet.

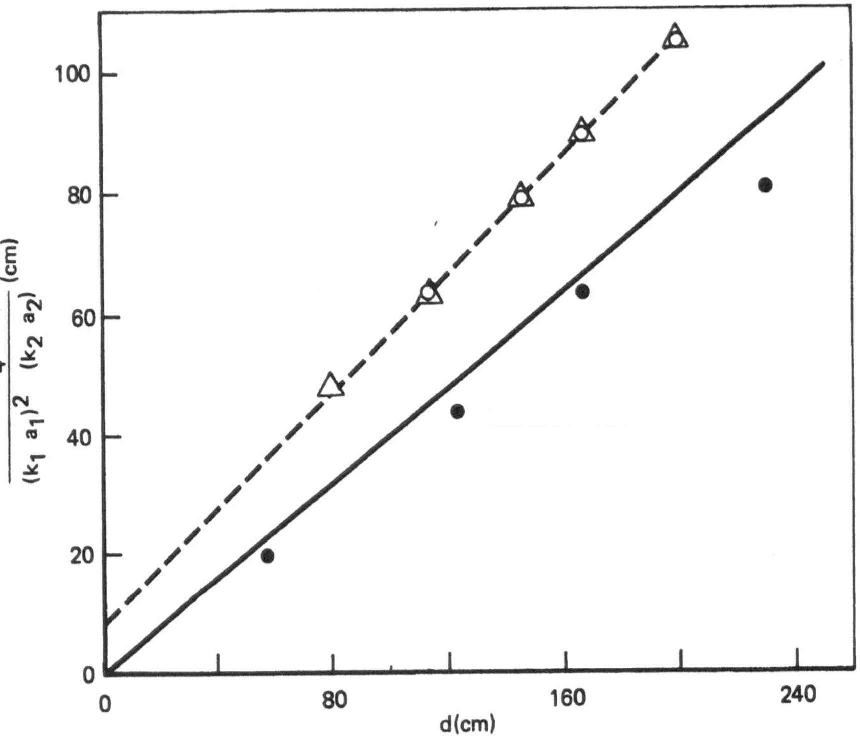

Figure 6.3 The growth of amplitude of the interaction product with distance d. The continuous line represents the predicted rate, A = 0.442d. The solid circles represent measurements by Longuet-Higgins and Smith. The open circles and triangles represent two different series of measurements by McGoldrick et al. in which the slopes of the primary waves were also varied.

7. Resonant Instabilities and the Nonlinear Schrödinger Equation

We have emphasized that *theoretically* the resonant interactions among gravity waves provide the dominant mechanism for energy transfer in the gravity wave field. In this section we examine the interaction of a few discrete gravity waves in various approximations and compare the analytic results with the direct numerical integration of (4.16). In particular, we discuss the instability of a Stokes wave as observed experimentally by Benjamin and Feir[40] and paraphrase their discussion of the instability mechanism. We also study the space-time evolution of a narrow spectrum of surface gravity waves which propagates as a stable wave form for a time period greater than the e-folding time of the Benjamin-Feir instability. Restricted analyses such as these are useful in outlining some of the general properties of the cubic nonlinearity in the mode rate equations.

In the experiment of Benjamin and Feir a finite amplitude gravity wave was mechanically generated and allowed to propagate on the fluid surface of a water tank. The spectral width of the generated wave was sufficiently narrow to consider it to be monochromatic. Low amplitude noise was injected into the wave and it was observed to lose its integrity at a distance far down the tank, i.e., the wave crests appeared randomized and could no longer be associated with a wavetrain. Analysis reveals that the noise modulates the carrier wave by means of a sideband instability. The nonlinear coupling of the noise to the sideband frequencies of the nonlinear carrier wave results in a preferential transfer of energy from the carrier wave to these sidebands. As the energy in the sidebands grows, the carrier wave is modulated more and more strongly until it appears to have lost its integrity. This interpretation of the experiment has persisted for over a decade and the analysis of the sideband instability is presented below. The more recent experiments of Yuen and Lake[68] and collaborators[46], however, indicate that this is not a true instability. Their experiments reveal that the apparent loss of integrity is reversed at an even greater distance along the tank so that the original state of the system is recaptured (almost!).[129] The analysis of their results is given later.

A direct calculation of the Benjamin-Feir instability can be made using the mode rate equation (4.16) in one dimension to determine how the energy initially confined to a central mode k_0 is redistributed to other modes in the wave field. We assume the wavenumbers to be spaced in increments Δk with the central mode given by $k_0 = N\Delta k$ and since $N >> 1$, N is taken to be an integer. As a first approximation to the energy transfer determined by (4.16) we neglect the nonresonant interactions, i.e., we set $T_{\underline{k}}^{(2)}(t)$ equal to zero and assume that only the nearest neighbors of k_0 are coupled to it. This approximation will be improved upon later. The approximate solution is then compared with the exact numerical solution to develop a feeling for how the more distant modes enter the interaction process.

We introduce the slope variable q_n which is indexed by the discrete mode number n as a deviation from the central mode number N such that

$$q_0(t) \equiv k_0 B_{k_0}(t) = N\Delta k B_{N\Delta k}(t) \tag{1}$$

and the approximate mode rate equations are given by

$$\dot{q}_1 \cong \frac{-i}{2}\omega_0[|q_0|^2 q_1 + q_0^2 q_{-1}^*]$$

$$\dot{q}_0 \cong 0$$

$$\dot{q}_{-1} \cong -\frac{i}{2}\omega_0 [|q_0|^2 q_{-1} + q_0^2 q_1^*] \tag{2}$$

where we have assumed the deviation in frequency from the central mode to be negligible, i.e., of $\mathcal{O}(n/N) = \mathcal{O}(\Delta k/K)$

$$\omega_0 \cong \omega_1 \cong \omega_{-1} \cong \sqrt{gN\Delta k} \ . \tag{3}$$

The effect of the frequency deviations, neglected here, will be discussed shortly. In (2) we assume that the central mode slope $q_0(t)$ remains unchanged throughout the time of the interaction with its sidebands $(N \pm 1)\Delta k$. It is interesting to see how long the system evolves as if this assumption were valid.

From (2) we construct the rate equations for the "energy" in each of the sidebands just as in the preceding lecture. We introduce $E_n \equiv 1/2 \ |q_n|^2/k_n^2$ for the energy in mode k_n. Multiplying each of the equations in (2) by the appropriate factor and adding each of the expressions to its complex conjugate we obtain the rate equations

$$k_1^2 \frac{d}{dt} E_1(t) = k_{-1}^2 \frac{d}{dt} E_{-1}(t) = -\omega_0 k_0^2 E_0 \mathrm{Im}[q_1(t) q_{-1}(t)] \ . \tag{4}$$

The initial value problem for which we solve (4) is specified by the initial conditions $q_1(0) = q_{-1}(0) = q$ with q a real constant and $q_0(0) = q_0^*(0)$. The energy in each of the sidebands is then given by

$$k_1^2 E_1(t) = k_{-1}^2 E_{-1}(t) = q^2 \cosh\,[2\omega_0 k_0^2 E_0 t] \tag{5}$$

and the growth in the sideband amplitudes by

$$q_1(t) = q_{-1}(t) = q\ \{\cosh\,[\omega_0 k_0^2 E_0 t] - i\ \sinh\,[\omega_0 k_0^2 E_0 t]\} \ . \tag{6}$$

The solutions (5) and (6) clearly indicate an unbounded growth of the sidebands in a time interval determined by the initial energy of the central modes.

Obviously the growth of the sidebands will be limited by both the energy transferred to other modes in the wave field which have not been included in this analysis and by the depletion of energy from the central mode. The solutions (5) and (6) diverge because the central mode provides an infinite source of energy.

In Figure 7.1 we compare the solution (5) with the numerical integration of (4.16). In this test calculation we have taken the central mode to be k_0 = .02863 cm^{-1} with a wavenumber interval $\Delta k = 1.463 \times 10^{-3}$ cm^{-1}. The slope of the primary mode was selected to be q_0 = .071 and $|q_{\pm 1}|$ = .014 for the secondary modes which corresponds to wave amplitudes of 2.5 cm for the primary and .45 cm for the greater and .55 cm for the lesser of the secondary modes. In the numerical calculation we use nine modes to see how rapidly energy is transferred from the central three modes to the remainder of the system. The initial slope of these other six modes is taken to be 10^{-5} so as to simulate noise.

In this figure we see that the approximation of constant q_0 is satisfactory for early times but is progressively worse as the energy in the remainder of the system builds up. The linear approximation to the energy density of the secondary modes $k_{\pm 1}$ is seen to agree very well with the numerical results for early times. Dramatic changes occur at late times, however, when the analytic linear solution begins to diverge while the non-linear calculational result remains finite. The numerical results indicate that the energy transfer from the central to the other modes in the system inhibits the divergence of the secondary sidebands. Precisely how energy builds up in the system is shown in the next figure.

In Figure 7.2 the magnitudes of the nine modes in the calculation are shown at four different times. The snapshot at t = 0 indicates the initial state of the calculation. At later times the energy appears to be draining in an organized way from the central to the more remote modes. The process resembles that of diffusion produced by a cascading of energy out of the central mode. Unlike diffusion, however, this process is in principle reversible. In the absence of dissipation this energy transfer can be reversed [see West, et al.[45] and Bryant[130]].

To simulate the Benjamin-Feir experiment we establish the initial conditions shown at time t = 0 in Figure 7.3. There are nine modes in the calculation, all of which have a slope of 10^{-3} except the primary mode which has the experimental value of $q_0 = .12$. The primary wavenumber is $k_0 = .02863\ cm^{-1}$ with four modes on either side spaced at intervals $\Delta k = 1.4365 \times 10^{-3}\ cm^{-1}$. From the slope snapshots in Figure 7.3 it is clear that the local process of energy diffusion observed in Figure 7.2 does *not* dominate in the present interaction. It is apparent that energy is being preferentially transferred to modes centered about $k_0 \pm 3\Delta k$ which is in essential agreement with the perturbation analysis of Benjamin and Feir[40]. Benjamin[41] shows that the frequencies at $\omega_{\pm} = (1 \pm q_0)\omega_0$ would be amplified from out of the background noise in a preferential manner. The numerical calculations of West, et al.[45] indicate that this amplification occurs for a band of frequencies centered around the mode given by such a perturbation analysis.

A narrow spectrum of gravity wave such as those calculated above is often more appropriately discussed in terms of an envelope function. The monochromatic gravity wave generated in the Benjamin-Feir experiment transfers energy to wavenumbers adjacent to that of the carrier wave (central mode) by the nonlinear interactions. The growth of these carrier sidebands is associated with the *break-up* of the initial Stokes wave as it propagates down the tank, i.e., the surface displacement becomes randomized and cannot be *visually* associated with a wavetrain. A more quantitative measure of the loss of integrity of the carrier wave is provided by the one dimensional envelope function G(x,t).

The complex surface displacement in one dimension is given by (4.6) so that removing the linear frequency using (5.3) we have

$$Z(x,t) = \sum_k \sqrt{2/V_k}\, b_k(t)\, e^{i(kx-\omega_k t)} \quad . \tag{7}$$

Equation (7) can be factored by again writing the wavenumber and frequency of the carrier wave as (k_0,ω_0) and assuming the wavenumbers are integer multiples of Δk, i.e.,

$$Z(x,t) = e^{i(k_0x-\omega_0t)} \sum_n \sqrt{2/V_n}\, b_n\, e^{in\Delta k(x-V_g^{(0)}t)} \quad , \tag{8}$$

where

$$V_g^{(0)} \equiv (\omega_n - \omega_0)/n\Delta k \tag{9}$$

is the unperturbed group velocity of the carrier wave and the spatial components are indexed by n = 0, ±1, ±2, ... The envelope function is defined by (8) as the coefficient of the carrier plane wave, i.e.,

$$G(x,t) = \sum_n \sqrt{2/V_n}\, b_n\, e^{in\Delta k\xi} \quad ; \qquad \xi = x - V_g^{(0)}t \ . \tag{10}$$

Thus a point of constant phase of the envelope function propagates with the group velocity of the carrier wave. In Figure 7.4 the *magnitude* of G(x,t) along the water surface is depicted. The modulation is on a stretch of surface

200 times the primary wavelength (λ_0 = 2.19 m). In this calculation the initial phases of the nine modes were selected at random to avoid biasing the conclusions. Although different realizations of the initial phases changes the detailed appearance of the surface, an initial *bump* in the envelope function maintains its identity as the surface distortion increases. The surface is of course being viewed in a coordinate system translating at the group velocity of the carrier wave (phase velocity of the envelope) so that we are always viewing the same phase points of the envelope function.

The envelope clearly suppresses waves at some locations and enhances waves at others, these points being determined by the initial phases. Visually this strong modulation of the carrier gives the impression that the surface consists of a group of wavepackets rather than a single strongly modulated wave. The coherent nonlinear interaction between a finite amplitude wave and "noise" on the ambient surface, therefore, leads to a "break-up" of the wave into apparently randomized wavepackets. The characteristic time for this break-up process is dependent on the initial conditions as can be seen from (5), i.e., the e-folding time of the perturbation is the inverse of the square of the initial carrier wave slope $\tau \sim 1/q_0^2\omega_0$. We have viewed this break-up as a distortion of the surface wave envelope function induced by the coherent diffusion of energy from the primary (carrier) wave to the other modes in the system. As energy accumulates in these additional modes the surface distortion increases markedly, but the *pattern* of the distortion remains the same. In Figure 7.4 the distortion translates and grows but does not change. This suggests that the velocity of the pattern, which is the group velocity of the carrier wave, is unaffected by the development of the other modes in the system. The newly excited modes *lock onto* the initial surface distortion pattern and create a pattern stationary in the moving frame of the envelope.[45]

The stationary property of the distortion pattern provides us with a piece of information which was not apparent in the experiment of Benjamin and Feir and which we did not have before. In effect, the numerical calculations have provided us with a new piece of data to explain. Why are the modulation patterns stationary in the translating coordinate system? To answer this question, we examine the stability characteristics of the envelope function.

To determine the space-time evolution of the envelope function we generalize (10) to two dimensions and write

$$G(\underline{x},t) = \sum_{\Delta\underline{k}(\underline{k}_0)} \sqrt{2/V_{|\underline{k}_0+\Delta\underline{k}|}}\; b_{\underline{k}_0+\Delta\underline{k}}(t)\, e^{i(\Delta\underline{k}\cdot\underline{x}+\Delta\omega_k t)} \tag{11}$$

where the notation $\Delta\underline{k}(\underline{k}_0)$ is used to indicate that we sum modes in a localized spectral region about $\underline{k}_0$, i.e., $\Delta k << k_0$. Rewriting the mode rate equations (4.16) including only the cubic nonlinearities (the effect of the nonresonant terms act here only to modify the interaction coefficient of the cubic terms) we have at *exact* resonance

$$\dot{b}_{\underline{k}}(t) = i \sum_{\underline{\ell},\underline{m},\underline{n}} \delta_{\underline{k}+\underline{n}-\underline{\ell}-\underline{m}} \Gamma^{\underline{k}\underline{n}}_{\underline{\ell}\underline{m}}\, b_{\underline{\ell}}(t) b_{\underline{m}}(t) b^*_{\underline{n}}(t) \tag{12}$$

i.e., $\omega_k + \omega_n - \omega_\ell - \omega_m = 0$. Now taking the time derivative of (11) we obtain

$$\frac{\partial G}{\partial t} = \sum_{\Delta\underline{k}(\underline{k}_0)} [i\Delta\omega_k b_{\underline{k}_0+\Delta\underline{k}} + \dot{b}_{\underline{k}_0+\Delta\underline{k}}]\, e^{i(\Delta\underline{k}\cdot\underline{x}+\Delta\omega_k t)} [2/V_{|\underline{k}_0+\Delta\underline{k}|}]^{1/2} \tag{13}$$

where for the time derivative of the mode amplitudes in (13) we will substitute the rate equation with the cubic nonlinearities (12).

The linear part of (13) is of the form $\omega_{k_0} - \omega_{|\underline{k}_0+\Delta\underline{k}|} = -\Delta\omega_k$. Expanding the frequency $\omega_{|\underline{k}_0+\Delta\underline{k}|}$ about ω_0, which is valid since $\Delta k << k_0$, and assuming that the carrier wave is propagating in the k_x direction, we obtain

$$\omega_0 - \omega_{|\underline{k}_0+\Delta\underline{k}|} = -\frac{1}{2}\Delta k_x \frac{\omega_0}{k_0} - \frac{1}{8}(\Delta k_x)^2 \frac{\omega_0}{k_0^2} + \frac{1}{4}(\Delta k_y)^2 \frac{\omega_0}{k_0^2} + \ldots . \qquad (14)$$

Also, we have used

$$\omega_0' = \frac{1}{2}\frac{\omega_0}{k_0} ; \qquad \omega_0'' = -\frac{1}{4}\frac{\omega_0}{k_0^2} \qquad (15)$$

by using the linear dispersion relation for gravity waves. The frequency difference in (14) can be replaced in the equation of motion (13) by the derivatives

$$\frac{\partial G}{\partial x} = \sum_{\Delta\underline{k}(\underline{k}_0)} i\Delta k_x\, b_{\underline{k}_0+\Delta\underline{k}}\, e^{i(\Delta\underline{k}\cdot\underline{x}+\Delta\omega_k t)} \left[2/V_{|\underline{k}_0+\Delta\underline{k}|}\right]^{\frac{1}{2}} \qquad (16)$$

$$\frac{\partial^2 G}{\partial x^2} = \sum_{\Delta\underline{k}(\underline{k}_0)} -(\Delta k_x)^2\, b_{\underline{k}_0+\Delta\underline{k}}\, e^{i(\Delta\underline{k}\cdot\underline{x}+\Delta\omega_k t)} \left[2/V_{|\underline{k}_0+\Delta\underline{k}|}\right]^{\frac{1}{2}} \qquad (17)$$

with similar expression for derivatives in the y direction. Substituting the expansion for the frequency difference (14) into the rate of change of the envelope function (13) and replacing the wave number products by the derivatives in (16) and (17) yields

$$i\left[\frac{\partial}{\partial t} + \omega_0' \frac{\partial}{\partial x}\right] G = \frac{\omega_0}{8k_0^2}\left[\frac{\partial^2 G}{\partial x^2} - 2\frac{\partial^2 G}{\partial y^2}\right]$$

$$- \sum_{\Delta\underline{k}(\underline{k}_0)} \sum_{\underline{\ell},\underline{m},\underline{n}} \Gamma_{\underline{\ell}\underline{m}}^{\underline{k}_0+\Delta\underline{k},\underline{n}}\, e^{i(\Delta\underline{k}\cdot\underline{x}+\Delta\omega_k t)}\, \delta_{\underline{k}_0+\Delta\underline{k}+\underline{n}-\underline{\ell}-\underline{m}}\, b_{\underline{\ell}} b_{\underline{m}} b_{\underline{n}}^* . \qquad (18)$$

The character of (18) is now determined by the coupling coefficient $\Gamma_{\underline{\ell}\underline{m}}^{\underline{k}\underline{n}}$.

The cubic term in (18) can be simplified by inverting the defining equation for $G(\underline{x},t)$, (11), to obtain

$$b_{\underline{k}_0+\Delta\underline{k}}(t) = \frac{1}{(2\pi)^2}\int_{-\infty}^{\infty} d^2x\, e^{-i(\Delta\underline{k}\cdot\underline{x}+\Delta\omega_k t)}\, G(\underline{x},t)\, [V_{|\underline{k}_0+\Delta\underline{k}|}/2]^{1/2} \qquad (19)$$

Each of the mode amplitudes in (18) is of this form, since by assumption all interacting waves are in the same narrow spectral interval. In this interval we *assume* that the coupling coefficients $\Gamma^{\underline{k}\underline{n}}_{\underline{\ell}\underline{m}}$ are all equal. This probably over-estimates the coupling between the waves, but comparison with numerical experiments as well as with water tank experiments, (see, e.g., Cohen, Watson and West[130]) indicates this approximation is a satisfactory one. The coupling coefficient is determined from (4. 21) to be

$$\Gamma^{\underline{k}\underline{n}}_{\underline{\ell}\underline{m}} \simeq \Gamma^{\underline{k}_0\underline{k}_0}_{\underline{k}_0\underline{k}_0} = - \; k_0^3/\omega_0 V_{n_0} \tag{20}$$

Using this coupling coefficient and substituting (19) for the mode amplitudes into (18) and performing the indicated summations and integrations results in the expression

$$i\left(\frac{\partial}{\partial t} + v_g^{(0)}\frac{\partial}{\partial x}\right) G = \frac{\omega_0}{8k_0^2}\left[\frac{\partial^2 G}{\partial x^2} - 2\,\frac{\partial^2 G}{\partial y^2} + 4k_0^4|G|^2 G\right] . \tag{21}$$

Equation (21) is the much celebrated nonlinear Schrödinger equation in two dimensions.[130-133] Solitons are well known solutions of the one dimensional form of (21). The term "soliton" itself was coined by Zabusky and Kruskal[134] to label the stable, localized propagating disturbance they observed in their computer studies of nonlinear waves in plasmas. The equation of motion they studied was the Korteweg-de Vries equation developed in 1895 to describe the propagation of shallow water waves.[135] These coherent wave forms differ from ordinary wave packets or mounds of water in a number of ways: 1) solitons do not become diffuse as they propagate even though the medium supporting the motion must be dispersive; 2) solitons are stable against interactions with other solitons and 3) an arbitrary finite amplitude waveform will break up into one or more stable solitons as well as non-soliton wave forms. There is a rapidly growing literature on solitons and the interested student is referred to the excellent but already somewhat dated review article by Scott, Chu and McLaughlin[133]. To gain an appreciation for how ubiquitous this entity is, there are 267 references on solitons in the bibliography of this paper dating from its coinage in 1965. Because of this extended literature we make no attempt to review this fascinating topic, but restrict our limited considerations to a discrete representation of a soliton provided by the mode rate equation (4.16).

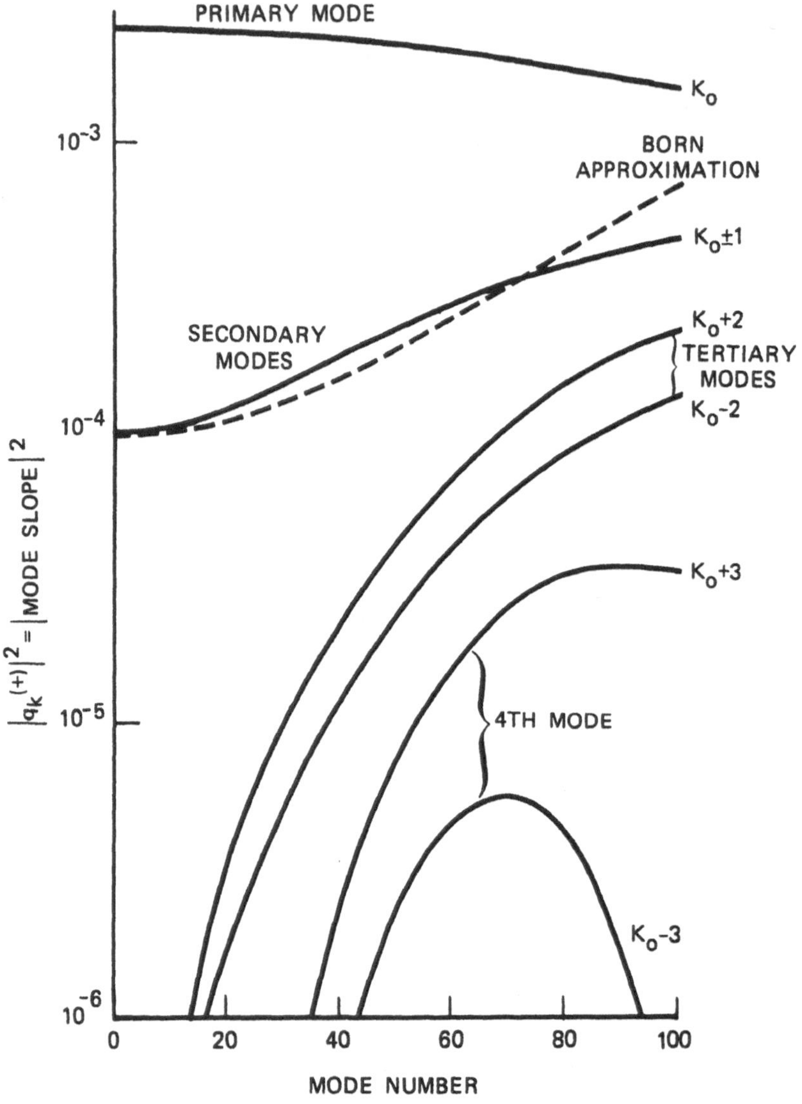

Figure 7.1: The transfer of energy in the nine mode test calculation is given as a continuous function of time. The secondary modes are compared to a calculation of their growth in a linear approximation [cf. Eq. (7.)] .

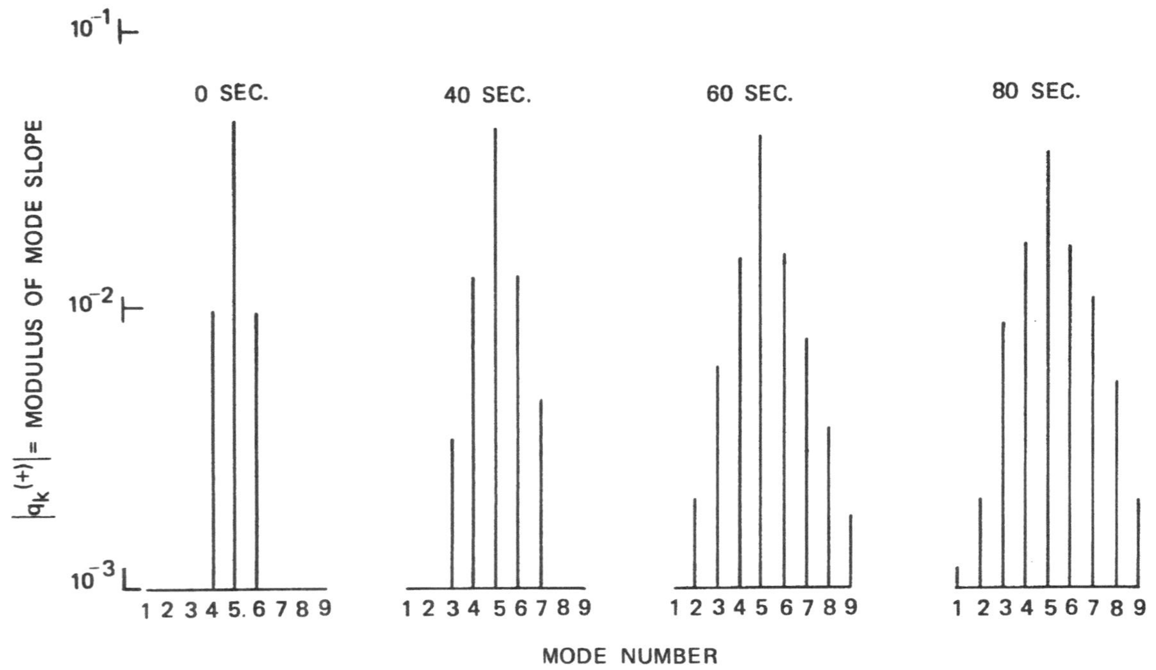

Figure 7.2: The energy distribution among the nine modes in the test calculation is shown at four different times. The central mode (5) is $k_o = .02863\ cm^{-1}$ and the step size (Δk) is $2.863 \times 10^{-3}\ cm^{-1}$.

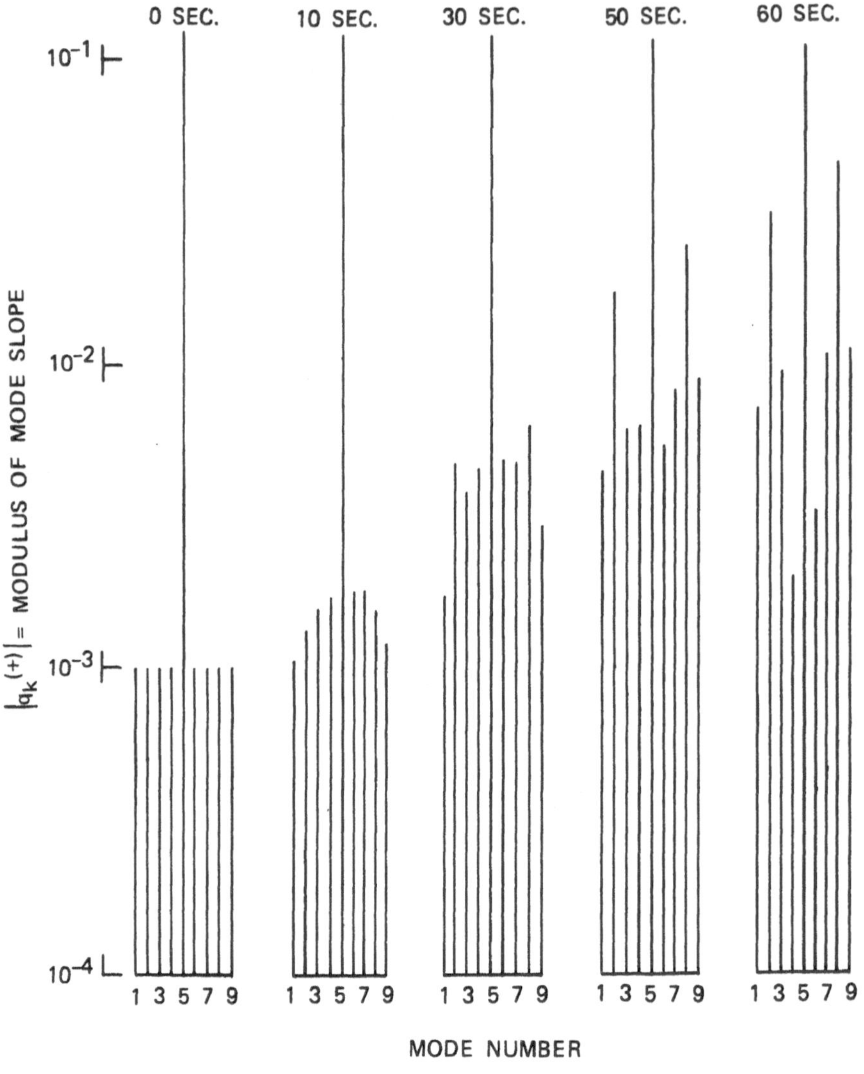

Figure 7.3: The energy distribution among the nine modes is depicted for the simulated Benjamin-Feir experiment at five different times. The center mode (5) is k_o = .02863 cm^{-1} and the step size is 2.86 x 10^{-3} cm^{-1}

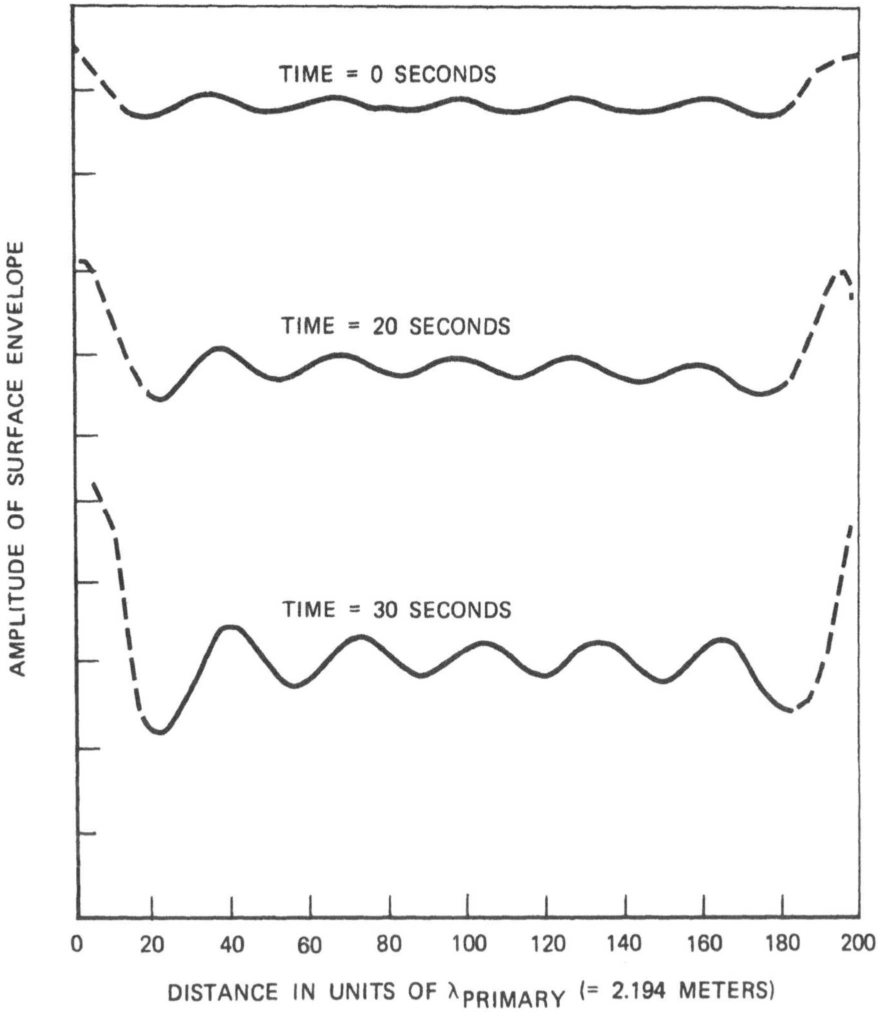

Figure 7.4: The surface envelope of the surface waves is shown for a 13-mode calculation of the Benjamin-Feir experiment.

8. Unstable Finite Amplitude Waves and Envelope Functions

The stability of the Stokes wave was long thought to be absolute. The idea had such universal acceptance that when Whitham obtained a result which implied that a Stokes wave was unstable, he assumed (according to his own words[136]) that his analysis was "probably wrong" and set it aside. It was not until Lighthill,[137] using the method of Whitham,[66,67] examined the problem in detail that it was seen that indeed a Stokes wave could not be stable. The stability argument is straightforward, but rather lengthy, so we present an argument based on the nonlinear Schrödinger equation (7.15) to demonstrate the instability. This will also enable us to contrast the unstable solution with that of a coherent soliton. The relation between this and a simplified version of the original stability argument is then presented.

Consider the stability of the one dimensional envelope function generated by the interaction of two sinusoidal wavetrains. We use an envelope function of the form

$$G(x,t) = (A_0 + \lambda\delta A) \exp i[\kappa x - \Omega t + \lambda\delta\theta] \qquad (1)$$

where λ is a smallness parameter; A_0, κ and Ω are the unperturbed amplitude, wavenumber and frequency of the envelope respectively, and the perturbed amplitude δA and phase $\delta\theta$ are functions of x and t.

Taking the derivatives indicated in the nonlinear Schrödinger equation (7.21) of the assumed solution (1) yields

$$i\left(\frac{\partial}{\partial t} + \overset{(o)}{V}_g \frac{\partial}{\partial x}\right) G = -\left\{\left(\overset{(o)}{V}_g\kappa - \Omega\right)A_0 + \lambda\left(\frac{\partial\delta\theta}{\partial t} + \overset{(o)}{V}_g \frac{\partial\delta\theta}{\partial x}\right)A_0 - i\left(\frac{\partial\delta A}{\partial t} + \overset{(o)}{V}_g \frac{\partial\delta A}{\partial x}\right)\right\} \exp i[\kappa x - \Omega t + \lambda\delta\theta] \; ;$$

and

$$\frac{\partial^2 G}{\partial x^2} = \left\{-\left(\kappa^2+2\lambda\kappa\ \frac{\partial\delta\theta}{\partial x}\right)A_o\ +\lambda\frac{\partial^2\delta A}{\partial x^2} + i\lambda\ \frac{\partial^2\delta\theta}{\partial x^2}\ A_o\ + 2i\lambda\kappa\ \frac{\partial\delta A}{\delta x}\ -\ \lambda\kappa^2\delta A\right\} \exp\ i\ [\kappa x-\Omega t + \lambda\delta\theta] + O(\lambda^2) \tag{3}$$

to first order in λ. Substituting (2) and (3) into the one-dimensional form of equation (7.21) and separating real and imaginary parts of like powers of λ yields

$$\Omega\ \doteq\ \kappa\ V_g^{(o)} - \left(\frac{\kappa}{k_o}\right)^2 \frac{\omega_o}{8} + \frac{1}{2}\ \omega_o\ k_o^2\ A_o^{\ 2} \tag{4}$$

$$\frac{\partial}{\partial t}\ \delta A + \left(V_g^{(o)} - \frac{\omega_o}{4k_o^2}\ \kappa\right) \frac{\partial\delta A}{\partial x} = \frac{\omega_o}{8k_o^2}\ A_o\ \frac{\partial^2\delta\theta}{\partial x^2} \tag{5}$$

$$\frac{\partial\delta\theta}{\partial t} + \left(V_g^{(o)} - \frac{\omega_o}{4k_o^2}\ \kappa\right) \frac{\partial\delta\theta}{\partial x} = -\ \frac{\omega_o}{A_o 8k_o^2}\ \frac{\partial^2\delta A}{\partial x^2} - \omega_o\ k_o^2 A_o \delta A\quad . \tag{6}$$

Equation (4) is the nonlinear dispersion relation for the envelope function and the other two expressions, (5) and (6), describe the linear evolution of the amplitude and phase perturbations, respectively.

In a coordinate system translating with the velocity obtained from (4), i.e., $d\Omega/d\kappa = V_g^{(1)}$,

$$V_g^{(1)} = V_g^{(o)} - \kappa\omega_o^2/4k_o^2 \tag{7}$$

so that

$$\xi = x - V_g^{(1)} t \quad , \tag{8}$$

(5) and (6) can be written as

$$\frac{\partial \delta A}{\partial t} - \frac{\omega_o}{8k_o^2} \frac{\partial^2(\delta\theta A_o)}{\partial \xi^2} = 0 \quad , \tag{9}$$

$$\frac{\partial}{\partial t}(A_o \delta\theta) + \frac{\omega_o}{8k_o^2}\frac{\partial^2 \delta A}{\partial \xi^2} + k_o{}^2 \, \omega_o \, A_o{}^2 \, \delta A = 0 \quad . \tag{10}$$

The stability of G(x,t) can now be determined from the solutions to (9) and (10).

We substitute (10) into the time derivative of (9) to obtain the linear fourth order equation for δA,

$$\frac{\partial^2 \delta A}{\partial t^2} + \left(\frac{\omega_o}{8k_o^2}\right)^2 \frac{\partial^4 \delta A}{\partial \xi^4} + \frac{1}{8}\omega_o^2 \, A_o{}^2 \frac{\partial^2 \delta A}{\partial \xi^2} = 0 \quad . \tag{11}$$

Assuming a solution for the amplitude perturbation of the form

$$\delta A = \text{Re}\,\{\exp\,(i\ell\xi + \gamma t)\} \tag{12}$$

we obtain the expression for the growth rate γ,

$$\gamma^2 + \left(\frac{\omega_o}{8k_o^2}\right)^2 \ell^4 - \frac{1}{8}\omega_o^2 \, A_o{}^2 \, \ell^2 = 0 \quad . \tag{13}$$

The solution to (13) yields the amplitude dependent dispersion relation

$$\gamma = \pm \frac{\omega_o}{8k_o^2} \ell(8k_o^4 A_o^2 - \ell^2)^{1/2} \quad . \tag{14}$$

The perturbation solutions (12) are unstable when $\gamma > 0$, i.e., when

$$8k_o^2 m^2 > \ell^2 \quad ; \quad m = k_o A_o \tag{15}$$

for the positive root of (14). This is the condition obtained by Benjamin and Feir for the stability of a finite amplitude wave.[50] The amplitude of the sideband (δA) grows like

$$\delta A \sim e^{\gamma t} \quad . \tag{16}$$

Introducing the fractional change in wave number induced by the nonlinearity, i.e., $\ell = 2k_o\delta$, we have from (14)

$$\gamma = \omega_o \, [2m^2 - \delta^2]^{1/2} \, \frac{\delta}{2} \quad . \tag{17}$$

The maximum growth rate is determined by the condition that the variation in γ with respect to δ vanish, resulting in

$$\delta_{max} = k_o A_o \tag{18}$$

which is the slope of the envelope function. In terms of the wavenumber ℓ we have from the definition of δ above and (18) that

$$\ell = 2k_o^2 A_o = 2mk_o \quad . \tag{19}$$

The position of the sidebands receiving the maximum amplification have wavenumbers $k_{\pm} = k_o(1 \pm 3m)$ and frequencies $\omega_{\pm} = \omega_o(1 \pm m)$. At these wavenumbers the growth rate from (14) is

$$\gamma = \frac{1}{2} m^2 \omega_o \tag{20}$$

with the corresponding growth time

$$\tau_{BF} = 2/(m^2 \omega_o) \tag{21}$$

which gives the e-folding time for the Benjamin - Feir interaction. Reference to the nonlinear term in (7.21) suggests that τ_{BF} is the characteristic time scale for nonlinear interactions to develop, providing that the wavenumber separation between discrete Fourier modes is less than that given by (19), i.e., $\Delta k = 2mk_o$. The characteristic time for dispersive spreading of a soliton is also τ_{BF}. The balance between this dispersive spreading and the *peaking* produced by the nonlinear terms leads to the steady solutions discussed in the next lecture.

An alternate viewpoint of the above instability is obtained if one assumes the nonlinear wave does not become too nonlinear and therefore obeys the "energy" transport equation[136]

$$\frac{\partial}{\partial t} a^2 + \frac{\partial}{\partial x}\left[\omega_o' \, a^2\right] = 0 \quad . \tag{22}$$

The "energy" is here given by $\frac{1}{2}a^2$ and the nonlinear wave propagates at the speed $V_g^{(o)}(k) \equiv \omega_o'(k)$. The hydrodynamic equation of continuity expresses the fact that the number of wave crests entering a unit interval of space per unit time matches the frequency with which the wave crests impinge on the border of the spatial interval, i.e., in one dimension

$$\frac{\partial k}{\partial t} + \frac{\partial \omega}{\partial x} = 0 \quad . \tag{23}$$

The dispersion relation for a nonlinear wave relates the frequency to both the wave amplitude a and wavenumber k,

$$\omega = \Omega(k,a) \quad . \tag{24}$$

To have a concrete example for discussion we assume a dispersion relation of the form

$$\Omega = \omega_0(k) + \omega_1(k)a^2 \tag{25}$$

in the one dimensional case. Inserting these expressions into the hydrodynamic equation of continuity (23) we obtain

$$\frac{\partial k}{\partial t} + \frac{\partial}{\partial x}\left[\omega_0(k) + \omega_1(k)a^2\right] = 0 \quad . \tag{26}$$

Equation (26) can be combined with the characteristic curve for the group velocity in (22) to determine the characteristic speed for both the wavenumber and energy of the nonlinear wave.

We multiply (22) and (26) by the constants C_1 and C_2, respectively, and add the resulting equations to obtain

$$C_1 \left[\frac{\partial k}{\partial t} + \omega_0' \frac{\partial k}{\partial x} + 2a\omega_1 \frac{\partial a}{\partial x}\right] + C_2\left[\frac{\partial a}{\partial x} + \omega_0' \frac{\partial a}{\partial x} + \frac{1}{2} \omega_0'' \quad a \frac{\partial k}{\partial x}\right] = 0 \quad . \tag{27}$$

Here we have omitted a term of the form $\omega_1' a^2 \; \partial k/\partial x$ which is of higher order and whose inclusion would not modify any of the qualitative features discussed below. Equation (27) can be put in the characteristic form

$$C_1\left\{\frac{\partial k}{\partial t} + V_g \frac{\partial k}{\partial x}\right\} + C_2 \left\{\frac{\partial a}{\partial t} + V_g \frac{\partial a}{\partial x}\right\} = 0 \tag{28}$$

where V_g is the characteristic speed if C_1 and C_2 satisfy the equations

$$C_1V_g = C_1 \; V_g^{(o)} + C_2\omega_0'' \quad a/2$$

$$C_2V_g = C_1 2a \; \omega_1 + C_2 \; V_g^{(o)} \quad . \tag{29}$$

Substituting C_2 from the second into the first expression in (29) yields

$$(V_g - V_g^{(o)}) \; C_1 = \omega_0'' \quad a \frac{a\omega_1}{V_g - V_g^{(o)}} \; C_1$$

resulting in the group velocity

$$V_g = V_g^{(o)} \pm a\sqrt{\omega_0'' \; \omega_1} \quad . \tag{30}$$

That (30) is indeed the characteristic speed can be seen by substituting the values of C_1 and C_2 obtained from (29) into (28) to obtain the characteristic form of the equation

$$\frac{1}{2}\left(\frac{\omega_0''\ (k)}{\omega_1(k)}\right)^{1/2} dk \pm da = 0 \tag{31}$$

on the characteristics

$$\frac{dx}{dt} = V_g^{(o)} \pm a[\omega_0''\ \omega_1]^{1/2} \quad . \tag{32}$$

The character of the nonlinear system is determined by the sign of $\omega_0''\ \omega_1$. For $\omega_0''\ \omega_1 > 0$ the characteristics are real and the system is hyperbolic; for $\omega_0''\ \omega_1 < 0$ the characteristics are imaginary and the system is elliptic. In the former case we have a splitting of the linear group velocity into two values by the nonlinear perturbation. This is the nonlinear generalization of the concept of a group velocity. As pointed out by Whitham[66] (Sect. 14.2) an initial disturbance will introduce perturbations in both families of characteristics. For example, a uniform wave train which is disturbed locally will eventually split the resulting modulation in two groups as prescribed by (32). This is distinct from linear theory where the pattern of modulation would distort due to the k dependence of $V_g^{(o)}(k)$, but would not split.

In the elliptic case the nonlinear wave is *unstable*. Small perturbations in frequency and amplitude in (28) have the harmonic solution

$$e^{i\ell(x-V_g t)} \quad . \tag{33}$$

In the elliptic case ($\omega_0'' \; \omega_1 < 0$) the characteristic speed given by (32) is complex and the modulations (33) grow exponentially. Since the Stokes wave has the nonlinear dispersion relation [cf. (6.15)]

$$\omega = (1 + \frac{1}{2} k^2 a^2) \sqrt{gk} \tag{34}$$

we obtain by comparing with the assumed form of the dispersion relation (25)

$$\omega_0(k) = \sqrt{gk} \quad ; \qquad \omega_1(k) = \frac{1}{2} \sqrt{g}\, k^{5/2} \tag{35}$$

and $\omega_0''(k)\omega_1(k) < 0$. Therefore the Stokes wave in deep water is unstable.

This instability of a finite amplitude water wave is just the Benjamin-Feir sideband instability. The growth rate of the instability is given by $\gamma = \ell a \sqrt{\omega_0'' \; \omega_1}$ from (33) so that using (34) and (35), and assuming that the location of the wavenumber ℓ is given by (19) we obtain $\gamma = \frac{1}{2} m^2 \omega_0$ as the maximum growth rate. Therefore, the Stokes wave instability indicates that the nonlinear interaction among gravity waves is weak and is due to the resonant third order term in (4.16).

9. Stability vs. Instability

We here contrast the instability discussed in the previous lecture with the stable soliton solution of the nonlinear Schrödinger equation. Feir[138] conducted an experiment in which he mechanically generated a high frequency water wave modulated by a low frequency driving force on the surface of a water tank. One such pulse generated in this manner was found to propagate without change in form along the water surface, but with an overall attenuation of the pulse amplitude. Feir noted that if one changed the amplitude of the pulse, the integrity of the pulse shape was lost as it propagated down the tank. He was not able to *explain* the phenomenon at that time and presented his results as a curiosity. Feir had created a deep water envelope soliton on the surface of his water tank. The first explanation of his experimental results in terms of solitons was made by Chu and Mei.[43] They used a nonlinear WKB method for solving the equations of motion, including dispersion, to study the propagation of various pulse shapes including those experimentally generated by Feir. General discussion of envelope solitons in the present context can be found in a number of places.[131-133]

We now consider the solution to the one-dimensional form of the nonlinear Schrödinger equation which is stable in space and time. We assume a solution of the form

$$G(x,t) = A(x,t)e^{i\Theta(x,t)} \tag{1}$$

with both A and Θ real. As in the preceeding lecture we substitute (1) into the one dimensional from of (7.21) and equating real and imaginary parts obtain

$$\frac{\partial^2 A}{\partial x^2} - \left(\frac{\partial\Theta}{\partial x}\right)^2 A + \frac{8k_o^2}{\omega_o}\left[\frac{\partial\Theta}{\partial t} + V_g^{(o)}\frac{\partial\Theta}{\partial x} + \frac{1}{2}\omega_o k_o^2 A^2\right]A = 0 \tag{2a}$$

$$\frac{\partial}{\partial x}\left(\frac{\partial\Theta}{\partial x}A^2\right) - \frac{8k_o^2}{\omega_o}\left(\frac{\partial A}{\partial t} + V_g^{(o)}\frac{\partial A}{\partial x}\right)A = 0 \quad . \tag{2b}$$

Identifying the derivatives of the phase function θ with a wavenumber κ and frequency Ω, as done in the wavepacket picture, i.e.,

$$\kappa \equiv \frac{\partial\Theta}{\partial x} \quad ; \quad \Omega \equiv -\frac{\partial\Theta}{\partial t} \tag{3}$$

we obtain from (2)

$$A_{xx} - \kappa^2 A + \frac{8k_o^2}{\omega_o}\left(\frac{1}{2}\omega_o k_o^2 A^2 + \kappa V_g^{(o)} - \Omega\right)A = 0 \tag{4a}$$

$$(\kappa A^2)_x - \frac{8k_o^2}{\omega_o}\left(A_t + V_g^{(o)} A_x\right)A = 0 \tag{4b}$$

and we have used the subscript notation for derivatives, i.e., $A_x = \partial A/\partial x$, etc.

Assuming that the wavenumber κ and frequency Ω are independent of x and t allows us by multiplying (4a) by A_x to directly integrate (4a) and obtain

$$A_x^2 = -\frac{8k_o^2}{\omega_o}\left[-\Omega + \kappa V_g^{(o)} - \frac{1}{8}\left(\frac{\kappa}{k_o}\right)^2\omega_o + \frac{1}{4}\omega_o k_o^2 A^2\right]A^2 \ . \tag{5}$$

Integration of (5) yields

$$A = A_o \operatorname{sech} \sqrt{2}\, k_o m\left(x - V_g^{(1)} t\right) \tag{6}$$

with A_o a constant amplitude; m the slope (=k_oA_o); $V_g^{(1)}$ the shifted group velocity and Ω the frequency given by

$$\Omega = \kappa V_g^{(o)} - \frac{1}{8}\left(\frac{\kappa}{k_o}\right)^2 \omega_o + \frac{1}{2}\,\omega_o k_o^2 A_o{}^2 \tag{7}$$

as found in the preceeding lecture, [cf., Eq. (8.4)].

By rearranging (4b) we obtain the transport equation

$$\frac{\partial A}{\partial t} + \left(V_g^{(o)} - \frac{1}{4}\,\kappa\,\omega_o k_o^{-2}\right)\frac{\partial A}{\partial x} = 0 \tag{8}$$

so that the amplitude A is constant along the trajectory

$$V_g^{(1)} \equiv \frac{dx}{dt} = V_g^{(o)} - \frac{1}{4}\,\kappa\,\omega_o/k_o^2 \quad . \tag{9}$$

A solution to (7.21) in one dimension is therefore

$$G(x,t) = A_o \operatorname{sech}\left[\sqrt{2}\, m k_o\left(x - x_o - V_g^{(1)} t\right)\right] \exp i[\kappa x - \Omega t + \phi_o] \tag{10}$$

where κ, ϕ_o, x_o and A_o are real parameters. With $\kappa = 0$ we have the *centered solution*

$$G(\xi,t) = A_o \operatorname{sech}(\sqrt{2}\, m k_o \xi) \exp\left[-i m^2 \omega_o t/2 + i\phi_o\right] \tag{11}$$

in the translating coordinate system

$$\xi = x - x_o - V_g^{(1)} t \ , \tag{12}$$

which is both localized and stationary.

In Figure (9.1) is depicted a realization of the pulse shape given by the envelope soliton solution (11) for $k_o = .2516 \text{ cm}^{-1}$, $m = .064$ and $\phi_o = 0$. In part (b) of this figure the actual displacement of the fluid surface is shown. Both (a) and (b) have been normalized to unity. It is clear that if one chooses the initial state of the system to be that of a soliton, then the nonlinear interactions in the mode rate equations should not modify the pulse shape; at least to the degree that the approximations leading from the mode rate equations (4.16) to the nonlinear Schrödinger equation (7.21) are valid.

The selection of the appropriate initial mode amplitude for a soliton solution are made using the Fourier Transform of (11). If we specify periodic boundary conditions over a distance L at $x \pm L/2$, then the initial mode amplitudes are

$$b_n(0) = \frac{1}{L} \int_{-L/2}^{L/2} G(\xi,0) e^{i(k_o - k_n)\xi} d\xi \tag{13}$$

where the mode spacing is $\Delta k = 2\pi/L$ and $k_n = n\Delta k$. Integrating and normalizing (13) we obtain the mode amplitudes for the discrete representation of a soliton

$$b_n(0) = b_N(0) \operatorname{sech} \left[\frac{(N-n)\pi\Delta k}{\sqrt{8}\, mk_o} \right] . \tag{14}$$

In the calculations done by Cohen et al,[130] (14) was used to represent the initial soliton by a finite number of modes. The envelope depicted in Figure (9.1) is determined by the absolute value of

$$G(x,t) = \sum_n \sqrt{2/V_n}\; b_n(t) \exp[i(N-n)\Delta k(x-V_g^{(0)}t)] \tag{15}$$

where

$$V_g^{(0)} \equiv \frac{(\omega_n-\omega_o)}{(N-n)\Delta k} \tag{16}$$

is the group velocity of the primary mode. The parameters for this example calculation were selected to correspond to the wave pulse experiments of Feir[138]. The initial amplitude of the central mode in (15) is obtained from the experiment using the expression

$$\left(\frac{2}{V_o}\right)^{\frac{1}{2}} b_o(0) = G(x=0)\left\{\sum_n \text{sech}\,[\pi\Delta k(n-N)/\sqrt{8}\; mk_o]\right\}^{-1} \tag{17}$$

where m = 0.064 and G(x=0) = 0.254 cm for the first trace of Figure 3 of Feir[138], which yields $\left(\frac{2}{V_o}\right)^{\frac{1}{2}} b_o(0) = 0.056$ cm. The slope of the soliton is given by

$$m = k_o \sum_n (2/V_n)^{\frac{1}{2}} b_n(0) \tag{18}$$

which is the central wavenumber times the maximum surface displacement.

It was observed that the stable pulse generated in Feir's experiments, although it maintained its form did decrease in amplitude. After propagating a distance of 7.3 m the pulse was observed to be one half its initially measured amplitude. This dampling of the pulse is simulated in the present calculation using a phenomenological viscosity coefficient in the rate equations. The linear amplitude damping coefficient yields

$$G(\xi,t) = G(\xi,o)e^{-\alpha t} \tag{19}$$

where t is given by the ratio of the distance traveled to the group velocity. The decay rate is given by $\alpha = 0.03 \text{ sec}^{-1}$ or in terms of a viscosity coefficient $\nu = \alpha/k_0^2 = 0.47 \text{ cm}^2/\text{sec}$. The shape of the initial pulse indicated in Figure (9.1) is virtually unchanged after propagation 7.3 m, however, its amplitude is calculated to decrease by a factor of .45 consistent with experiment.

Feir described the propagation of six pulse shapes with increasing amplitude, but with the same carrier wavenumber k_0 and frequency of modulation ω_0. Since the pulse shape already discussed was that of soliton, the other pulses, with their changes in amplitude, must *not* be solitons. From analysis then, these waveforms must be unstable and they should be observed to breakup. This was indeed the observation. In Figure (9.2) a simulation of one of the latter runs is depicted. The initial pulse in (a) is seen to lose its integrity 24 seconds later as indicated in (b). A more detailed comparison with Feir's experiments is made by Chu and Mei[143] by direct numerical integration of the nonlinear Schrödinger equation.

We have discussed two descriptions of the soliton and its stability in one dimension; 1) the exact form as the solution to the nonlinear Schrödinger equation, and 2) the approximate form as a numerical solution to a finite number of modes using the mode rate equations. For a local perturbation as in the Feir experiment the modulation of the soliton shape can be calculated by a numerical integration of either system. The mode rate equations, however, allow one to study the behavior of the soliton embedded in a more general system. This was done by Cohen <u>et al</u>[130] and rather than reproducing their numerical results here we merely list some of their conclusions.

The soliton interacting with two high frequency gravity waves is found to be unstable. The mode rate equations include the dynamics of the high frequency waves as well as the dynamics of the soliton. These waves are non-linear and can interact with the low frequency soliton by means of both the quadratic and cubic nonlinearities. If we examine the cubic nonlinearities in (4.16), we see that one of the modes in the interaction can be the carrier wave of the soliton, while the two others are the high frequency perturbing waves. In physical space this quadratic interaction of the perturbing waves generates sum and difference waves. The difference waves have a low frequency and long wavelength and can interact with the soliton for a long time - thereby inducing an instability.

The soliton interacting with a single low frequency, long wavelength gravity wave is also found to be unstable, but on a much longer time scale than when interacting with the high frequency water waves. The soliton couples to the surface current generated by the long waves and it is found that since the current is periodic, so too is the distortion of the soliton. After a complete cycle of the perturbation the soliton <u>almost</u> returns to its initial form. There is some residual distortion after each cycle however, and the effect is cumulative so that eventually the soliton does lose its integrity.

No systematic study of the stability properties of solitons in the presence of an ambient wave field has been made to date (1980). The existence of solitons in geophysical water wave fields would modify the coherence properties of the measured correlation functions and power spectral densities of the waves. In particular, the predicted modulation of the water waves in a broad band system with random initial phases has not been studied in any detail. Such studies are crucial for extracting the properties of water waves from the ambient ocean data.

We now seem to be presented with two contradictory physical phenomena. The first is the instability of a Stokes wave (finite amplitude water wave) which for so long was thought to be stable. The second is the absolute stability of the isolated finite amplutude soliton wavepacket. The resolution of these apparently incompatible modes of evolution for a nonlinear water wave field is made by examining the way the separate calculations were done and the corresponding experiments performed.

The Stokes wave instability is determined by a *linear* stability analysis with the bifurcation of the group velocity discussed in lecture 8 resulting from the calculation of a second derivative using the linear dispersion relation. The soliton, on the other hand, is an analytic solution of a nonlinear equation and is absolutely stable (in one dimension). To make these two results compatible we assert that the dispersion relation must change (formally) between the Stokes wave and soliton analysis. This is found to be the case in the presentation of the average Lagrangian approach used by Yuen and Lake[68]. The dispersion relation is modified to include the curvature of the envelope function and thus changes the equation of motion from elliptic in the case of the Stokes wave back to hyperbolic for the soliton. This theoretical result is supported by their experiments and those of their collabortors[68,46,129], which have shown that the Benjamin-Feir instability is not absolute, but is in fact reversible. The implications of their experiments will be discussed after we review more of the properties of envelope solitons.

The dispersion relation for a finite amplitude wave was determined by Yuen and Lake[68] to be

$$\omega \cong \omega_0 \left[1 + \frac{1}{2}k_o^2 a^2 + \frac{1}{8k_o^2 a} \frac{\partial^2 a}{\partial x^2}\right] . \tag{20}$$

The spatial derivative term in (20) is analogous to the diffraction term in the Fresnel theory of diffraction in optics. This is *not* the dispersion relation which leads to the instability of a Stokes wave predicted by Lighthill[137,140] and Whitham[136]. The elliptic character of the energy transport equation (8.22) disappears when the curvature of the nonlinear wave is properly accounted for in the dispersion relation.

We have constructed the single soliton solution to the nonlinear Schrödinger equation. Zakharov and Shabat[141] in discussing the same problem showed, using the inverse scattering technique,[83] that the nonlinear Schrödinger equation can be solved exactly for initial conditions that approach zero sufficiently rapidly as $x \to \pm\infty$. Such an initial condition corresponds to a pulse. A summary of the characteristics of the predicted soliton in the water wave context is given by Yuen and Lake. A number of these points have already been emphasized in the previous discussions, but it is useful to review them before we leave this topic:

(i) As mentioned in conjunction with the Benjamin-Feir experiments, an initial envelope pulse of arbitrary shape will eventually disintegrate into a number of solitons and an oscillatory tail. The initial conditions completely determine the number and structure of the solitons and the character of the tail.

(ii) For an initial condition corresponding to a pulse, the tail is relatively unimportant. It will disperse linearly and therefore the amplitude will decay as $t^{-1/2}$.

(iii) Each soliton, indexed with an integer n, is a permanent progressive wave solution of (7.21) of the form

$$G_n = A_n \operatorname{sech}\left[\sqrt{2}\, k_o m_n (x - X_n - (V_g^{(n)} + V_g^{(o)})t\right]$$
$$\times \exp\left\{-m_n^2 \omega_o t/2 - 2ik_o \frac{V_g^{(n)}}{V_g^{(o)}}\left[x - X_n - (V_g^{(n)} + V_g^{(o)})t + \phi_n\right]\right\} \qquad (21)$$

where A_n and $V_g^{(n)}$ characterize the amplitude and speed [relative to $V_g^{(o)} \equiv \omega_o/(2k_o)$] of the nth soliton; X_n and ϕ_n represent its initial position and phase and m_n is the slope of the nth soliton. Note that A_n and $V_g^{(n)}$ are independent quantities in (21).

(iv) Solitons can interact strongly in overlapping regions of space, but they are asymptotically stable in the sense that no permanent change in form is induced by such interaction. At most there results a shift in position and phase, i.e., in X_n and ϕ_n, from such an interaction.

(v) The time scale for the formation of solitons from an initial data set is directly proportional to the length of the pulse and inversely proportional to the amplitude. The single parameter which measures this ratio is the slope of the soliton.

(vi) For an initial pulse with a narrow spectrum, i.e., small $\delta\omega$ and δk, we can estimate the number of solitons N_s contained in the profile by the formula

$$N_s \simeq \frac{1}{\pi} \int \sqrt{2}\, k_o m\, f(x) dx$$

where the pulse profile is given by $A(x,0) = a_o f(x)$ with $0 \leq f(x) \leq 1$, and $m = k_o a_o$.

Yuen and Lake along with collaborators[46,68] did a number of interesting water tank experiments using a wavemaker to generate prescribed modulations of high frequency water waves. The evolution of *pulses* was recorded at a number of stations along the tank using capacitance wire gauges, i.e., every five feet up to thirty feet. Figures (9.3) - (9.6) indicate three sequence of pulses measured in some of their experiments.

Figure (9.3) shows three pulses of approximately the same effective duration but with different initial profiles. Case (a) is a soliton profile. Case (b) is a sech profile whose amplitude is twice that of the soliton profile. Case (c) is a sine profile with the same amplitude as the soliton profile. Case (c) and the first two measurements in case (b) are similar to results obtained in the pulse experiments performed by Feir as discussed earlier. In cases (b) and (c) the initial profiles break up into solitons;

while in case (a) where the initial profile is already that of a soliton profile, no *apparent* interaction occurs.

Figure (9.4) shows two successive pulses of different carrier frequencies, with the 3 Hz pulse followed by the 1.5 Hz pulse. The second pulse is therefore twice as fast as the first. At 30 ft. they have passed through one another and emerged relatively unaffected by the interaction.

Figure (9.5) shows the head-on collision of two pulses where the first pulse has been backscattered off a vertical end wall before sending in the second. Again the pulses show little effect of interaction after the collision.

From these experiments estimates for the formation time of the solitons as well as the number of solitons have been verified.

The theory as interpreted above does not adequately describe the most recent experimental results of Lake and Yuen. The theoretical prediction of the Benjamin-Feir sideband instability is that of a breakup of a nonlinear wave with the energy eventually appearing in a number of small amplitude randomized waves. With the identification of the soliton it was conjectured that the final state of the unstable Stokes wave would be one or more asymptotically stable solitons. The coherence in the initial nonlinear wave would be partially preserved in the solitons. It now appears that this is not the case experimentally.

Lake et al[46] have found that the initial Benjamin-Feir instability of a wavetrain does not induce the breakup of the train into either of the above final states. What is observed is a periodic return of the wavetrain to the initial state of the system. The modulation induced by the instability periodically increases and decreases in a way that is reminiscent of the Fermi-Pasta-Ulam recurrance phenomenon discussed earlier. In a similar fashion Lake et al find that even though energy is transferred to distant modes in the spectrum, once the magnitude of the modulation has reached some critical value, that there is no loss of coherence. The initial state will reclaim this distant energy with a definite reclamation frequency.

A numerical integration of the nonlinear Schrödinger equation indicates that this experimental result is consistent with this description with the appropriate initial conditions. To quote from the *conclusion* of Lake et al;[46] "*The results of this investigation provide evidence that the end state of the evolution of a non-dissipative nonlinear continuous wave train is neither steady nor random, but is instead a series of periodically recurring states. The nonlinear resonant interactions which occur during the evolution of a nonlinear wave train on deep water do not lead to an irreversible spread of energy over all spectral components; in fact, coherence is retained throughout the process of evolution. In realistic environments coherence will be lost ultimately not as a result of the nonlinearity of the wave trains but because of the longtime effects of dissipative mechanisms such as surface viscous dissipation, generation of capillary waves and wave breaking.*"

The analysis of Lake, Yuen and collaborators is based on the nonlinear Schrödinger equation description of the water wave field. A direct analysis of the mode coupled equations has been made by Bryant[142] and again the recurrance phenomenon is observed. Bryant's discussion is based in large part on the results of numerically integrating the mode coupled equations of evolution (4.16). He does however show that these equations have an analytic solution for a system of four interacting modes. The interaction is described by (6.10) where modes 1 and 4, say, have wave number k_0 and modes 2 and 3 have wave numbers $k_0 \pm K$ respectively and $K << k_0$. The solution to this system of equations is shown by Bryant to be an elliptic integral of the first kind in the energy of the central mode, i.e., $|B_1|^2$. The energy is periodic in the time variable m^2t where m is the slope of the central mode, indicating the period over which return to the initial state ocurrs. The conclusions reached based on the analytic calculation using the four mode model are consistent with those based on the more extensive numerical calculations using up to 18 additional modes.

The stability of the soliton to modulation distortion was also observed in the calculations of Cohen, Watson and West[130] in which the envelope soliton with carrier wavenumber k_0 was coupled to a gravity wave of wavenumber K, with $k_0 \gg K$. The distortion of the soliton pulse produced by the interaction with the orbital current of the gravity wave was found to be reversible, i.e., the pulse returned to *nearly* its initial state with each period of the long gravity wave. The coupling of the same pulse to very short gravity waves, however, did induce an instability so that the soliton distortion is not stable in this latter case. The presence of such high frequency waves in a wind wave spectrum such as present on the ambient ocean would seem to rule out the persistence of envelope solitons in naturally occuring water wave fields. Their effect, i.e., that of the nonlinear interaction in a narrow spectral band, might however appear in the coherence measurements of such fields. The question of solitons occuring in wind generated water wave fields and how one would determine their existence from a measurement record is still open. In later lectures, we consider the fluctuation properties of the wind wave field and hopefully gain some appreciation of the difficulty of determining such coherence effects.

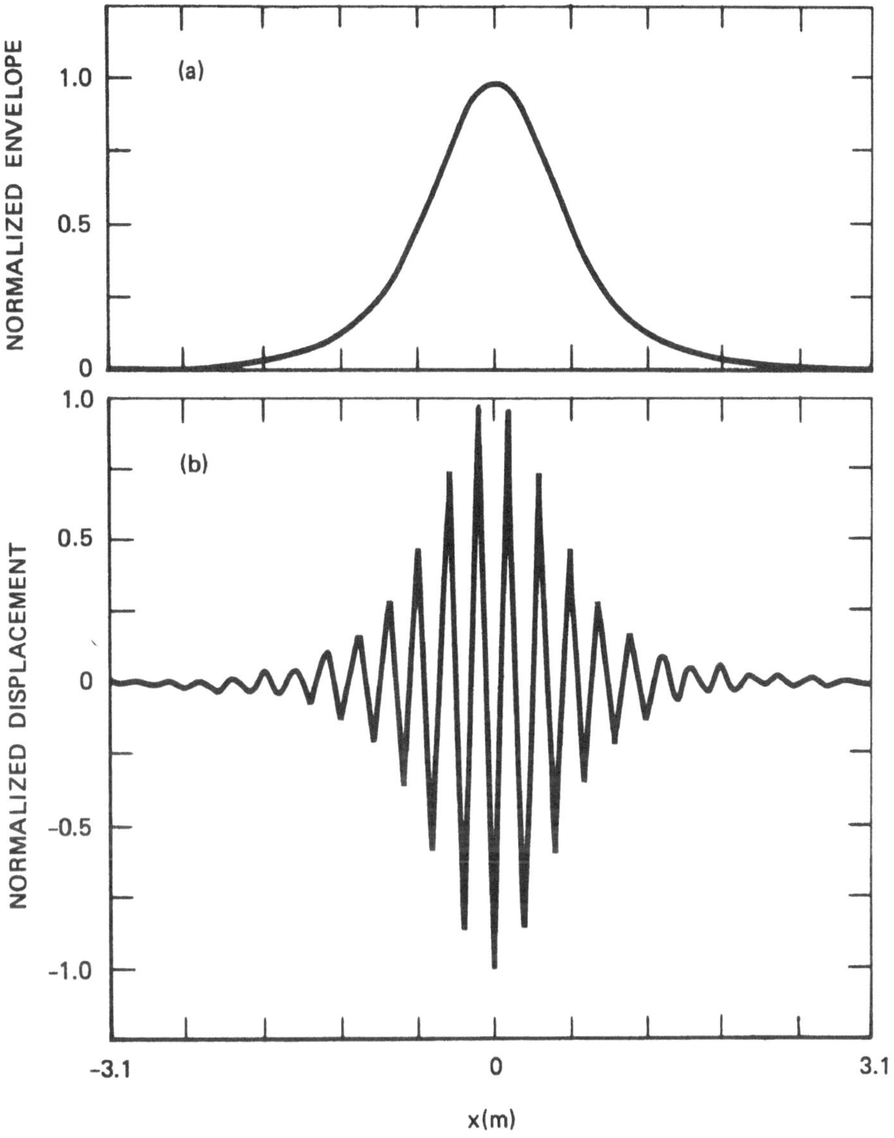

Figure 9.1: The soliton specified by the slope amplitude given in Cohen, Watson and West is shown at t = 0. (a) The Envelope G and (b) the wave displacement ζ are normalized to .251 cm.

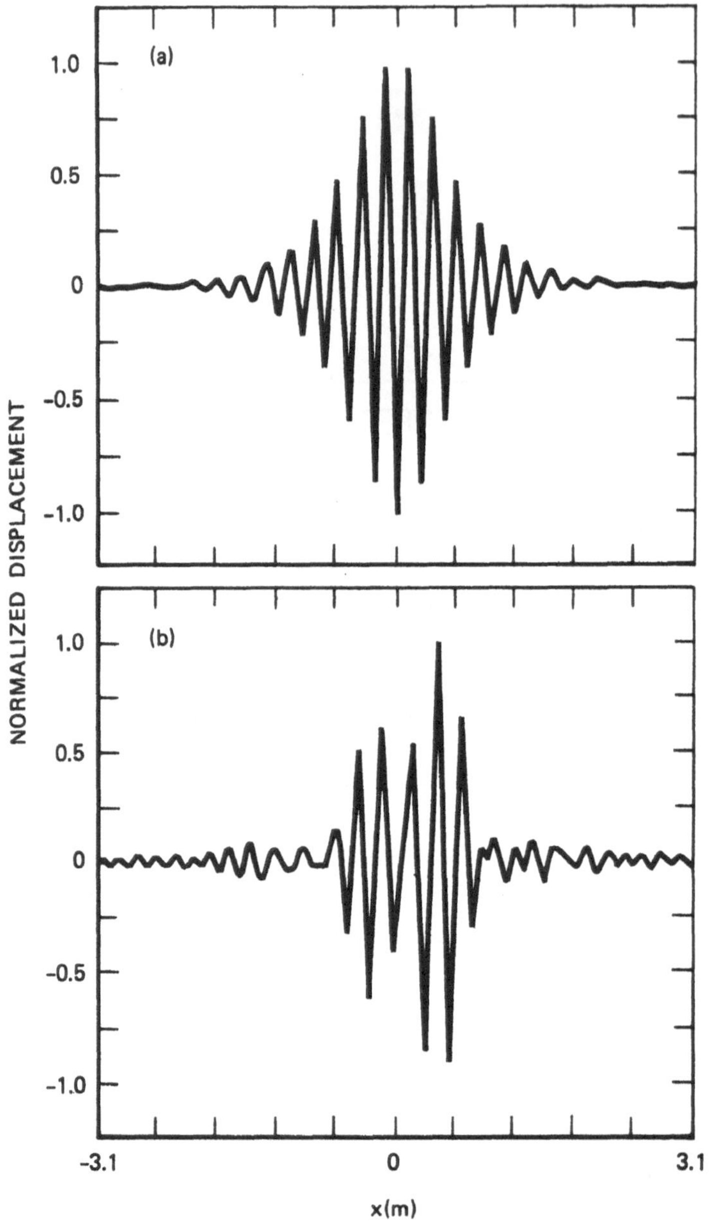

Figure 9.2: Wavepacket with slope and amplitudes given in Cohen, Watson and West is shown in (a) at t = o. It is shown at t = 24 sec in (b).

Figure 9.3: Evolution of wave pulses. Case A, initial pulse with soliton profile, ω_0 = 2 Hz, initial $(ka)_{max} \simeq 0.14$. Case B, initial pulse with sech profile and amplitude twice that for soliton profile, ω_0 = 2 Hz, amplitude scale of traces reduced by factor of 2.5 compared with Cases A and C. Case C, initial pulse with sine profile and amplitude equal to that for soliton profile, ω_0 = 2 Hz, initial $(ka)_{max} \simeq 0.14$. In all figures, time increases from left-to-right and the x = 0 ft traces show initial wave pulse inputs to the wavemaker.

Figure 9.4: One wave pulse overtaking and passing through another wave pulse. Left-hand trace: first pulse along, ω_0 = 1.5 Hz, initial $(ka)_{max} \simeq 0.10$, six-cycle pulse. Center trace: second pulse alone, ω_0 = 3 Hz, initial $(ka)_{max} \simeq 0.2$, 12-cycle pulse which disintegrates into two solitons. Right-hand traces: interaction of the two pulses.

Figure 9.5: Head-on collision of two wave pulses. The initial conditions for each pulse are identical to those for the pulse in Figure 9.4. Noted on this figure are the propagation distances for wave pulses A and B during the process.

For Figs.9.3-5 see pp.118-120

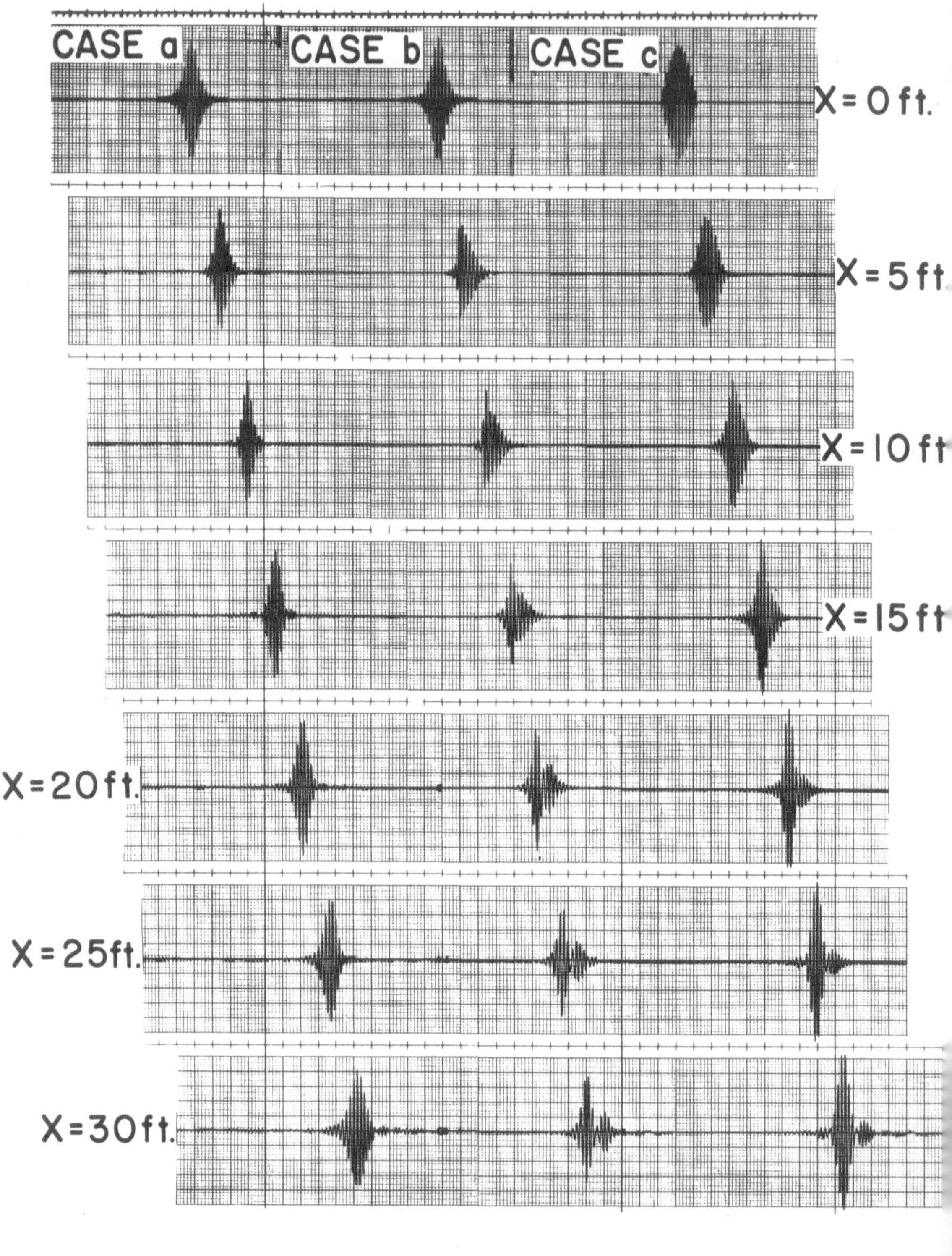
CASE a
CASE b
CASE c
X = 0 ft.
X = 5 ft.
X = 10 ft
X = 15 ft
X = 20 ft.
X = 25ft.
X = 30 ft.

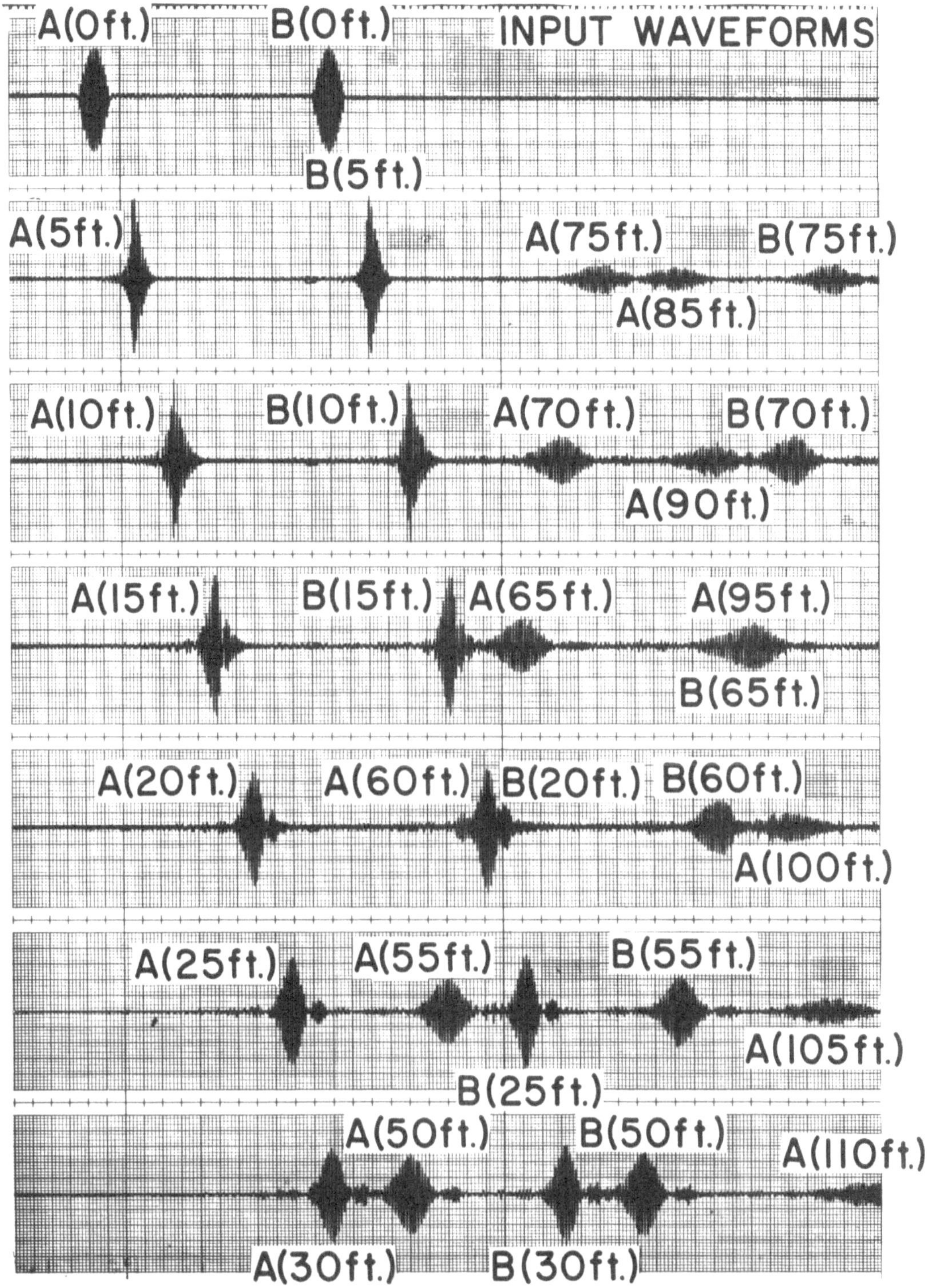
A(0ft.)
B(0ft.)
INPUT WAVEFORMS
B(5ft.)
A(5ft.)
A(75ft.)
B(75ft.)
A(85ft.)
A(10ft.)
B(10ft.)
A(70ft.)
B(70ft.)
A(90ft.)
A(15ft.)
B(15ft.)
A(65ft.)
A(95ft.)
B(65ft.)
A(20ft.)
A(60ft.)
B(20ft.)
B(60ft.)
A(100ft.)
A(25ft.)
A(55ft.)
B(55ft.)
A(105ft.)
B(25ft.)
A(50ft.)
B(50ft.)
A(110ft.)
A(30ft.)
B(30ft.)

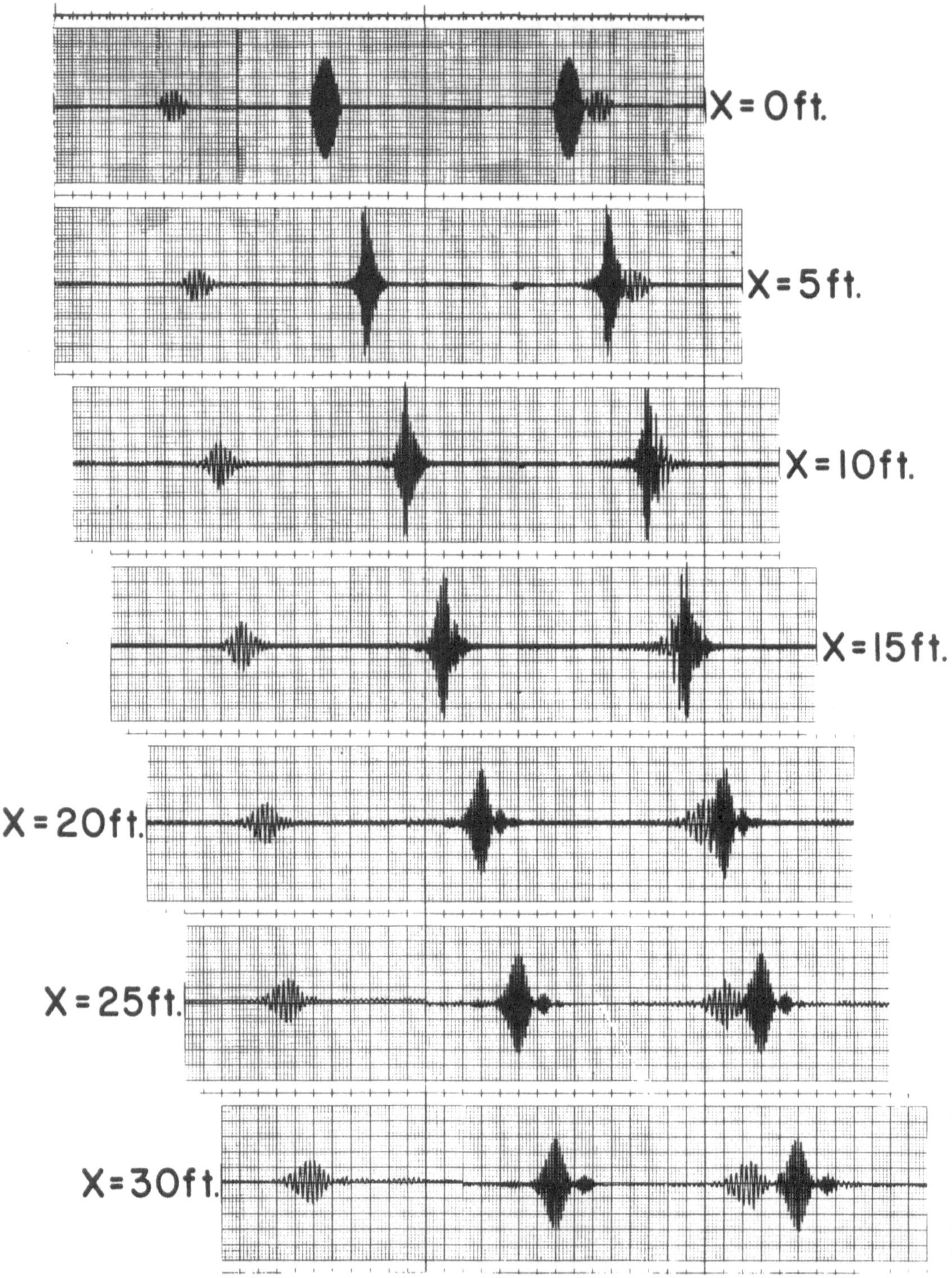
X = 0 ft.
X = 5 ft.
X = 10 ft.
X = 15 ft.
X = 20 ft.
X = 25 ft.
X = 30 ft.

10. Kolmogorov's Average Hamiltonian Technique

In the examination of the stability properties of the water wave field, using the mode rate equation (4.16), we concentrated on the dynamics of narrow band spectral processes. Restricted solutions have been obtained for such processes when the mode coupled equations reduce to the nonlinear Schrödinger equation to describe the envelope of the wave field.[64,143] This narrow band approximation is often made to describe the interaction between long waves and short waves[74,75], e.g., the interaction between gravity and capillary waves is usually made in two scale approximation[144,145] as in the formation of parasitic capillary waves at the crests of long waves.[146] However, the full mode rate equation, including both the quadratic and cubic interactions, have not yielded up any closed form solutions for a broad band spectrum of water waves. In this lecture we consider a perturbation approach developed by Kolmogorov[147] and recently discussed by Chirikov[34] which focuses attention on the generator of the motion, i.e., the Hamiltonian, rather than on the dynamic equations. This technique is to be contrasted with the classical perturbation technique discussed in the fourth lecture.

The technique for solving the mode rate equation (4.16) most familiar to physicists is a straightforward expansion of the mode amplitudes in terms of a smallness parameter. The steps involved in this approach are to: 1) express the mode amplitudes in the rate equations in a power series of the smallness parameter λ, i.e.,

$$B_{\underline{k}}(t) = B_{\underline{k}}^{(0)}(t) + \lambda B_{\underline{k}}^{(1)}(t) + \lambda^2 B_{\underline{k}}^{(2)}(t) + \dots ,$$

2) substitute this expression into the rate equations (4.16) and equate the coefficients of equal powers of the parameter λ to form the hierarchy of

of equations

$$\lambda^0: \quad \dot{B}^{(0)}_{\underline{k}}(t) + i\omega_k B^{(0)}_{\underline{k}}(t) = 0$$

$$\lambda^1: \quad \dot{B}^{(1)}_{\underline{k}}(t) + i\omega_k B^{(1)}_{\underline{k}}(t) = T^{(2)}_{\underline{k}}(\underline{B}^{(0)},\underline{B}^{(0)})$$

$$\lambda^2: \quad \dot{B}^{(2)}_{\underline{k}}(t) + i\omega_k B^{(2)}_{\underline{k}}(t) = T^{(2)}_{\underline{k}}(\underline{B}^{(0)},\underline{B}^{(1)}) + T^{(3)}_{\underline{k}}(\underline{B}^{(0)},\underline{B}^{(0)},\underline{B}^{(0)})$$

$$\vdots \qquad \vdots$$

where $T^{(2)}_{\underline{k}}$ and $T^{(3)}_{\underline{k}}$ are defined in (4.17) and (4.18) and the arguments indicate the order of B's included in the interaction; 3) integrate the $\mathcal{O}(\lambda^0)$ equations to obtain $B^{(0)}_{\underline{k}}(t)$ which is then inserted into the right hand side of the $\mathcal{O}(\lambda)$ equations and integrated to obtain $B^{(1)}_{\underline{k}}(t)$, etc. As discussed in the fourth lecture one cannot be assured of the convergence properties of this perturbation series so that the method must be applied with caution. An improvement on this method has been made by Weber and Barrick[106,107] by expanding the frequency as $\omega_k = \omega^{(0)}_k + \lambda\omega^{(1)}_k + \lambda^2\omega^{(2)}_k + \ldots$ and Fourier transforming the mode amplitudes in time. This analysis is an extension of the work of Stokes and was discussed earlier. The technique has not yet been applied to the wave-wave interaction process, however.

A method superior to the above perturbation approach[34] is based on the repeated applications of a phase averaging procedure in conjunction with canonically transforming the system variables.[147] To discuss this successive transformation technique of Kolmogorov we express the mode amplitudes for the water wave field $B_{\underline{k}}(t)$ in *polar coordinates*, i.e.,

$$B_{\underline{k}}(t) \equiv \sqrt{I_{\underline{k}}(t)}\, e^{-i\psi_{\underline{k}}(t)} \, . \qquad (1)$$

In the absence of the mode mixing terms, i.e., $H_3 = H_4 = \ldots = 0$ in (3.34) the set $(\underline{I},\underline{\psi})$ are the action-angle variables for the gravity wave field. These quantities are *not* the action-angle variables for the nonlinear wave field, however, because the interaction potential $(V = H_3 + H_4)$ is not independent of the angles $\underline{\psi}$. In polar coordinates the Hamiltonian for an isolated field of gravity waves (3.18) is

$$H_g = \sum_{\underline{\ell}} \omega_\ell I_{\underline{\ell}} + \lambda \sum_{\underline{\ell},\underline{m},\underline{p}} [I_{\underline{\ell}} I_{\underline{m}} I_{\underline{p}}]^{1/2} \Big\{\delta_{\underline{\ell}-\underline{m}-\underline{p}} V_3(\underline{\ell}\underline{m}\underline{p}) \sin[\psi_{\underline{\ell}} - \psi_{\underline{m}} - \psi_{\underline{p}}]$$

$$+ \delta_{\underline{\ell}+\underline{m}+\underline{p}} V_3'(\underline{\ell}\underline{m}\underline{p}) \sin[\psi_{\underline{\ell}} + \psi_{\underline{m}} + \psi_{\underline{p}}]\Big\}$$

$$+ \lambda^2 \sum_{\underline{\ell},\underline{m},\underline{p},\underline{g}} [I_{\underline{\ell}} I_{\underline{m}} I_{\underline{p}} I_{\underline{g}}]^{1/2} \Big\{\delta_{\underline{\ell}+\underline{m}-\underline{p}-\underline{g}} V_4(\underline{\ell}\underline{m}\underline{p}\underline{g}) \cos[\psi_{\underline{\ell}} + \psi_{\underline{m}} - \psi_{\underline{p}} - \psi_{\underline{g}}]$$

$$+ \delta_{\underline{\ell}+\underline{m}+\underline{p}+\underline{g}} V_4'(\underline{\ell}\underline{m}\underline{p}\underline{g}) \cos[\psi_{\underline{\ell}} + \psi_{\underline{m}} + \psi_{\underline{p}} + \psi_{\underline{g}}]$$

$$+ \delta_{\underline{\ell}-\underline{m}-\underline{p}-\underline{g}} V_4''(\underline{\ell}\underline{m}\underline{p}\underline{g}) \cos[\psi_{\underline{\ell}}-\psi_{\underline{m}}-\psi_{\underline{p}}-\psi_{\underline{g}}]\Big\} + O(\lambda^3) \qquad (2)$$

with the relation between these real coupling coefficients and those in (4.20) given by

$$V_3(\underline{\ell}\underline{m}\underline{p}) = 2iV_{\underline{m}\underline{p}}^{\underline{\ell}} = -2iV_{\underline{p}}^{\underline{\ell}\underline{m}} \; ; \qquad V_4'(\underline{\ell}\underline{m}\underline{p}\underline{g}) = 2V^{\underline{p}\underline{g}\underline{\ell}\underline{m}}$$

$$V_3'(\underline{\ell}\underline{m}\underline{p}) = 2iV^{\underline{\ell}\underline{m}\underline{p}} = =2iV_{\underline{\ell}\underline{m}\underline{p}} \; ; \qquad V_4''(\underline{\ell}\underline{m}\underline{p}\underline{g}) = 2V_{\underline{\ell}}^{\underline{p}\underline{g}\underline{m}}$$

$$V_4(\underline{\ell}\underline{m}\underline{p}\underline{g}) = 2V_{\underline{\ell}\underline{m}}^{\underline{p}\underline{g}} \, . \qquad (3)$$

The equations of motion in terms of the canonical polar coordinates are

$$\dot{\underline{I}}_{\underline{k}} = -\frac{\partial H_g}{\partial \underline{\psi}_{\underline{k}}} \; ; \qquad \dot{\underline{\psi}}_{\underline{k}} = \frac{\partial H_g}{\partial \underline{I}_{\underline{k}}} \; . \tag{4}$$

Kolmogorov's procedure is to first average the Hamiltonian $H_g(\underline{I},\underline{\psi})$ over the phase angle $\underline{\psi}$ to define an average Hamiltonian $K_g(\underline{I})$, i.e.,

$$K_g(\underline{I}) \equiv \langle H_g(\underline{I},\underline{\psi})\rangle_{\underline{\psi}} \; . \tag{5}$$

The average Hamiltonian K_g is used as a new reference from which the system can be perturbed, so that formally introducing an ordering parameter (interaction strength) λ we write

$$H_g(\underline{I},\underline{\psi}) = K_g(\underline{I}) + \lambda[H_g(\underline{I},\underline{\psi}) - \langle H_g(\underline{I},\underline{\psi})\rangle_{\underline{\psi}}] \; . \tag{6}$$

This parameter can be set equal to unity at the end of the analysis, however it is a useful way to index the contributions to (6) during the intermediate steps of analysis. The second step of the procedure is to introduce a canonical transformation to eliminate the perturbation at $O(\lambda)$ by going to a new set of canonical variables $(\underline{I}',\underline{\psi}')$. This new set of variables will have amplitude dependent frequencies and an explicit perturbation of $O(\lambda^2)$. Repeated application of these two steps shifts the nonlinearity to successively higher orders in λ. In classical perturbation theory the successive transformations $(\underline{I},\underline{\psi}) \rightarrow (\underline{I}',\underline{\psi}') \rightarrow (\underline{I}'',\underline{\psi}'') \ldots \rightarrow (\underline{J},\underline{\theta})$ are selected in such a way that at each level of this sequence the perturbation is reduced by one power of λ. Such series are known to be divergent, see e.g., Poincaré[148], so that their application to physical systems is only valid for short times. Kolmogorov's method, however, is one in which the canonical transformations is chosen such that at each step in the above sequence the perturbation is reduced by an order of the square of the preceding one. After n such transformations the classical approach has a perturbation of $\mathcal{O}(\lambda^n)$ and Kolmogorov's perturbation is of $\mathcal{O}(\lambda^{2^n})$.

Although Kolmogorov's procedure is an attractive one for treating nonlinear mechanical systems because of the rapid convergence in λ, there still exists the difficulty of the resonant interactions. The theorem of Kolmogorov, Arnold and Moser[33] establishes that if the unperturbed frequencies are linearly independent, the motion of the system is constrained to be on a N-dimensional torus imbedded in a 2 N-dimensional phase space for the system given that the system has N-degrees of freedom. The nonlinear perturbation terms, which maintain the linear independence condition, will distort these tori resulting in a modification of the linear frequencies and amplitudes, but the dynamic picture will be stable. In the Hamiltonian (1) the terms which satisfy this condition would not be expected to modify the system dynamics drastically. On the other hand, terms which violate the independence condition, i.e., resonance terms, are expected to destroy these nicely behaved tori and lead to a drastically altered evolution of the wave field, see e.g., the review papers, References 35 and 37.

Meiss and Watson[149] have indicated how one can use Kolmogorov's method for practical calculations when resonances are present in the wave field. Although they were concerned with the resonant triad interaction of two surface waves with one internal wave, the suggestion they make is still useful in the present study. They use the expediant of avoiding the singularities in the canonical transformation by omitting those terms from the transformation which are *close to resonance* (defined in a suitable way). For the gravity wave field there are no resonances at $O(\lambda)$ so that the terms in H_3 of (3.34) could be removed to $O(\lambda^2)$ by means of a canonical transformation. However, resonances do occur at $O(\lambda^2)$ some directly from the H_4 terms and others are induced by the transformation of the H_3 terms, see e.g., Watson and West[103]. The application of this modified averaging technique to remove the resulting divergences from the canonical transformation is crucial if one intends to study the exact dynamics of the gravity wave field.

The phase averaged Hamiltonian, obtained by averaging (2) over all $\underline{\psi}$, is

$$K_g(\underline{I}) = \sum_{\underline{\ell}} \omega_{\ell} I_{\underline{\ell}} + \lambda^2 \sum_{\underline{\ell m p g}}^{E} \delta_{\underline{\ell}+\underline{m}-\underline{p}-\underline{g}} [I_{\underline{\ell}} I_{\underline{m}} I_{\underline{p}} I_{\underline{g}}]^{1/2} V_4(\underline{\ell m p g}) \qquad (7)$$

where the E on the summation indicates that all *exactly resonant* four-wave interactions are included in (7). We have used the symmetry properties of the interaction coefficients discussed in the fourth lecture to write (7) in this simple form. The phase average of the three-wave interactions in H_3 vanish because they are of odd order in sines and cosines. The quartic nonlinearities have non-zero averages when the phases are *on-resonance*. The perturbation can now be defined as

$$U \equiv H_g - K_g = \lambda U_3 + \lambda^2 U_4 + \ldots \qquad (8)$$

with

$$U_j = H_j - \langle H_j \rangle_{\underline{\psi}} \; ; \qquad j = 3,4 \ldots \qquad (9)$$

so that by construction the phase average of U vanishes, i.e., $\langle U \rangle_{\underline{\psi}} = 0$. The total Hamiltonian for the isolated gravity wave field is now

$$H_g(\underline{I},\underline{\psi}) = K_g(\underline{I}) + \lambda \sum_{\underline{\ell}\underline{m}\underline{p}} [I_{\underline{\ell}} I_{\underline{m}} I_{\underline{p}}]^{1/2} \Big\{ \delta_{\underline{\ell}-\underline{m}-\underline{p}} V_3(\underline{\ell}\underline{m}\underline{p}) \sin [\psi_{\underline{\ell}} - \psi_{\underline{m}} - \psi_{\underline{p}}]$$

$$+ \delta_{\underline{\ell}+\underline{m}+\underline{p}} V_3'(\underline{\ell}\underline{m}\underline{p}) \sin [\psi_{\underline{\ell}} + \psi_{\underline{m}} + \psi_{\underline{p}}] \Big\} + \lambda^2 U_4 + \ldots \tag{10}$$

If the explicit λ dependence in (10) is neglected, the equations of motion given by (4) become

$$\dot{\underline{I}} = -\nabla_{\underline{\psi}} K_g = 0 \; ; \qquad \dot{\underline{\psi}} = \nabla_{\underline{I}} K_g = \underline{\Omega}(\underline{I}) \tag{11}$$

where $\underline{\Omega}(\underline{I})$ is the vector of perturbed frequencies $[\Omega_{\underline{k}_1}(\underline{I}),\ldots,\Omega_{\underline{k}_N}(\underline{I})]$. In the phase averaged, nonlinear system the action of each mode is seen to be a constant of the motion, i.e., K_g is independent of $\underline{\psi}$ by construction. The new frequencies $\Omega_{\underline{k}}(\underline{I})$ are, therefore, determined by the average effect of the non-linearities

$$\Omega_{\underline{k}}(\underline{I}) = \omega_k + \lambda^2 \sum_{\underline{\ell}} 2I_{\underline{\ell}} V_4(\underline{\ell}\underline{k}\underline{\ell}\underline{k}) \tag{12}$$

This amplitude dependence of the frequency is the same as we determined by the approximate integration of (6.15), only now we have a self-consistent method for including the shifted phases in the equations of motion. These frequencies were also obtained by Weber and Barrick by a different procedure. Having included this effect in the $(\underline{I},\underline{\psi})$ representation, we next remove the harmonic nonlinearity at $O(\lambda)$. As discussed in the fifth lecture, Hamilton-Jacobi theory prescribes how to do this by means of a canonical transformation.

A generating function for transforming the set of variables $(\underline{I},\underline{\psi})$ to a new canonical set $(\underline{I}',\underline{\psi}')$ is given by

$$S(\underline{I}',\underline{\psi}) = \underline{I}'\cdot\underline{\psi} + \lambda S_1(\underline{I}',\underline{\psi}) \; . \tag{13}$$

Here the function $S_1(\underline{I}',\underline{\psi})$ is specified by the condition that the $O(\lambda)$ perturbations are eliminated from the Hamiltonian (10). The prescription for the relation between $(\underline{I},\underline{\psi})$ and $(\underline{I}',\underline{\psi}')$ is given by (5.21) to be

$$\underline{I} = \nabla_{\psi} S(\underline{I}',\underline{\psi}) = \underline{I}' + \lambda \nabla_{\psi} S_1(\underline{I}',\underline{\psi}) \ ,$$

$$\underline{\psi}' = \nabla_{\underline{I}'} S(\underline{I}',\underline{\psi}) = \underline{\psi} + \lambda \nabla_{\underline{I}'} S_1(\underline{I}',\underline{\psi}) \ . \tag{14}$$

As discussed in Lecture 5, we substitute the transformed variables from (14) into the Hamiltonian (10) which we then expand in powers of λ. Equating the coefficients of equal powers of λ yields a partial differential equation for $S_1(\underline{I}',\underline{\psi})$ which we integrate to obtain

$$S_1(\underline{I}',\underline{\psi}) = \sum_{\underline{\ell m p}} \delta_{\underline{\ell}-\underline{m}-\underline{p}} [I'_{\underline{\ell}} I'_{\underline{m}} I'_{\underline{p}}]^{1/2} \left\{ \frac{V_3(\underline{\ell m p})}{\omega_\ell - \omega_m - \omega_p} \cos [\psi_{\underline{\ell}} - \psi_{\underline{m}} - \psi_{\underline{p}}] \right.$$
$$\left. + \frac{V'_3(\underline{\ell m p})}{\omega_\ell + \omega_m + \omega_p} \cos [\psi_{\underline{\ell}} + \psi_{\underline{m}} + \psi_{\underline{p}}] \right\} \tag{15}$$

to lowest order in λ.

The Hamiltonian (2) can now be written in the $(\underline{I}',\underline{\psi}')$ representation as

$$H_g(\underline{I}',\underline{\psi}') = K_g(\underline{I}') + \lambda^2 \sum_{\underline{\ell m p g}}^{\overline{E}} [I'_{\underline{\ell}} I'_{\underline{m}} I'_{\underline{p}} I'_{\underline{g}}]^{1/2} \left\{ \delta_{\underline{\ell}+\underline{m}-\underline{p}-\underline{g}} V_4(\underline{\ell m p g}) \right.$$
$$\times \cos (\psi'_{\underline{\ell}} + \psi'_{\underline{m}} - \psi'_{\underline{p}} - \psi'_{\underline{g}}) + \delta_{\underline{\ell}+\underline{m}+\underline{p}+\underline{g}} V'_4(\underline{\ell m p g}) \cos [\psi'_{\underline{\ell}} + \psi'_{\underline{m}} + \psi'_{\underline{p}} + \psi'_{\underline{g}}]$$
$$\left. + \delta_{\underline{\ell}-\underline{m}-\underline{p}-\underline{g}} V''_4(\underline{\ell m p g}) \cos [\psi'_{\underline{\ell}} - \psi'_{\underline{m}} - \psi'_{\underline{p}} - \psi'_{\underline{g}}] \right\}$$
$$+ \frac{1}{2} \lambda^2 \sum_{\underline{\ell m p}} [I'_{\underline{\ell}} I'_{\underline{m}} I'_{\underline{p}}]^{1/2} \left\{ \frac{1}{I'_{\underline{\ell}}} \frac{\partial S_1}{\partial \psi'_{\underline{\ell}}} + \frac{1}{I'_{\underline{m}}} \frac{\partial S_1}{\partial \psi'_{\underline{m}}} + \frac{1}{I'_{\underline{p}}} \frac{\partial S_1}{\partial \psi'_{\underline{p}}} \right\} F_3(\underline{\ell m p})$$

(16)

with

$$F_3(\underline{\ell m p}) = \delta_{\underline{\ell}-\underline{m}-\underline{p}} V_3(\underline{\ell m p}) \sin [\psi'_{\underline{\ell}} - \psi'_{\underline{m}} - \psi'_{\underline{p}}] + \delta_{\underline{\ell}+\underline{m}+\underline{p}} V'_3(\underline{\ell m p}) \sin [\psi'_{\underline{\ell}} + \psi'_{\underline{m}} + \psi'_{\underline{p}}] . \tag{17}$$

The Σ on the summation in (16) indicates that only the *exactly* non-resonant terms are *excluded* from the sum since they already contribute to the average Hamiltonian $K_g(\underline{I}')$. In the process of eliminating the $O(\lambda)$ interactions, however, we have generated additional resonant interactions in the new $O(\lambda^2)$ term of (16). Note that we have replaced $\underline{\psi}$ in (15) by $\underline{\psi}'$ since the difference in phases contribute to the Hamiltonian at $O(\lambda^3)$. These new terms arise from the substitution of the transformation of $\underline{I}$ in terms of $\underline{I}'$ in H_3 and does not get eliminated by $S_1(\underline{I}',\underline{\psi})$.

We now begin the procedure again. The generating function $S_1(\underline{I}',\underline{\psi})$ is well defined since there are no resonant triad interactions among gravity waves. However, in eliminating the $O(\lambda^2)$ terms in (16) by the canonical transformation $(\underline{I}',\underline{\psi}') \to (\underline{J},\underline{\theta})$ we encounter resonant quartet interactions so that the generating function for this transformation must be selected with care. We again phase average the Hamiltonian to obtain

$$\begin{aligned} K'_g(\underline{I}') &\equiv \langle H_g(\underline{I}',\underline{\psi}')\rangle_{\underline{\psi}'} \\ &= K_g(\underline{I}') + \frac{1}{2}\lambda^2 \sum_{\underline{\ell m p}} [I'_{\underline{\ell}} I'_{\underline{m}} I'_{\underline{p}}]^{1/2} \left\{ \frac{1}{I'_{\underline{\ell}}} \langle \frac{S_1}{\partial\psi'_{\underline{\ell}}} F_3\rangle_{\underline{\psi}'} + \frac{1}{I'_{\underline{m}}} \langle \frac{\partial S_1}{\partial\psi'_{\underline{m}}} F_3\rangle_{\underline{\psi}'} \right. \\ &\quad \left. + \frac{1}{I'_{\underline{p}}} \langle \frac{\partial S_1}{\partial\psi'_{\underline{p}}} F_3\rangle_{\underline{\psi}'} \right\} . \end{aligned} \tag{18}$$

Evaluating the phase averages in (18) and using (7) we obtain for the terms exactly on-resonance

$$K'_g(\underline{I}') = \sum_{\underline{\ell}} \omega_\ell I'_{\underline{\ell}} + \lambda^2 \sum_{\underline{\ell m p g}}^{\overline{E}} \delta_{\underline{\ell}+\underline{m}-\underline{p}-\underline{g}} [I'_{\underline{\ell}} I'_{\underline{m}} I'_{\underline{p}} I'_{\underline{g}}]^{1/2} T^{\underline{pg}}_{\underline{\ell m}} \tag{19}$$

where the new coupling coefficient is calculated to be

$$T^{\underline{pg}}_{\underline{\ell m}} = V_4(\underline{\ell m p g}) - \left\{ \frac{V_3(\underline{\ell},\underline{\ell}-\underline{g},\underline{g})V_3(\underline{p},\underline{p}-\underline{m},\underline{m})}{\omega_p - \omega_{|\underline{\ell}-\underline{g}|} - \omega_m} + \frac{1}{4}\frac{V_3(\underline{\ell}+\underline{m},\underline{\ell},\underline{m})V_3(\underline{\ell}+\underline{m},\underline{p},\underline{g})}{\omega_{|\underline{\ell}+\underline{m}|} - \omega_p - \omega_q} - \frac{1}{2}\frac{V_3(\underline{\ell},-(\underline{\ell}+\underline{m}),\underline{m})V'_3(\underline{p},-(\underline{\ell}+\underline{m}),\underline{g})}{\omega_p + \omega_{|\underline{\ell}+\underline{m}|} + \omega_q} - \frac{1}{4}\frac{V'_3(-(\underline{\ell}+\underline{m}),\underline{\ell},\underline{m})V'_3(-(\underline{\ell}+\underline{m}),\underline{p},\underline{g})}{\omega_{|\underline{\ell}+\underline{m}|} + \omega_p + \omega_q} - \frac{1}{2}\frac{V_3(\underline{\ell}+\underline{m},\underline{\ell},\underline{m})V_3(\underline{\ell}+\underline{m},\underline{p},\underline{g})}{\omega_p - \omega_{|\underline{\ell}+\underline{m}|} - \omega_q} \right\} . \tag{20}$$

The shifted Hamiltonian can be written as

$$H_g(\underline{I}',\underline{\psi}') = K'_g(\underline{I}') + \lambda^2 U'_4(\underline{I}',\underline{\psi}') + O(\lambda^3) \tag{21}$$

where again the resonant terms have been shifted from U_4 to K'_g by defining $U'_4 \equiv U_4 - \langle H_g(\underline{I}',\underline{\psi}')\rangle_{\underline{\psi}'}$.

We now introduce the second canonical transformation

$$S(\underline{J},\underline{\psi}') \equiv \underline{J} \cdot \underline{\psi}' + \lambda^2 S_2(\underline{J},\underline{\psi}') \tag{22}$$

which we insist remove all *non-resonant* interactions at $O(\lambda^2)$ in (21). As done earlier we have the new canonical coordinates

$$\underline{I}' \equiv \nabla_{\underline{\psi}'} S(\underline{J},\underline{\psi}') = \underline{J} + \lambda^2 \nabla_{\underline{\psi}'} S_2(\underline{J},\underline{\psi}')$$

$$\underline{\theta} \equiv \nabla_{\underline{J}} S(\underline{J},\underline{\psi}') = \underline{\psi}' + \lambda^2 \nabla_{\underline{J}} S_2(\underline{J},\underline{\psi}') . \tag{23}$$

In the $(\underline{J},\underline{\theta})$ coordinate representation the Hamiltonian for an isolated field of gravity waves is

$$H_g(\underline{J},\underline{\theta}) = K'_g(\underline{J}) + \sum_{\underline{\ell m p g}}^{R} \delta_{\underline{\ell}+\underline{m}-\underline{p}-\underline{g}}[J_{\underline{\ell}}J_{\underline{m}}J_{\underline{p}}J_{\underline{g}}]^{1/2}\, T^{\underline{pg}}_{\underline{\ell m}} \cos[\theta_{\underline{\ell}}+\theta_{\underline{m}}-\theta_{\underline{p}}-\theta_{\underline{g}}] \tag{24}$$

where the superscript R on the series indicates that all terms *near resonance* are to be explicitly included in the perturbation and are excluded from the generating function (22). We have both the exact and near resonance terms in (24), the former in $K'_g(\underline{J})$ and the latter indicated explicitly. The generating function $S_2(\underline{J},\underline{\psi}')$ obtained from (21) is

$$S_2(\underline{J},\underline{\psi}') = -\sum_{\underline{\ell m p g}}^{\bar{R}} [J_{\underline{\ell}}J_{\underline{m}}J_{\underline{p}}J_{\underline{g}}]^{1/2} \Bigg\{ \delta_{\underline{\ell}+\underline{m}-\underline{p}-\underline{g}} \frac{V_4(\underline{\ell m p g})}{\omega_\ell+\omega_m-\omega_p-\omega_q} \cos[\psi'_{\underline{\ell}}+\psi'_{\underline{m}}-\psi'_{\underline{p}}-\psi'_{\underline{g}}]$$
$$+ \delta_{\underline{\ell}+\underline{m}+\underline{p}+\underline{g}} \frac{V'_4(\underline{\ell m p g})}{\omega_\ell+\omega_m+\omega_p+\omega_q} \cos[\psi'_{\underline{\ell}}+\psi'_{\underline{m}}+\psi'_{\underline{p}}+\psi'_{\underline{g}}]$$
$$+ \delta_{\underline{\ell}-\underline{m}-\underline{p}-\underline{g}} \frac{V''_4(\underline{\ell m p g})}{\omega_\ell-\omega_m-\omega_p-\omega_q} \cos[\psi'_{\underline{\ell}}-\psi'_{\underline{m}}-\psi'_{\underline{p}}-\psi'_{\underline{g}}] \Bigg\} \tag{25}$$

where the superscript $\bar{R}$ indicates that all terms at or near the four-wave resonance band have been excluded from the summation.

The equations of motion in the $(\underline{J},\underline{\theta})$ repsentation are

$$\dot{J}_{\underline{k}} = -\sum_{\underline{\ell m p}}^{R} 4\delta_{\underline{\ell}+\underline{m}-\underline{p}-\underline{k}}\, T^{\underline{kp}}_{\underline{\ell m}} [J_{\underline{\ell}}J_{\underline{m}}J_{\underline{p}}J_{\underline{k}}]^{1/2} \sin[\theta_{\underline{\ell}}+\theta_{\underline{m}}-\theta_{\underline{p}}-\theta_{\underline{k}}] \tag{26a}$$

$$\dot{\theta}_{\underline{k}} = \Omega_{\underline{k}} - \sum_{\underline{\ell m p}}^{R} 2\delta_{\underline{\ell}+\underline{m}-\underline{p}-\underline{k}}\, T^{\underline{kp}}_{\underline{\ell m}} \left(\frac{J_{\underline{\ell}}J_{\underline{m}}J_{\underline{p}}}{J_{\underline{k}}}\right)^{1/2} \cos[\theta_{\underline{\ell}}+\theta_{\underline{m}}-\theta_{\underline{p}}-\theta_{\underline{k}}] \tag{26b}$$

where the set of terms in the resonance region is presumed to be much smaller than that in the original Hamiltonian (2). The canonical transformations have carried us through $(\underline{I},\underline{\psi}) \xrightarrow{S_1} (\underline{I}',\underline{\psi}') \xrightarrow{S_2} (\underline{J},\underline{\theta})$ so that the original $\underline{I}$ variables are of $O(\underline{J}^3)$ and we have indeed included high order nonlinearities in the system of

Eqs. (26). The utility of using the system (26) to make predictions on the evolution of a field of gravity waves has not yet been assessed and, therefore, the determination of the value of Kolmogorov's perturbation technique as a calculational tool in the present context is still uncertain.

11. Modification of the Hamiltonian to Couple with the Wind

The evolution of the gravity wave field on the ocean surface in the absence of external driving forces is modeled by (10.26). For a wind generated spectrum of gravity waves there are at least three physical mechanisms which modify the form of this relation. The first is the pressure term in Bernoulli's equation which models the air-sea coupling;[5] the second is the drift current generated at the ocean surface by the coupling of the water surface to the wind field[25] and finally the interaction with the spectrum of short waves generated by the wind prior to the development of the more energetic gravity waves.[144,145,150] Note that we are excluding from consideration external current sources such as internal waves, etc. These three external forcing terms, i.e., external to the gravity wave field, induce fluctuations in the gravity waves because of the existence of turbulent eddies in the air flow at the air-sea interface. The fluctuations in the atmosphere induce a fluctuating response of the water surface, by direct interactions at the gravity wave frequencies and wavelengths and by indirect processes through the wind generated gravity-capillary waves and the wind drift current. To describe the effect of the wind on the gravity wave field we must have a model of this coupling in the Hamiltonian.

We adopt the linear air-sea coupling model of Miles[86,87] and Phillips[85] to describe the initial growth of the water waves. The mean shear flow instability in the air and the turbulent eddies in the air flow act in concert to generate the fluctuating water wave field. Van Dorn[151] experimentally observes that there is a significant reduction of the momentum flux to the surface wave field from the wind in the absence of short waves. Dobson[88] in his measurements of the induced pressure perturbations in the air by low frequency water waves, experimentally determined that a large fraction of the momentum flux from the air to the sea goes intially into the water wave field rather than into a surface drift current. He further observes that the rates of growth of these long waves exceeds that predicted by the Miles'[86] inviscid laminar theory by a factor of between 5 and 8. Dobson resolved an apparent inconsistency between his data and those of Van Dorn by conjecturing that the high frequency short waves act as a catalyst to the "low frequency" growth mechanism.

Although our interest here is in the gravity wave field, it is clear that we cannot ignore the ambient environment in which these waves are generated and evolve. The physical effects of surface tension and viscosity make the evolution of gravity-capillary waves, as driven by the turbulent wind, much different from the simultaneous evolution of the long wavelength gravity waves. For a specified wind the short waves reach their steady state level at a given fetch, the longer waves equilibrating at greater fetches for greater wavelengths. The steady state spectrum of high frequency waves is, therefore, present on the surface of the evolving low frequency waves and can supply energy to the long waves as discussed by Longuet-Higgens[152] or contrarily, extract energy from the long waves as discussed by Hasselmann[153]. Valenzuela and Wright[145] point out that when the correlations between the energy influx from the atmosphere and the large-scale water flow (these were assumed to be uncorrelated by Hasselmann) are included in the energy balance equation that the modulated, short-wave, surface stress supplies energy flux to maintain the growth of large-scale gravity waves. It is presumed that this energy transfer mechanism enhances the gravity wave growth rate above that predicted by the Miles model,[86] and in part explains the discrepancy of the linear theory with observations.

Dynamic models, such as those proposed by Miles and Phillips provide a description of the initial states of the growth of high frequency water waves which is essentially correct. This description, however, does not yield a steady state spectral energy density for the water waves. The steady state observed in the data, e.g., Kitaigorodskii[7], is a result of the nonlinear interactions among surface waves which quench the wind induced instability. West[154], using the experimental conclusions of Plant and Wright[144] that "*the longest wave which grows primarily by the direct influx (of energy) from the air flow is probably about 10 cm in wavelength*", proposes a dynamic model for the gravity-capillary waves including the average effect of the nonlinear interactions. The result of West's analysis is a steady state energy spectral density for the high frequency surface waves which depends parametrically on the average nonlinear interaction and the spectrum of fluctuations in the wind field.

The arguments of van Dorn[151], Dobson[88], Valenzuela and Wright[145] and Plant and Wright[144] among others[9,155-157] suggest that the existence of the steady state spectrum of gravity-capillary waves is more important to the growth of gravity waves than the direct influx of energy from the air flow. We argue that the short waves are also responsible for the observed fluctuations in the gravity wave field and exclude the fluctuations in the gravity waves induced by the direct coupling to the wind field from consideration here. There is no doubt that the air-sea coupling mechanism is in part responsible for the observed stochastic behavior of the wind generated gravity waves, but to keep these mechanisms separate we postpone discussion of the direct air-sea coupling until later.

For a sufficiently large surface area Σ_0 the spectrum of modes we are considering is dense and our description of the dynamics should be accurate for water waves whose wavelengths are much less than the smaller of the lateral dimensions of our model ocean. Restricting our analysis to the short fetch situation in which only the gravity-capillary waves are generated by the wind and not the gravity waves, we modify the form of the Hamiltonian density obtained in the fourth lecture. The existence of capillary waves is a consequence of the fact that the interface between two immiscible fluids is in a state of uniform tension. The stress along the surface depends only on the nature of the fluid and the temperature. This *surface tension* is usually denoted by T and the free surface of water, i.e., the air-water interface, at 20°C has the value 75 dynes/cm. The surface tension contributes to the free energy of the fluid and therefore contributes to the Hamiltonian (3.18).

To incorporate this state of tension of the surface into the equations of motion we must modify our earlier discussion and construct the equations of motion through a variation of action argument for the water wave system. In such a variational formulation the state of tension may be treated by means of a Lagrange multiplier.[96] Introducing the Lagrangian L by the integral relation

$$L \equiv \int d^2x \, \phi_s(\underset{\sim}{x},t) \, \frac{\partial \zeta}{\partial t} \, (\underset{\sim}{x},t) - H_g \tag{1}$$

we express the action I for the gravity wave field in isolation as

$$I_g = \int_{t_1}^{t_2} dt \left\{ \int d^2x \, \phi_s(\underset{\sim}{x},t) \, \frac{\partial \zeta}{\partial t} \, (\underset{\sim}{x},t) - H_g \right\} . \tag{2}$$

For fixed end points the variation of (2) is most readily obtained by using Parseval's theorem and noting the series expansions for $\phi_s(\underset{\sim}{x},t)$ and $\zeta(\underset{\sim}{x},t)$ given by (4.7), i.e.,

$$\phi_s(\underline{x},t) = \sum_{\underline{\ell}} \sqrt{\frac{V_\ell}{2}}\,[B_{\underline{\ell}}(t)\, e^{i\underline{\ell}\cdot\underline{x}} + B^*_{\underline{\ell}}(t)\, e^{-i\underline{\ell}\cdot\underline{x}}]$$
$$\zeta(\underline{x},t) = i \sum_{\underline{\ell}} \frac{1}{\sqrt{2V_\ell}}\,[B_{\underline{\ell}}(t)\, e^{i\underline{\ell}\cdot\underline{x}} - B^*_{\underline{\ell}}(t)\, e^{-i\underline{\ell}\cdot\underline{x}}] \tag{3}$$

the action integral (2) becomes

$$I_g = \int_{t_1}^{t_2} dt \left\{ i \sum_{\underline{\ell}} (B_{\underline{\ell}} + B^*_{-\underline{\ell}})(\dot{B}_{-\underline{\ell}} - \dot{B}^*_{\underline{\ell}}) - H_g \right\} . \tag{4}$$

The equations of motion obtained in Lecture 4 result from the variational condition $\delta I_g = 0$, yielding after some rearrangement

$$\delta I_g = 0 = \int_{t_1}^{t_2} dt \left\{ i \sum_{\underline{\ell}} [-\delta B_{\underline{\ell}} \dot{B}^*_{\underline{\ell}} + \delta B^*_{\underline{\ell}} \dot{B}_{\underline{\ell}}] - \frac{\partial H_g}{\partial B_{\underline{\ell}}} \delta B_{\underline{\ell}} - \frac{\partial H_g}{\partial B^*_{\underline{\ell}}} \delta B^*_{\underline{\ell}} \right\} . \tag{5}$$

The variations in $B_{\underline{\ell}}$ and $B^*_{\underline{\ell}}$ are mutually independent and the end points are held fixed so that equating coefficients of $\delta B_{\underline{\ell}}$ and $\delta B^*_{\underline{\ell}}$ in (5) we obtain Hamilton's equations of motion

$$\dot{B}_{\underline{\ell}} = -i \frac{\partial H_g}{\partial B^*_{\underline{\ell}}} ; \qquad \dot{B}^*_{\underline{\ell}} = i \frac{\partial H_g}{\partial B^*_{\underline{\ell}}} \tag{6}$$

which, of course, is just (4.16).

The contribution of the surface tension to the energy of the system does in turn, through the Hamiltonian, alter the motion. Introducing the Lagrange multiplier T we consider the new action integral I', i.e.,

$$I' = I_g + T \left(\Sigma_0 - \int_{\Sigma_0} \frac{d\sigma}{\rho_w} \right) \tag{7}$$

with I_g given by (3), Σ_0 the surface area of interest, ρ_w the fluid density and $d\sigma$ an element of surface area. The original equations of motion (4.18) condition imposed by (7) is that the variation in surface area vanish, i.e., the total surface area Σ_0 remains constant during the fluid motion

$$\delta \int_{\Sigma_0} \frac{d\sigma}{\rho_w} = 0 \ . \tag{8}$$

The Lagrange multiplier T provides the force of reaction necessary to maintain the equation of constraint (8). This implies that there is a new mechanical force in the equations of motion generated by

$$\delta I' = \delta I_g - T\,\delta \int_{\Sigma_0} \frac{d\sigma}{\rho_w} = 0 \ . \tag{9}$$

The variation of δI_g again yields the seven terms in (4.16), but now we have an additional force to consider. The area of the fluid surface is given in terms of the fixed coordinate $\underline{x}$ by the projection from the free surface $z = \zeta(\underline{x},t)$ by

$$d\sigma = \sqrt{1 + (\nabla_{\underline{x}}\zeta)^2}\; d^2x \ . \tag{10}$$

Equation (10) results from the scalar product of the unit normal to the free surface $\underline{\hat{n}} = (-\nabla_{\underline{x}}\zeta, 1)/\sqrt{1 + (\nabla_{\underline{x}}\zeta)^2}$ and the unit normal to the (x,y) plane (0, 0, 1). Using (10), assuming a constant fluid density $\rho_w = \rho_0$ and defining the new Lagrange multiplier $\gamma \equiv T/\rho_0$ yields for the variation (8)

$$\begin{aligned} T\delta \int_{\Sigma_0} \frac{d\sigma}{\rho_w} &= \gamma\delta \int_{\Sigma_0} \sqrt{1 + (\nabla_{\underline{x}}\zeta)^2}\; d^2x \\ &= \gamma \int_{\Sigma_0} d^2x \; \frac{\nabla_{\underline{x}}\zeta \cdot \nabla_{\underline{x}}\delta\zeta}{\sqrt{1 + (\nabla_{\underline{x}}\zeta)^2}} \ . \end{aligned} \tag{11}$$

Expanding the denominator in the integrand of (11) and neglecting terms higher than cubic in ζ we obtain by again applying Parsavel's theorem

$$\gamma\delta \int_{\Sigma_0} \sqrt{1 + (\nabla_{\underline{x}}\zeta)^2}\, d^2x = \gamma \sum_{\underline{\ell}} \frac{\ell^2}{2V_\ell} [B_{\underline{\ell}}\delta B^*_{\underline{\ell}} + B^*_{\underline{\ell}}\delta B_{\underline{\ell}}]$$

$$+ \sum_{\underline{\ell m p g}} \Big\{ \delta_{\underline{\ell}+\underline{m}+\underline{p}+\underline{g}} \bar{\Gamma}_{\underline{\ell m p g}} B_{\underline{\ell}} B_{\underline{m}} B_{\underline{p}}$$

$$+ \delta_{\underline{\ell}+\underline{m}-\underline{p}+\underline{g}} \bar{\Gamma}^{\underline{p}}_{\underline{\ell m g}} B_{\underline{\ell}} B_{\underline{m}} B^*_{\underline{p}}$$

$$+ \delta_{\underline{\ell}-\underline{m}-\underline{p}+\underline{g}} \bar{\Gamma}^{\underline{mp}}_{\underline{\ell g}} B_{\underline{\ell}} B^*_{\underline{m}} B^*_{\underline{p}}$$

$$+ \delta_{-\underline{\ell}-\underline{m}-\underline{p}+\underline{g}} \bar{\Gamma}^{\underline{\ell m p}}_{\underline{g}} B^*_{\underline{\ell}} B^*_{\underline{m}} B^*_{\underline{p}} \Big\} \delta B_{\underline{g}} + cc \qquad (12)$$

where

$$\bar{\Gamma}_{\underline{\ell m p g}} = \bar{\Gamma}^{\underline{\ell m p}}_{\underline{g}} = \frac{\gamma}{4[V_\ell V_m V_p V_q]^{1/2}} (\underline{\ell}\cdot\underline{m})(\underline{p}\cdot\underline{g})$$

$$\bar{\Gamma}^{\underline{p}}_{\underline{\ell m g}} = \frac{\gamma}{4[V_\ell V_m V_p V_q]^{1/2}} [(\underline{\ell}\cdot\underline{m})(\underline{p}\cdot\underline{g}) + (\underline{\ell}\cdot\underline{q})(\underline{p}\cdot\underline{m}) + (\underline{\ell}\cdot\underline{p})(\underline{m}\cdot\underline{g})]$$

$$\bar{\Gamma}^{\underline{mp}}_{\underline{\ell g}} = \frac{\gamma}{4[V_\ell V_m V_p V_q]^{1/2}} [2(\underline{\ell}\cdot\underline{m})(\underline{p}\cdot\underline{g}) + (\underline{\ell}\cdot\underline{p})(\underline{g}\cdot\underline{m})] . \qquad (13)$$

By combining the first order terms from (12) and (4.8)we observe that for a given wave vector $\underline{\ell}$, that the phase velocity V_ℓ and the parameters g and γ occur in the combinations

$$\frac{1}{2}\left[\frac{\ell V_\ell^2 + g}{V_\ell} + \frac{\gamma\ell^2}{V_\ell}\right] = \frac{1}{2}\frac{\omega_\ell^2 + g + \gamma\ell^3}{\omega_\ell} . \qquad (14)$$

If instead of interpreting ω_ℓ to be $\sqrt{g\ell}$ in (14), we interpret it as $\sqrt{g\ell + \gamma\ell^3}$, then (14) reduces to ω_ℓ $(=\sqrt{g\ell + \gamma\ell^3})$, the frequency of a small amplitude, free gravity-capillary wave. Thus by interpreting the phase velocity in the complex amplitude $Z(\underline{x},t)$ [cf., (4.6)] as $V_\ell = \sqrt{g/\ell + \gamma\ell}$, we can include the effect of surface tension directly into the equations of motion for the water wave field. The higher order effects of surface tension may be included by modifying the interaction coefficients V_3, V_3' and V_4 to include the $\bar{\Gamma}$'s, but due to the small size of γ this is probably not necessary.

The final result of the above argument is that the variational equation (9) has an extremum at $\delta I' = 0$. The resulting equations of evolution subject to the surface constraint (8) is identically that obtained by modifying the canonical variables to include surface tension and then using Hamilton's equations (6). We will not indicate this change of variables, but merely note the new dispersion relation for the remaining discussion. We denote this new Hamiltonian by H_{sw} indicating that all surface waves are included.

The equations of motion (6) still do not provide for a source of wave energy from the wind field. We consider the flow of energy across the free surface due to a pressure distribution in the air $p_a(\underline{r},t)$ using an argument which in part is due to Stoker.[2] This pressure is discontinuous across the surface $z = \zeta(\underline{x},t)$, the discontinuity being determined by the surface tension, i.e., the pressure in the water at the free surface is related to the pressure in the air at the free surface by

$$p_a(\underline{x},t) = p(\underline{x},t) + \frac{\gamma\nabla^2_{\underline{x}}\zeta(\underline{x},t)}{\{1 + [\nabla_{\underline{x}}\zeta(\underline{x},t)]^2\}^{3/2}} \,. \tag{15}$$

Thus from Bernoulli's equation for the motion of the surface,

$$\frac{\partial\phi}{\partial t} + \frac{1}{2}\nabla\phi\cdot\nabla\phi + g\zeta = -\frac{p}{\rho_w} \qquad \text{on } \zeta(\underline{x},t) \tag{16}$$

we can write the energy of the surface wave field either as

$$E = \rho_o \int_V dV \left[\frac{1}{2}\nabla\phi\cdot\nabla\phi + g\zeta\right] \tag{17}$$

for a constant fluid density as in (3.12) or as

$$E = -\int_V dV \left[p + \rho_o \frac{\partial\phi}{\partial t}\right] . \tag{18}$$

The time rate of change of the surface wave energy can now be written using $E = \int_V dV\, f(\underline{r},t)$ as

$$\frac{dE}{dt} = \int_V dV \frac{\partial f}{\partial t} + \int_{\Sigma_o} f \frac{\partial\phi}{\partial n} d\sigma \tag{19}$$

where $\partial\phi/\partial n$ is the velocity of the fluid normal to the free surface and $d\sigma$ is an element of area on that surface. To evaluate (19) we use the definition of $f(\underline{r},t)$ implied by (17) in the first integral and that implied by (18) in the second integral to obtain

$$\frac{dE}{dt} = \rho_o \int_V dV\, \nabla\phi\cdot\nabla \frac{\partial\phi}{\partial t} - \int_{\Sigma_o} d\sigma \left[p + \rho_o \frac{\partial\phi}{\partial t}\right] \frac{\partial\phi}{\partial n} . \tag{20}$$

Using the fact that the velocity potential satisfies Laplace's equation $\nabla^2\phi$ we may apply Greens' formula to the first integral in (20) to obtain

$$\rho_o \int_{\Sigma_o} \frac{\partial\phi}{\partial t}\frac{\partial\phi}{\partial n} d\sigma \tag{21}$$

and thus express the time rate of change in the energy as

$$\frac{dE}{dt} = -\int_{\Sigma_o} d\sigma\, p \frac{\partial\phi}{\partial n} . \tag{22}$$

Using the pressure condition across the free surface given by (15) we obtain from (22)

$$\frac{dE}{dt} = -\int_{\Sigma_0} d\sigma p_a \frac{\partial\phi}{\partial n} + \gamma \int_{\Sigma_0} \frac{d\sigma\, \nabla^2\zeta}{[1 + (\nabla_{\underline{x}}\zeta)^2]^{3/2}} \frac{\partial\phi}{\partial n} \,. \tag{23}$$

However, noting that the velocity normal to the surface is given by

$$\frac{\partial\phi}{\partial n} = - \frac{1}{\sqrt{1 + (\nabla_{\underline{x}}\zeta)^2}} \frac{\partial\zeta}{\partial t} \tag{24}$$

and the increment of surface area is given by (10); we can rewrite (23) as

$$\begin{aligned} \frac{dE}{dt} &= -\int_{\Sigma_0} d\sigma p_a \frac{\partial\phi}{\partial n} + \frac{d}{dt} \gamma \int_{\Sigma_0} d\sigma \,, \\ &= -\int_{\Sigma_0} d^2x p_a \frac{\partial\zeta}{\partial t} + \frac{d}{dt} \gamma \int_{\Sigma_0} d\sigma \,. \end{aligned} \tag{25}$$

Thus the variation in surface wave energy is determined by the energy influx due to the pressure distribution in the wind field at the free surface $p_a(\underline{x},t)$ and the change in the energy associated with the surface tension acting to preserve the total surface area.

We have argued that the second term in (25) can be explicitly included in the Hamiltonian for the water wave field by using H_{sw} rather than H_g. If we now express the pressure distribution of the air flow at the free surface in the Fourier series

$$p_a(\underline{x},t) = \sum_{\underline{\ell}} p_{\underline{\ell}}(t)\, e^{i\underline{\ell}\cdot\underline{x}} \tag{23}$$

then the equations of motion for the water wave modes become

$$\dot{B}_{\underline{\ell}} = -i \frac{\partial H_{SW}}{\partial B^{*}_{\underline{\ell}}} - \frac{p_{\underline{\ell}}}{\rho_0 \sqrt{2V_{\ell}}} \tag{24}$$

$$\dot{B}^{*}_{\underline{\ell}} = i \frac{\partial H_{SW}}{\partial B_{\underline{\ell}}} + \frac{p^{*}_{\underline{\ell}}}{\rho_0 \sqrt{2V_{\ell}}} . \tag{25}$$

Since the pressure is a real quantity the coefficients have the property $p_{\underline{\ell}} = p^{*}_{-\underline{\ell}}$. Alternatively one could express the interaction of the pressure field with the fluid surface by means of the interaction potential

$$V_p \equiv \sum_{\underline{\ell}} \frac{i}{\sqrt{2V_{\ell}}} [B_{\underline{\ell}} p^{*}_{\underline{\ell}} - B^{*}_{\underline{\ell}} p_{\underline{\ell}}] . \tag{26}$$

The new Hamiltonian for the interactive system can be the written as

$$H_T \equiv H_{SW} + V_p \tag{27}$$

so that the $B_{\underline{\ell}}$ waves are now supplied with energy from the surface pressure field, e.g., from the wind generated surface stress.

The final force we consider to be acting on the surface is viscous damping and the arguments associated with the development of its functional form are generally constructed in analogy with the theory of the elasticity of solids. Although a variational argument is given by Rayleigh[158] for including dissipation in a Lagrangian formalism we will not reproduce that argument here. These arguments are avoided because they involve rotational flow and the generation of vorticity, see, e.g., Lamb[1] (p. 581), which would take us far afield of our present interest. We merely note that the dissipation function in irrotational flow results in an overall loss of energy at a rate $2\nu k^2$ for a wave of wave vector $\underline{k}$. We, therefore, introduce viscosity phenomenologically and write the rate equations (23) as

$$\dot{B}_{\underline{k}}(t) + (\nu k^2 + i\omega_k)B_{\underline{k}}(t) = T_{\underline{k}}^{(2)}(t) + T_{\underline{k}}^{(3)}(t) + f_{\underline{k}}(t) \tag{28}$$

and $f_{\underline{k}}(t)$ is given by

$$f_{\underline{k}}(t) \equiv -p_{\underline{k}}(t)/\sqrt{2V_k}\ \rho_0 \ . \tag{29}$$

The phenomenological parameter ν is often referred to as the "bare" viscosity to distinguish it from the effective damping measured in turbulent fluid flows. In later lectures we will also have occasion to make a similar distinction.

12. Lagrangian Formulation

In the preceding lecture we developed a set of mode rate equations to describe the dynamics of a field of wind driven water waves. In particular we were interested in how energy may be transferred from the short scale waves, which are presumed to be strongly coupled to the air flow, to the long scale waves which are only weakly coupled to the air flow. The presentation was based on a variational formulation of the surface properties making use of the Hamiltonian constructed earlier. It is worthwhile at this point to compare the mode rate equations developed there with an alternate derivation based on a different variational argument. The present argument is based in part on a stationary pressure condition at the surface of the fluid due to Luke[108]. We consider this alternative because it provides a straightforward method for modifying the linear eigenfunctions in which we are expanding the surface quantities. This modification is necessary to systematically include the physics of the high frequency waves in the equations of motion.

The equation of motion for the fluid, in terms of the velocity potential, is obtained by a variation of the integral

$$I = \int_{\Sigma_0} d^2x \, dt \int_{-B}^{\zeta(\underline{x},t)} \left\{ p(\underline{r},t) + \rho_0 \left[\dot{\phi}(\underline{r},t) + \frac{1}{2} \nabla\phi(\underline{r},t) \cdot \nabla\phi(\underline{r},t) + gz \right] \right\} dz \,. \tag{1}$$

Note that the integral over the vertical water column is from the great depth $z = -B$ to the free surface $z = \zeta(\underline{x},t)$. Thus the variation of I is done by varying $\phi(\underline{r},t)$ under the integral in (1) and $\zeta(\underline{x},t)$ at the limit of the inner-most integral to obtain

$$\delta I = \int_{\Sigma_0} d^2x \int_{t_1}^{t_2} dt \left\{ \left[\dot{\phi} + \frac{1}{2} (\nabla\phi)^2 + gz \right] \Big|_{z=\zeta(\underline{x},t)} \delta\zeta \right.$$
$$\left. + \int_{-B}^{\zeta(\underline{x},t)} (\delta\dot{\phi} + \nabla\phi \cdot \nabla\delta\phi) \, dz + \frac{p(\underline{r},t)}{\rho_0} \Big|_{z=\zeta(\underline{x},t)} \delta\zeta \right\} \rho_0 \,. \tag{2}$$

It is only recently that Luke[108] determined that the stationary value of (2) yields both the equations of motion for the water waves and the boundary conditions at both the free surface and the ocean floor (with p set to zero).

If we integrate the second expression within the brackets in (2) by parts, we obtain

$$\delta I = \int_{\Sigma_0} d^2x \int_{t_1}^{t_2} dt \left\{ \frac{\partial}{\partial t} \int_{-B}^{\zeta} \delta\phi dz + \nabla_{\underline{x}} \cdot \int_{-B}^{\zeta} \nabla_{\underline{x}}\phi \; \delta\phi dz - \int_{-B}^{\zeta} \left(\nabla_{\underline{x}}^2 \phi + \frac{\partial^2 \phi}{\partial z^2} \right) \delta\phi dz \right.$$

$$- \left[\dot{\zeta} + \nabla_{\underline{x}}\phi \cdot \nabla_{\underline{x}}\zeta - \frac{\partial \phi}{\partial z} \right]_{z=\zeta} \delta\phi + \left[\nabla_{\underline{x}}\phi \cdot \nabla_{\underline{x}} B + \frac{\partial \phi}{\partial z} \right]_{z=-B} \delta\phi$$

$$\left. + \left(\left[\dot{\phi} + \frac{1}{2} \nabla\phi \cdot \nabla\phi + gz \right]_{z=\zeta} + \left. \frac{p}{\rho_0} \right|_{z=\zeta} \right) \delta\zeta \right\} \rho_0 \tag{3}$$

The first two integrals in (3) are made to vanish by selecting $\delta\phi$ such that it vanishes at the boundaries of Σ_0 and is stationary at the time points t_1 and t_2. If the integral I is to be stationary, i.e., $\delta I = 0$, then the coefficients of $\delta\phi$ and $\delta\zeta$ must vanish independently:

(2nd Integral) $\quad \nabla^2 \phi = 0 \; ; \qquad \zeta(\underline{x},t) \geq z \geq -B \qquad$ (4a)

(3rd Integral) $\quad \frac{\partial \zeta}{\partial t} + \nabla_{\underline{x}}\phi \cdot \nabla_{\underline{x}}\zeta = \frac{\partial \phi}{\partial z} \; ; \qquad z = \zeta(\underline{x},t) \qquad$ (4b)

(4th Integral) $\quad \nabla_{\underline{x}}\phi \cdot \nabla_{\underline{x}} B = - \frac{\partial \phi}{\partial z} \; ; \qquad z = -B \qquad$ (4c)

(5th Integral) $\quad \frac{\partial \phi}{\partial t} + \frac{1}{2} \nabla\phi \cdot \nabla\phi + gz = -p/\rho_0 \quad z = \zeta(\underline{x},t) \; . \qquad$ (4d)

Equations (4a) to (4d) completely define the dynamics of the nonlinear water wave field. The velocity potential is seen to satisfy Laplace's equation (4a) throughout the fluid volume. Physically this is equivalent to having an incompressible fluid, i.e., $\nabla \cdot \underline{u} = 0$ since $\underline{u} = \nabla\phi$. The boundary condition at the free surface of the fluid (4b) matches the motion of the surface with the vertical component of the fluid velocity. The boundary condition at the ocean bottom (4c) similarly matches the vertical fluid velocity with the stationary bottom topography. The dynamic equation for the water wave field at the free surface is given by (4d) which is Bernoulli's equation. Note that all this information is a consequence of the change in Lagrangian density and the boundary conditions on the variation of the field variables.

The problem defined by (4) is much more extensive than we have elected to investigate. We, therefore, restrict our attention to water waves on the free surface of the ocean $z = \zeta(\underline{x},t)$ and assume B is very much greater than the longest wavelength of these surface waves. The two expressions for our model ocean system are, of course,

$$\left.\begin{aligned} &\frac{\partial\phi}{\partial t} + \frac{1}{2}\nabla\phi\cdot\nabla\phi + g\zeta = \frac{-p}{\rho_0} \\ &\frac{\partial\zeta}{\partial t} + \nabla_{\underline{x}}\phi\cdot\nabla_{\underline{x}}\zeta = \frac{\partial\phi}{\partial z} \end{aligned}\right\} \quad z = \zeta(\underline{x},t) \; . \tag{5}$$

To establish the relation between the configuration space partial differential equations (5) and the mode rate equations (4.16) we recall that a differential operator D acting on a function F may be written

$$DF_s(\underline{x},t) = DF(\underline{x},z,t)\Big|_{z=\zeta} + \frac{\partial F}{\partial z}\Big|_{z=\zeta} D\zeta \; . \tag{6}$$

The first term on the right hand side of (6) is the variation in F with z held constant and then evaluated at $z = \zeta(\underline{x},t)$; the second term is the variation in F with z due to the free surface being varied.

If we replace the operator D in (6) with $\partial/\partial t$ and $\nabla_{\underline{x}}$, and replace the function F with the surface velocity potential we obtain

$$\left.\frac{\partial\phi}{\partial t}\right|_{z=\zeta} = \frac{\partial\phi_s}{\partial t} - W\frac{\partial\zeta}{\partial t}$$

$$\left.\nabla_{\underline{x}}\phi\right|_{z=\zeta} = \nabla_{\underline{x}}\phi_s - W\nabla_{\underline{x}}\zeta \ . \qquad (7)$$

In (7) we have defined the vertical component of the fluid velocity at the free surface as,

$$W \equiv \left.\frac{\partial\phi}{\partial z}\right|_{z=\zeta} . \qquad (8)$$

Substituting (7) into (5) yields after some simplification

$$\frac{\partial\phi_s}{\partial t} + \frac{1}{2}\nabla_{\underline{x}}\phi_s\cdot\nabla_{\underline{x}}\phi_s + g\zeta = \frac{1}{2}W^2[1 + \nabla_{\underline{x}}\zeta\cdot\nabla_{\underline{x}}\zeta] - \frac{p}{\rho_0}$$

$$\frac{\partial\zeta}{\partial t} + \nabla_{\underline{x}}\phi_s\cdot\nabla_{\underline{x}}\zeta = [1 + \nabla_{\underline{x}}\zeta\cdot\nabla_{\underline{x}}\zeta]W \qquad (9)$$

where ϕ_s and ζ are surface quantities and W is the vertical fluid velocity at the surface. Series expressions for W in terms of ϕ_s were discussed in the third lecture; we do not recalculate its form here. We note that the two approaches result in the same equations of motion.

If we now use the relation between the pressure in the water $p(\underline{x},t)$ at $z = \zeta(\underline{x},t)$ with the pressure in the air $p_a(\underline{x},t)$ at $z = \zeta(\underline{x},t)$ given by (11.23) we can rewrite the equations of motion (9), again including viscosity phenomenologically, as

$$\frac{\partial\phi_s}{\partial t} + \frac{1}{2}\nabla_{\underline{x}}\phi_s\cdot\nabla_{\underline{x}}\phi_s + g\zeta + \frac{1}{2}[1 + \nabla_{\underline{x}}\zeta\cdot\nabla_{\underline{x}}\zeta]W^2 - \gamma\frac{\nabla_{\underline{x}}^2\zeta}{[1 + \nabla_{\underline{x}}\zeta\cdot\nabla_{\underline{x}}\zeta]^{3/2}} - 2\nu\nabla_{\underline{x}}^2\phi = -p_a/\rho_0 ,$$

$$\frac{\partial\zeta}{\partial t} + \nabla_{\underline{x}}\phi_s\cdot\nabla_{\underline{x}}\zeta = [1 + \nabla_{\underline{x}}\zeta\cdot\nabla_{x}\zeta]\ W , \qquad (10)$$

evaluated at $z = \zeta(\underline{x},t)$.

To generate waves on the ocean surface as described by (10) we use the Phillips-Miles model to couple the sea surface to the turbulent wind field. This model consists of separating the pressure field into the incoherent pressure fluctuations $p_0(\underline{x},t)$ and the in-phase pressure variations

$$p_1(\underline{x},t) = 2\mu_\kappa \rho_0 V_\kappa \dot{\zeta}(\underline{x},t) \ , \tag{11}$$

yielding the total pressure field

$$p_a(\underline{x},t) = p_0(\underline{x},t) + P_1(\underline{x},t) \ . \tag{12}$$

The parameter μ_κ is the fractional increase in the surface energy per radian for which we use the wavenumber dependent quantity evaluated by Miles.[86]

We are now in a position to contrast the normal mode expansion for (10) with the modifications we made in the Hamiltonian in the previous lecture to accommodate surface tension and the surface pressure field. We expand the surface deflection and surface velocity potential in the Fourier series

$$\zeta(\underline{x},t) = \sum_{\underline{k}} i\zeta_{\underline{k}}(t)\, e^{i\underline{k}\cdot\underline{x}} \ , \tag{13}$$

$$\phi_s(\underline{x},t) = \sum_{\underline{k}} \phi_{\underline{k}}(t)\, e^{i\underline{k}\cdot\underline{x}} \ , \tag{14}$$

and substitute them into the form of (1) which results when the series expansion for the vertical surface velocity W is used. The equations of motion are

$$\frac{\partial \phi_{\underline{k}}}{\partial t} + 2(\nu k^2 - \mu_k \omega_k)\phi_{\underline{k}} + (g + \gamma k^2)\zeta_{\underline{k}} = -p_{\underline{k}} + \bar{F}_\phi(\underline{k})$$

$$\frac{\partial \zeta_{\underline{k}}}{\partial t} - k\phi_{\underline{k}} = \bar{F}_\zeta(\underline{k}) \ . \tag{15}$$

Here ω_k is the angular frequency $\sqrt{gk + \gamma k^3}$ and the functions $F_\phi(\underline{k})$ and $F_\zeta(\underline{k})$ are defined by the spatial Fourier transform of the nonlinear terms resulting from (10). These terms are

$$F_\phi(\underline{x},t) = \frac{1}{2}\,[(\kappa\phi_s)^2 - (\nabla_{\underline{x}}\phi_s)^2] + \kappa\phi_s[\zeta\kappa^2\phi_s - \kappa(\zeta\kappa\phi_s)] - 2\mu_\kappa V_\kappa F_\zeta$$

$$- \frac{3}{2}\,\gamma(\nabla_{\underline{x}}\zeta)^2\nabla_{\underline{x}}^2\zeta - 2\nu\nabla_{\underline{x}}^2\zeta[\kappa\phi_s - \kappa\zeta\kappa\phi_s + \zeta\kappa^2\phi_s] \; ; \qquad (16a)$$

$$F_\zeta(\underline{x},t) = -\nabla_{\underline{x}}\phi_s\cdot\nabla_{\underline{x}}\zeta - \kappa\zeta\kappa\phi_s + \zeta\kappa^2\phi_s + \kappa\phi_s(\nabla_{\underline{x}}\zeta)^2 + \kappa\zeta\kappa\zeta\kappa\phi_s$$

$$- \frac{1}{2}\,\zeta^2\kappa^2\phi_s - \zeta\kappa^2\zeta\kappa\phi_s + \frac{1}{2}\,\zeta^2\kappa^3\phi_s \; ; \qquad (16b)$$

from which we see that the original linear models of viscous dissipation and the in-phase coupling to the mean air flow give rise to nonlinear terms at the free surface, i.e., those terms proportional to ν and μ_k in (16). Here $p_{\underline{k}}(t)$ is the Fourier transform of the fluctuating pressure field $p_0(\underline{x},t)$.

Equation (15) is cast in the form of a system of coupled mode amplitudes by diagonalizing its linear part. The homogeneous linear system of equations is given by

$$\frac{\partial}{\partial t}\begin{pmatrix}\phi_{\underline{k}}(t)\\ \zeta_{\underline{k}}(t)\end{pmatrix} + \begin{pmatrix}2(\nu k^2 - \mu\omega_k) & (g + \gamma k^2)\\ -k & 0\end{pmatrix}\begin{pmatrix}\phi_{\underline{k}}(t)\\ \zeta_{\underline{k}}(t)\end{pmatrix} = 0 \;, \qquad (17)$$

which can be diagonalized to yield

$$\frac{\partial}{\partial t}\begin{pmatrix} C_{\underline{k}}(t) \\ C^*_{\underline{k}}(t) \end{pmatrix} + \begin{pmatrix} \lambda_k & 0 \\ 0 & \lambda^*_k \end{pmatrix}\begin{pmatrix} C_{\underline{k}}(t) \\ C^*_{\underline{k}}(t) \end{pmatrix} = 0 \; ; \tag{18}$$

where λ_k is the complex eigenvalue

$$\lambda_k = \nu k^2 - \mu_k\omega_k - i\omega_1(k) \; , \tag{19}$$

with

$$\omega_1(k) \equiv \sqrt{\omega_k^2 - (\nu k^2 - \mu_k\omega_k)^2} \; . \tag{20}$$

Here $\omega_1(k)$ is the frequency of a linear gravity-capillary wave shifted by the effect of viscous dissipation νk^2 and the linear coupling to mean air flow $\mu_k\omega_k$. The similarity transformation which diagonalizes (17) is

$$\underline{\underline{S}} = \begin{pmatrix} V_k^{-1} & -i \\ V_k^{*-1} & i \end{pmatrix} \tag{21}$$

where the eigenvalues of (17) are used to define the "complex velocity" V_k, i.e.,

$$V_k \equiv -i\lambda_k/k \tag{22}$$

whose magnitude $|V_k|$ is the ordinary phase velocity $V_k = \sqrt{g/k + \gamma k^2}$ of a small amplitude gravity-capillary wave. In the absence of wind ($\mu = 0$) and viscosity ($\nu = 0$) the complex velocity V_k again reduces to V_k.

Transforming (15) using (21) yields the set of coupled mode rate equations

$$\dot{C}_{\underline{k}}(t) + \lambda_k\, C_{\underline{k}}(t) = T^{(2)}_{\underline{k}}(\underline{C}) + T^{(3)}_{\underline{k}}(\underline{C}) + f_{\underline{k}}(t) \tag{23}$$

where the nonlinear functions $T_{\underline{k}}^{(2)}(\underline{C})$ and $T_{\underline{k}}^{(3)}(\underline{C})$ have the forms of the quadratic and cubic interactions given in (4.18). However, the coupling coefficients are quite different and different in a way not suggested by the analysis of the previous lecture. The coupling coefficients in the $T_{\underline{k}}^{(2)}$ terms are complex due to the inclusion of viscosity and the air-sea coupling parameters in the linear eigenvalues, i.e.,

$$\bar{\Gamma}^{\underline{k}}_{\underline{\ell}\underline{m}} = \frac{1}{8}\left[\frac{V_\ell V_m}{V_k}(\ell m + \underline{\ell}\cdot\underline{m}) - V_\ell(\ell k - \underline{\ell}\cdot\underline{k}) - V_m(mk - \underline{m}\cdot\underline{k})\right]$$

$$+ \frac{i\nu}{4V_k}[\ell^2 m V_m + m^2\ell V_\ell] - \frac{i}{4}\mu_k[(\ell k - \ell\cdot k)V_\ell + (mk - \underline{m}\cdot\underline{k})V_m]$$

$$\bar{\Gamma}^{-\underline{k},\underline{m}}_{\underline{\ell}} = \frac{1}{4}\left[\frac{V_\ell V^*_m}{V_k}(\ell m - \underline{\ell}\cdot\underline{m}) + V_\ell(\ell k - \underline{\ell}\cdot\underline{k}) - V^*_m(mk + \underline{m}\cdot\underline{k})\right]$$

$$+ \frac{i\nu}{4V_k}[\ell^2 m(V^*_m - V_m) + m^2\ell(V^*_\ell - V_\ell)] + \frac{i\mu_k}{4}$$

$$\times [(\ell k - \underline{\ell}\cdot\underline{k})V_\ell - (km + \underline{k}\cdot\underline{m})V^*_m]$$

$$\bar{\Gamma}^{-\underline{k},\underline{\ell},\underline{m}} = \frac{1}{8}\left[\frac{V^*_\ell V^*_m}{V_k}(\ell m + \underline{\ell}\cdot\underline{m}) + V^*_\ell(\ell k + \underline{\ell}\cdot\underline{k}) + V^*_m(mk + \underline{m}\cdot\underline{k})\right]$$

$$- \frac{i}{4}\frac{\nu}{V_k}[m\ell^2 V^*_m + \ell m^2 V^*_\ell] + \frac{i}{4}\mu_k[(k\ell + \underline{k}\cdot\underline{\ell})V^*_\ell + (km + \underline{k}\cdot\underline{m})V^*_m] \tag{24}$$

where the complex velocity is explicitly given by

$$V_k = \sqrt{V_k^2 - (\mu_k V_k - \nu k)^2} + i(\mu_k V_k - \nu k)\ . \tag{25}$$

These coefficients differ markedly from those obtained in the studies of eg. Valenzuela and Laing[159] and Holliday.[109]

From the arguments in Lecture 10 we keep only the resonant term at third order and approximate $\bar{\Gamma}^{kn}_{\ell m}$ by the *form* of $\Gamma^{kn}_{\ell m}$ given by (4.18) with the phase velocities V_k replaced by $\mathcal{V}_k$. This approximation neglects cubic interactions proportional to γ, ν, and μ_k which are expected to be very small.

The function $f_{\underline{k}}(t)$ in (23) models the turbulent eddies in the wind field and is a fluctuating function of time, i.e.,

$$f_{\underline{k}}(t) = i \frac{p_{\underline{k}}(t)}{\rho_0 \mathcal{V}_k} \quad , \tag{26}$$

so that the expression (23) is a *nonlinear stochastic differential equation* . The properties of the solutions to the stochastic equation (23) can only be obtained when the statistics of the fluctuating force (26) are specified. The statistical properties of the solution are determined by averaging powers of the mode amplitudes over the fluctuations in the wind field. However, because we have not as yet prepared the groundwork for solving a stochastic differential equation, we postpone discussion of (23) until the next lecture.

We emphasize here that the lowest order equation of motion (23), even in the absence of wind ($\mu_k = 0$), is different from (11.24). First of all the frequency is shifted from its inviscid value ω_k to $\omega_1(k)$, which is significant at high wave numbers. Second of all, the interaction coefficients are all complex, both due to the additive effects calculated in (11.11) and due to the complex velocities arising from the eigenvalues of the linear problem. The difference in the dynamics predicted by the two systems has not yet been fully explored.

13. Linear Stochastic Differential Equation (Langevin Model)

We now have two distinct equations of motion for the description of wind driven waves of the surface of the ocean. In the Hamiltonian formulation in lecture (11) the instantaneous pressure couples linearly with the surface displacement to provide an additive flux of energy from the wind to the wave field. In the Langrangian formulation in lecture (12) the pressure field separates into an average part determined by Miles' air-sea coupling parameter and a fluctuating part determined by Phillips' incoherent additive flux due to the turbulent eddies in the wind field. In the latter formulation the diagonalization of the linear homogeneous part of the equations of motion produce a number of differences between the two approaches even though the the final *form* of the equations of evolution (11.28) and (12.23) is the same in both approaches. The most significant difference between these equations is the dependence of the coupling coefficients in (12.23) on the air-sea coupling parameter. This dependence implies that the interaction between waves changes as a function of the wind speed.

The salient result from both lectures (11) and (12) is that we obtain a *nonlinear stochastic differential equation* to describe the evolution of the water wave field. Thus it is necessary to review some of the known properties of stochastic equations and how they arise in physical systems. First of all we note that there is ample experimental support for modeling the evolution of the wind wave field by a system of stochastic differential equations, [see eg. refs. (4-3,12,155)]. Observations on the open ocean indicate that a measurement of a set of mode amplitudes $\{B_{\underline{k}}\}$ at time t is not sufficient to determine this set at time $t + \Delta t$. The observed fluctuations in wave amplitudes make it impossible to predict the instantaneous properties of the vertical displacement and fluid velocity near the surface, i.e., the instantaneous canonical field variables. For this reason the evolution of wind generated sea waves is often described in terms of transport equations rather than dynamic equations.[155] An example of a transport equation is the space-time growth of the energy spectral density of the water waves. For a homogeneous fluid such a description has been given by Hasselmann,[11] and for a nonhomogeneous fluid by Watson and West[103] and Willebrand.[160]

The fluid transport properties are often obtained by injecting fluctuations into the deterministic equations of motion in some prescribed way and then averaging the response of the surface over these fluctuations. The resulting averaged equations are presumed to determine the transport or average evolution of some surface property of the water wave field. These transport properties are the moments of the velocity and surface displacements and require for their determination a knowledge of the statistics of the surface wave fluctuation. Various hypotheses about the statistics of the surface wave field may be tested by observing oceanographic data. A primary piece of evidence concerning wave statistics is the distribution of wave slopes deduced from sunglint experiments by Cox and Munk.[161] The distribution of slopes was determined to be approximately Gaussian in these observations and since the slope and displacement are related by a linear operation, i.e., the former is the spatial derivative of the latter, the distribution of surface heights is often naively assumed to be Gaussian also. This information has traditionally been incorporated into the equations of evolution by simply assuming that the mode amplitudes in (4.7) are Gaussian random variables; [see e.g., Hassleman (11,29,162) and Watson and West (103)]. In the two preceding lectures we developed models which include fluctuations explicitly in the hydrodynamic field equations. Here we intend to calculate rather than postulate the statistical properties of the field variables.

Both the transient and static (steady state) statistical properties of water waves have been investigated. The strategy for determining the evolution of the statistics has previously been to assume that the statistics of the linear wave field are known. The linear waves are those of sufficiently small (infinitesimal) amplitude that the nonlinear interactions can be neglected. As the amplitudes of the water waves increase due to the input of energy from the wind, for example, the nonlinear interactions cease to be negligible and these interactions alter the assumed statistical properties of the field. In order to determine the effect of the nonlinearities on the statistics one must solve a set of nonlinear stochastic differential equations for the field variables, eg. (11.28) or (12.23). We have expressed the surface properties as a superposition of interacting modes, however, the evolution of the wave field is often described by a set of transport equations. This set of transport equations for nonlinear systems forms an infinite heirarchy of moment rate equations[163,164], i.e., the evolution of the average mode amplitude is coupled to the second moment of the mode amplitude, the evolution of the second moment is coupled to the third moment, etc. Truncating this hierarchy of equations at a given order is referred to as *closure* and is usually done by expressing the transport equations in terms of cumulant averages rather than moments.[165-167] The so-called quasi-Gaussian approximation used, e.g., by Hasselmann,[11] Valenzuela and Laing[159] and others[103,109,160], to construct transport equations for the energy spectral density of the water wave field is a second order closure, i.e., it assumes that all cumulants above the second vanish identically. The general validity of this particular approximation has been argued for in the limit of weak nonlinearity by Benney and Saffman[165] and in the asymptotic limit ($t \rightarrow \infty$) by Benney and Newell[166] and reviewed by Newell.[167]

In later lectures we adopt the viewpoint employed in statistical mechanics that the *source* of fluctuations in physical measurements is a result of the finite resolution possible with any macroscopic measuring device. In a physical system with many degrees of freedom (infinite in the case of a field) there is always some loss of information due to the dynamic range of the measurements, i.e., the degrees of freedom which are not resolved are lost. However, the unresolved dynamics of the system are manifest in apparently erratic or irregular changes in the measured results, i.e., in fluctuations.[164] In classical statistical mechnaics it is assumed that a large number of degrees of freedom are required for a system to exhibit stochastic behavior. This assumption is necessary because it is precisely the degree of freedom that one is *not* able to resolve in an experiment which gives rise to observable fluctuations. Consider a closed system having a number of integrals of the motion given by a set of constants $\underline{C}$. These are classically conserved quantities, e.g., mass, energy, momentum, etc. If the system is now opened to the environment the composite system has many more degrees of freedom and the integrals of the motion may become approximations, i.e., the components of $\underline{C}$ become slowly varying functions of time. The other degrees of freedom vary rapidly on the time scale of the components of $\underline{C}$, so that in the equations of motion for these degrees of freedom $\underline{C}$ may still be treated as constant. In the equations of motion for $\underline{C}$ the other degrees of freedom (the fast ones) are therefore eliminated and the evolution of $\underline{C}$ is determined by $\underline{C}$ alone. Such a procedure for constructing the equations for $\underline{C}$ is known as adiabatic elimination and yields the macroscopic equations of motion, i.e., transport equations, for the slowly varying few degrees of freedom[91] as mentioned earlier.

The concept that all but a few degrees of freedom in a mechanical system vary on a short time scale dates back to Boltzmann.[168] It was Langevin,[58] however, who conceived of directly modeling this separation in time scales by means of a differential equation. He considered the momentum of a massive particle suspended in a fluid of much lighter particles to have a slowly varying component (viscous dissipation) and a fluctuating component (collisions of the ambient particles with the more massive Brownian particle). The dissipation parameter, the power spectral density of the fluctuations, and the temperature of the ambient fluid were then inter-related by the fluctuation-dissipation theorem using the thermodynamic properties of a simple fluid[50,53,54,164] A model for the relaxation of a given internal wave due to the nonlinear interaction with a spectrum of ambient internal waves has been developed by Meiss, Pomphrey and Watson[164] using generalizations of Langevin's ideas. The dissipa tion parameter in this latter case is related to the ambient equilibrium spectrum of internal waves given by Garrett and Munk.[170] The point we stress is that fluctuations are a consequence of the large time scale separation for the dynamics of the approximate integrals of the motion and the other degrees of freedom in a system.

In this and the next lecture we restrict our analysis to the dynamics of the high frequency region of the water wave spectrum, i.e., that region in which surface tension and viscosity are important. These two physical effects make the evolution of these waves, as driven by the turbulent wind, much different from the simultaneous evolution of the longer wavelength gravity waves; equations (11.28) and (12.23) become the same for long waves. For a specified wind the short waves reach their steady state level at a given fetch, the longer waves equilibrate at greater fetches for greater wavelengths. The steady state spectrum of high frequency waves provides a modulated, short-wave surface stress which as mentioned in lecture 11, supplies the energy flux to maintain the growth of large scale gravity waves. This dynamic transfer of energy is as complicated as the air-sea interaction (at least as modeled here) and evolves on a time scale much longer than that required for the high frequency waves to come to a steady state with the wind field and viscous dissipation.

Our model of the evolution of the wind driven surface is that the wind is strongly coupled to the short wavelength waves on the sea[88,151,144]. These high frequency waves develop to their steady state level in a short fetch before the long wavelength gravity waves have attained appreciable amplitude. The short waves act as a catalyst to the growth of the longer gravity waves,[88] a model which we will develop fully in the next few lectures. The observed rates of growth for long waves exceeds those predicted by Miles' inviscid laminar theory of wave instability by a factor of between 5 and 8[9,88,171,172]. The energy cascade mechanism by which the gravity-capillary waves transmit energy from the air flow to long wavelength gravity waves requires that we first demonstrate that the gravity-capillary waves come to an asymptotic steady state with the wind field.

To establish familiarity with the nomenclature and the physical concepts involved with the dynamics of a system driven by fluctuations, let us consider the motion of a heavy particle immersed in a fluid of lighter particles as orginally considered by Langevin. This process is referred to as Brownian motion since it was the English botonist Robert Brown who first observed the erratic motion of pollen in water.[173] It in noteworthy that a full fifty years *prior* to Browns' observation the Dutch physician Ingenhausz studied the chaotic motion of powdered charcoal suspended in alcohol.[174] Fortunately for the English speaking world the process is named after Brown and not Ingenhausz. The forces acting on the Brownian particle, representing the momentum of the heavy particle by $\underline{P}(t)$, are

$$\frac{d\underline{P}(t)}{dt} = -\lambda\underline{P}(t) + \underline{f}(t) \tag{1}$$

where λ is the scalar phenomenological dissipation parameter and $\underline{f}(t)$ is a random function modeling the rapid and independent impulses due to collisions with the lighter ambient particles pushing the heavy particle about.

Equation (1) was the first modeling of the connection between the microscopic and macroscopic world by means of a stochastic differential equation. Part of the success of the description was based on the fact that it is a linear equation and that the fluctuations are independent of the motion of the Brownian particle itself. Thus the equation can be solved analytically. The solution to (1) is

$$\underline{P}(t) = \underline{P}(o)e^{-\lambda t} + \int_0^t e^{-\lambda(t-t')}\underline{f}(t')dt' \tag{2}$$

which, because $\underline{f}(t)$ is a fluctuating quantity, the momentum $\underline{P}(t)$ is also a fluctuating quantity. Further, because (2) is a linear equation, the statistics of $\underline{P}(t)$ are the same as the statistics of the driving force $\underline{f}(t)$. Therefore, to determine the properties of the solution we must specify the statistics of $\underline{f}(t)$.

The work of Einstein[175] on diffusion, provided the groundwork for the specification of the statistical properties of $\underline{f}(t)$ made by Langevin. The distribution of ambient fluid particles is assumed to be isotropic, i.e., the fluid is both homogeneous and macroscopically at rest, therefore, the collisions with the Brownian particle are also distributed isotropically. Indicating by an overbar the average over a time period much longer than the interaction time during a collision, but much shorter than the response time of the Brownian particle to a collision, we have

$$\overline{\underline{f}(t)} = 0 \quad . \tag{3}$$

The Brownian particle does not respond to a single collision, but in any instant there is a net imbalance in the number of collisions on one side of the Brownian particle with the other side. This instantaneous imbalance moves the heavy particle about. The zero time average (3) indicates that the distribution of impulsive forces acting on the Brownian particle is indeed isotropic so that over a finite interval of time there is no *net* force acting. Taking the time average of (2) and using (3) we obtain

$$\overline{\underline{P}(t)} = \underline{P}(o)\ e^{-\lambda t} \tag{4}$$

so that the average momentum of the Brownian particle decays exponentially in time from its initial value $\underline{P}(o)$. The parameter λ models the absorption of momentum by the ambient fluid.

The second assumption that is usually made in studies of Brownian motion concern the correlations over time between collisions. The mean free time between collisions of individual ambient particles with other ambient particles is long compared with the collision time of fluid particle with Brownian particle. Therefore, any two scatterings between ambient and Brownian particles are uncorrelated in time. Further, since the motion of the heavy particle results from an imbalance in a large number of collisions at time t, i.e., the fluctuating force $\underline{f}(t)$, such an imbalance would be uncorrelated with a similar imbalance at time $t + \tau$, i.e., $\underline{f}(t + \tau)$. The mathematical representation of this physical picture is that the fluctuations are delta correlated in time, i.e.,

$$\overline{\underline{f}(t)\ \underline{f}(t - \tau)} = 2\underline{\underline{D}}\delta(\tau) \qquad (5)$$

where $\underline{\underline{D}}$ is the matrix of the mean square strength of the components of the fluctuations and $\delta(\tau)$ specifies that the fluctuations are only correlated at time $\tau = 0$. A number of representations for $\delta(\tau)$ can be used, i.e., $\delta(\tau) = \lim_{\tau_c \to 0} \frac{1}{\sqrt{2\pi\tau_c}} \exp[-\tau^2/2\tau_c]$ where τ_c is the correlation time between successive fluctuations in $\underline{f}(t)$ taken to be zero in the limiting case.

Taking the product of $\underline{P}(t)$ with itself and using both the properties (3) and (5) we obtain

$$\overline{\underline{P}^2(t)} = \underline{P}^2(o)\ e^{-2\lambda t} + \int_0^t dt_1 \int_0^t dt_2\ 2\underline{\underline{D}}\delta(t_1 - t_2)e^{-\lambda(2t - t_1 - t_2)} \qquad (6)$$

which integrates to

$$\overline{\underline{P}(t)^2} = \underline{P}^2(o)\, e^{-2\lambda t} + \frac{\underline{\underline{D}}}{\lambda}\left[1 - e^{-2\lambda t}\right] \quad . \tag{7}$$

We observe that the mean square momentum (energy) of the Brownian particle does not vanish asymptotically as does the mean momentum. The energy lost by visious dissipation is replenished by the fluctuations until an equilbrium balance is struck. This balance is obtained from (7) by taking the long time limit

$$\lim_{t\to\infty} \overline{\underline{P}(t)^2} = \frac{\underline{\underline{D}}}{\lambda} \tag{8}$$

and is called the fluctuation-dissipation theorem of the first kind. Equation (8) relates the rate of dissipation λ to the mean square level of excitation of the fluctuating driving force $\underline{\underline{D}}$. The proportionality coefficient is the equilibrium energy of the Brownian particle. Einstein[175] established that if (8) is equated to the equi-partition value for the energy of the ambient fluid, i.e., $\frac{1}{2}k_BT$ for each degree of freedom of the heavy particle, where k_B is Boltzmanns' constant and T is the fluid temperature, then λ, $\underline{\underline{D}}$ and T are interrelated. Einstein's theory of random motion of "large" particles in fluids was used as a proof of the existence of molecules, and Perrin's[176] and Svedberg's[177] quantitative measurements of particle displacements as a function of time yielded a numerical value for Avogodro's number.

We now have some preparation for examining the equations of motion of the wind generated waves on a fluid surface.

14. Steady-State Gravity-Capillary Spectrum

We select (12.23) as the more accurate description of the dynamics of the wind generated gravity-capillary waves. At early times before the wave amplitudes have had time to develop, the nonlinear terms $T_{\underline{k}}^{(2)}$ and $T_{\underline{k}}^{(3)}$ are negligible and the equation of motion is

$$\dot{C}_{\underline{k}}(t) + \lambda_k C_{\underline{k}}(t) = f_{\underline{k}}(t) \ . \tag{1}$$

We adopt this as our linear model of the growth of the high frequency wave field. Because we are neglecting the nonlinear terms we may apply the ideas discussed in the last lecture pertaining to the Langevin equation. Equation (1) contains the average coupling term of Miles[86], $\mu_k\omega_k$, in the complex eigenvalue $\lambda_k = \nu k^2 - \mu_k\omega_k + i\omega_1(k)$ which models the in-phase coupling of the air flow to the fluid surface. It also contains the incoherent coupling term of Phillips[85], $f_{\underline{k}}(t)$, due to fluctuations in the air flow. Equation (1) is in fact the mode coupled form of the Miles-Phillips model[87] and appears to be similar to the Langevin equation (13.1). However, there are as many differences between these two equations as there are similarities.

The first distinction between (1) and (13.1) is that although $f_{\underline{k}}(t)$ is a fluctuating quantity, the source of these fluctuations is not due to microscopic processes. The fluctuations are due to turbulent eddies in the wind field which we can identify but do not understand. The physical mechanism for the random motion of the air is shrouded in confusion. We, therefore, do not trace the "cause" of the fluctuations beyond the function $f_{\underline{k}}(t)$ which we obtain from experiments and field observations. Thus we distinguish between microscopic dynamics which generate the fluctuations in (13.1) and the macroscopically irregular air flow which has a component at the surface scale

$2\pi/k$. We refer to these later fluctuations as mesoscopic since their time scale is very much longer than that of any molecular process, but is still significantly shorter than the macroscopic surface response time. There are, therefore, microscopic processes which are assumed to be negligible in the generation of waves; mesoscopic processes which generate the statistics of the wave field [at least in a linear generation process like (1)] and finally there are the macroscopic processes which determine the average evolution of the wave field.

The second distinction between the Miles-Phillips model (1) and the Langevin equation (13.1) is that the real part of the complex eigenvalue λ_k is negative instead of positive. Therefore, instead of the mean coupling to the ambient environment providing a dissipation as it does in the Langevin equation, it instead provides an additional source of energy to generate waves. This term in fact leads to an instability in the linear model which we discuss below. The parameter λ_k is, therefore, not related to the fluctuations by a fluctuation-dissipation relation and is in fact independent of the fluctuations.

We integrate (1) directly to obtain

$$C_{\underline{k}}(t) = C_{\underline{k}}(0)\, e^{-\lambda_k t} + \int_0^t e^{-\lambda_k(t-t')}\, f_{\underline{k}}(t')dt' \qquad (2)$$

where again the statistics of the mode amplitude $C_{\underline{k}}(t)$ are given by those of the wind fluctuations since (2) is linear. If we assume that the average influx of energy from the wind is adequately modeled by the air-sea coupling parameter, then the turbulent fluctuations in the air flow can be assumed to be zero-centered, i.e.,

$$\overline{f_{\underline{k}}(t)} = 0 \tag{3}$$

and the average mode amplitude is

$$\overline{C_{\underline{k}}(t)} = C_{\underline{k}}(0)\, e^{-\lambda_k t} \,. \tag{4}$$

However, because Re $\lambda_k < 0$ the mean amplitude diverges asymptotically. This instability in the mode amplitude is presumed to provide the secondary growth of water waves since $C_{\underline{k}}(0) = 0$, so that $f_{\underline{k}}$ must generate the waves which are to become unstable. The nonlinear interactions that we have neglected are assumed to quench this instability and provide a finite asymptotic value for the wave amplitudes.[4-6]

The second moment quantity of interest is $\overline{|C_{\underline{k}}(t)|^2}$ which using (2) we determine to be

$$\overline{|C_{\underline{k}}(t)|^2} = |C_{\underline{k}}(0)|^2\, e^{-2\mathrm{Re}\lambda_k t} + \int_0^t dt_1 \int_0^t dt_2\, \overline{f_{\underline{k}}(t_1) f^*_{\underline{k}}(t_2)}\, e^{-2\mathrm{Re}\lambda_k t} \times e^{\lambda_k t_1 + \lambda^*_k t_2} \,. \tag{5}$$

The second moment of the fluctuating flux $f_{\underline{k}}(t)$, using (12.26) and recalling that $|V_k| = V_k$, is

$$\overline{f_{\underline{k}}(t_1) f^*_{\underline{k}'}(t_2)} = \frac{2\delta_{\underline{k}-\underline{k}'}}{\rho_0^2 V_k^2}\, \Phi(\underline{k}, t_1 - t_2) \tag{6}$$

where the strength of the correlations in the air flow on the spatial scale $2\pi/k$ over a time interval τ is given by $\Phi(\underline{k},\tau)$ and the fluctuations are spatially homogeneous. The time Fourier transform of the correlation function, i.e., $\bar{\Phi}(\underline{k},\omega)$, is the three-dimensional space-time spectral density of the pressure field

$$\Phi(\underline{k},\omega) \equiv \int_{-\infty}^{\infty} e^{-i\omega\tau} \overline{p_{\underline{k}}(t)p_{\underline{k}}^*(t-\tau)}\, d\tau \,. \tag{7}$$

We introduce the descrete power spectrum for the surface deflection by

$$F_{\underline{k}}(t) \equiv \frac{1}{2} \overline{|C_{\underline{k}}(t)^2|} \tag{8}$$

which is normalized such that the mean square surface displacement is given by

$$\langle\zeta^2\rangle = \sum_{\underline{k}} F_{\underline{k}}(t) \tag{9}$$

and using (6) we rewrite (5) as

$$F_{\underline{k}}(t) = F_{\underline{k}}(0)\, e^{-2\mathrm{Re}\lambda_k t} + \int_0^t dt_1 \int_0^t dt_2 \frac{\Phi(\underline{k},t_1-t_2)}{\rho_0^2 V_k^2} e^{-2\mathrm{Re}\lambda_k t+\lambda_k t_1+\lambda_k^* t_2} \,. \tag{5'}$$

If the spectrum of the pressure field fluctuations is symmetric in time, i.e., $\Phi(\underline{k},\tau) = \Phi(\underline{k},-\tau)$, we can take the time dirivative of (5') and write the transport equation

$$\frac{\partial F_{\underline{k}}(t)}{\partial t} + 2\lambda_R(k)F_{\underline{k}}(t) = \frac{2}{\rho_0^2 V_k^2} \int_0^t e^{-\lambda_R(k)\tau}\, \Phi(\underline{k},\tau) \cos[\lambda_I(k)\tau]\, d\tau \tag{10}$$

where $\lambda_k \equiv \lambda_R(k) + i\lambda_I(k)$ and λ_R and λ_I are real quantities.

We are here primarily interested in the growth of the gravity-capillary waves. One might guess, therefore, that the correlation time of the fluctuations in the air flow at the water surface, i.e., τ_L, is very much longer than the characteristic times of these waves. As argued by Phillips[85] this might be true at very early times and over time intervals in which $\Phi(\underline{k},\tau)$ is essentially constant. During these intervals $\Phi(\underline{k},\tau)$ may be removed from the integral in (10) so that the initial response of the water surface is given by

$$F(\underset{-}{k},t) = \frac{\Phi(\underset{-}{k},0)}{\rho_w^2 V_k^2} \frac{1}{|\lambda_k|^2} \left\{ \frac{1 + e^{-2\lambda_R t}}{2} - e^{-\lambda_R t} \cos (\lambda_I t) \right\} . \tag{11}$$

Since (11) is only true initially, i.e., over time intervals in which $\Phi(\underset{-}{k},t)$ is constant, we assume $|\lambda_R t| << 1$ and expand the exponential functions to obtain

$$F(\underset{-}{k},t) \cong \frac{\Phi(\underset{-}{k},0)}{\rho_w^2 V_k^2} t^2 \tag{12}$$

which has the quadratic time dependence observed by Phillips.[85]

If we do not assume $\Phi(\underset{-}{k},\tau)$ is frozen in the integral (10), but that the correlation time between fluctuations is substantially shorter than the integration time, i.e., $t >> \tau_c$, then introducing

$$\bar{\Phi}(\underset{-}{k},\lambda) \equiv \int_0^\infty e^{-\lambda_R(\underset{-}{k})\tau} \Phi(\underset{-}{k},\tau) \cos [\lambda_I(k)\tau] \, d\tau \tag{13}$$

we obtain under the short fetch condition the asymptotic inhomogeneous transport equation

$$\frac{\partial F(\underset{-}{k},t)}{\partial t} + 2\lambda_R(k)F(\underset{-}{k},t) \underset{=}{\sim} \frac{\bar{\Phi}(\underset{-}{k},\lambda)}{\rho_w^2 V_k^2} . \tag{14}$$

The solution to (14) is given by

$$F(\underset{-}{k},t) = F(\underset{-}{k},0)\, e^{-2\lambda_R(k)t} + \frac{\bar{\Phi}(\underset{-}{k},\lambda)}{\rho_w^2 V_k^2} \frac{1 - e^{-2\lambda_R(k)t}}{2\lambda_R(k)} . \tag{15}$$

Since the initial state of the ocean is a glassy surface $F(\underset{-}{k},0) \equiv 0$, early in the evolution $|\lambda_R t| << 1$ we obtain by expanding the exponential the linear growth

$$F(\underline{k},t) \cong \frac{\bar{\Phi}(\underline{k},\lambda)}{\rho_w^2 V_k^2}\, t \tag{16}$$

in contrast to (12) which grows quadratically with time.

The most significant feature of the solution (15) is that it predicts an exponentially growing surface wave energy spectrum. This, of course, does not persist indefinitely, at some spectral level the nonlinear interactions begin to transfer energy out of the spectral interval which is extracting energy from the wind. This leads to a growth of long waves and an increased level of high frequency waves to be viscously damped. West[154] provides a model for including the average effect of the nonlinear interaction in the dynamic equations (1). We examine this technique for including the function $T_{\underline{k}}$ in the equation of motion (1). This approximation scheme is unlike normal perturbation theory in that it is valid at late times rather than early times. The asymptotic method involves replacing the nonlinear function $T_{\underline{k}}(t)$ with a "statistically equivalent" function no higher than linear in the mode amplitude $C_{\underline{k}}(t)$. The method is referred to as statistical linearization and has been favorably compared with higher order perturbation theories by West, Lindenberg and Shuler[178] and also by Budgor and West[179]. The method reproduces the steady state average of $C_{\underline{k}}(t)$ and yields a self-consistent expression for the steady state second moments, i.e., the energy spectral density. Although the first two moments are certainly an inadequate description of the distribution function for the system (unless it just happens to be Gaussian), they can provide a good estimate of the steady state second-order statistics, i.e., variances, correlation functions, and spectral densities.[180] These are precisely the properties which interest us in this lecture so the approximation is felt to be adequate.

To linearize (12.23) we replace the nonlinear function $T_{\underline{k}}(t)$ by the term $h_{\underline{k}}C_{\underline{k}}(t) + \beta_{\underline{k}}$ and adjust $h_{\underline{k}}$ and $\beta_{\underline{k}}$ such that the error $\varepsilon_{\underline{k}}^2$,

$$\varepsilon_{\underline{k}}^2 \equiv \lim_{T\to\infty} \frac{1}{2T}\int_{-T}^{T} \left|T_{\underline{k}}(t) - h_{\underline{k}}C_{\underline{k}}(t) - \beta_{\underline{k}}\right|^2 dt \ , \tag{17}$$

induced by this replacement is a minimum. The initial conditions in (17) are taken to be $C_{\underline{k}}(0) = 0$. The error minimization condition is contained in the expressions for the independent variations of $h_{\underline{k}}$ and $\beta_{\underline{k}}$,

$$\frac{\delta\varepsilon_{\underline{k}}^2}{\delta h_{\underline{k}}} = 0 \ ; \qquad \frac{\delta\varepsilon_{\underline{k}}^2}{\delta\beta_{\underline{k}}} = 0 \ ; \tag{18}$$

which from (17) yields

$$h_{\underline{k}} = \frac{\langle\hat{C}_{\underline{k}}^*(t)T_{\underline{k}}(t)\rangle_T}{\langle|\hat{C}_{\underline{k}}|^2\rangle_T} \ ; \tag{19a}$$

$$\beta_{\underline{k}} = \langle T_{\underline{k}}(t)\rangle_T - \langle C_{\underline{k}}\rangle_T \ \frac{\langle\hat{C}_{\underline{k}}^*(t)T_{\underline{k}}(t)\rangle_T}{\langle|\hat{C}_{\underline{k}}|^2\rangle_T} \ . \tag{19b}$$

The quantity $\hat{C}_{\underline{k}}$ ($\equiv C_{\underline{k}} - \langle C_{\underline{k}}\rangle_T$) is the fluctuating component of the mode amplitude away from its time averaged value indicated by the T subscript on the averaging brackets.

The linearized equations of motion are now

$$\frac{dC_{\underline{k}}(t)}{dt} + \alpha(\underline{k})C_{\underline{k}}(t) = \beta_{\underline{k}} + f_{\underline{k}}(t) \ , \tag{20}$$

where the complex coefficient $\alpha(\underline{k})$ [$\equiv \lambda_k - h_{\underline{k}}$] consists of the shifted frequency

$$\alpha_I(\underline{k}) \equiv \mathrm{Imag}\,[\alpha(\underline{k})] = \omega_1(\underline{k}) - \mathrm{Imag}\,[h_{\underline{k}}] \ , \tag{21}$$

and shifted "dissipation parameter"

$$\alpha_R(\underline{k}) \equiv \text{Real}\ [\alpha(\underline{k})] = \nu k^2 - \mu_k - \text{Real}\ [h_{\underline{k}}] \ . \tag{22}$$

Recalling that the mean value of the wind fluctuation is zero, the average of (20) is

$$\frac{d\langle C_{\underline{k}}\rangle_{\underline{f}}}{dt} + \alpha(\underline{k})\ \langle C_{\underline{k}}(t)\rangle_{\underline{f}} = \beta_{\underline{k}} \ , \tag{23}$$

which using the definitions of the variational parameters $h_{\underline{k}}$ and $\beta_{\underline{k}}$, with the time averages replaced by an average over an ensemble of realizations of the fluctuating flux $\underline{f}(t)$, reduces to

$$\frac{d}{dt} \langle C_{\underline{k}}\rangle_{\underline{f}} + \lambda_k \langle C_{\underline{k}}\rangle_{\underline{f}} = \langle T_{\underline{k}}\rangle_{\underline{f}} \tag{24}$$

Thus, by construction, (20) preserves the evolution of the mean mode amplitude. We justify replacing the time averages in (19a) and (19b) by steady state ensemble averages over fluctuations in the air flow because of the $T \to \infty$ limit taken in (17). This limit implies that the variational parameters are only valid in the asymptotic region where presumably the surface response has reached a steady state.

Using the definition of $\beta_{\underline{k}}$ we can rewrite (20) in terms of the variation of the mode amplitude from its average behavior, i.e.,

$$\frac{d}{dt} \hat{C}_{\underline{k}}(t) + \alpha(\underline{k})\ \hat{C}_{\underline{k}}(t) = f_{\underline{k}}(t) \ . \tag{25}$$

Equation (25) more closely corresponds to a Langevin equation for a complex variable since $\alpha_R > 0$ for $\alpha(\underline{k}) \equiv \alpha_R(\underline{k}) + i\alpha_I(\underline{k})$, with both α_R and α_I real. For the initial condition $\hat{C}_{\underline{k}}(t=0) = 0$, (25) has the solution

$$\hat{C}_{\underline{k}}(t) = \int_0^t e^{-\alpha(\underline{k})(t-t')} f_{\underline{k}}(t')\, dt . \tag{26}$$

The energy spectral density of the amplitude fluctuations

$$F(\underline{k},t) \equiv \frac{1}{2} \langle |\hat{C}_{\underline{k}}(t)|^2 \rangle_{\underline{f}} , \tag{27}$$

resulting from (26) is given by

$$F(\underline{k},t) = \frac{\bar{\Phi}(\underline{k},\alpha)}{\rho_w^2 V_k^2} \frac{1 - e^{-2\alpha_R(\underline{k})t}}{2\alpha_R(\underline{k})} , \tag{28}$$

where $\bar{\Phi}(\underline{k},\alpha)$ is the *damped* cosine transform of the power spectral density of the fluctuations in the wind field, since $\alpha_R > 0$, i.e.,

$$\bar{\Phi}(\underline{k},\alpha) \equiv \int_0^\infty e^{-\alpha_R(\underline{k})\tau}\, \Phi(\underline{k},\tau) \cos [\alpha_I(\underline{k})\tau] . \tag{29}$$

West[154] has shown that for a reasonable phenomenological model of the wind field fluctuations $\bar{\Phi}(\underline{k},\alpha)$ that the parameter $\alpha_R(\underline{k})$ is greater than zero throughout the gravity-capillary region of the spectrum so that (29) and (28) are consistent.

The steady state power spectrum density for the surface deflection due to gravity-capillary waves is defined by

$$F_{ss}(\underline{k}) \equiv \lim_{t\to\infty} F(\underline{k},t) = \frac{\bar{\Phi}(\underline{k},\alpha)}{\rho_w^2 V_k^2} \frac{1}{2\alpha_R(\underline{k})} , \tag{30}$$

which depends on the nonlinear interactions through the variational parameter $h_{\underline{k}}$ entering through $\alpha_R(\underline{k})$. West shows that a self-consistent calculation of $h_{\underline{k}}$ and $F(\underline{k})$ are required by reducing (19a) to the expression

$$h_{\underline{k}} = 2 \frac{\alpha_R^2(\underline{k})}{|\alpha(\underline{k})|^2} \tilde{C}_{\underline{k}\underline{k}}^{kk} F_{ss}(\underline{k}) + 4 \sum_{\underline{\ell}} \tilde{C}_{\underline{k}\underline{\ell}}^{k\ell} F_{ss}(\underline{\ell}) , \tag{31}$$

where the coupling coefficients are not of particular interest here. Further discussion of the properties of the gravity-capillary system in this model can be found in West[154]. We have developed the concepts to a sufficient degree that they can now be implemented in the discussion of the growth of gravity waves.

To summarize the linear growth model for the power spectral density of the surface wave field, we indicate the four primary regions of wave growth in Figure (14.1). For early times, region I, the waves grow linearly by Phillips pressure fluctuation mechanism, with a growth rate proportional to the power spectral density of the turbulent fluctuations in the air flow $\Phi(\underline{k},\alpha')$. The prime on α' is used here as a reminder that perturbation theory must be used to determine the early time behavior of the mode amplitudes. For somewhat later times there is a region of exponential growth as prescribed by Miles' instability mechanism, as indicated in region II. The asymptotic, or long time region, gives the steady state or saturation spectrum $F_{ss}(\underline{k})$ in region IV. The steady state spectral level is determined by $\Phi(\underline{k},\alpha)$, $\alpha(\underline{k})$ and $\boldsymbol{\nu}_k$, all of which depend on the average nonlinear hydrodynamic interactions through the parameter $h_{\underline{k}}$. The connection between regions II and IV in Figure 14.1 is left tenous because in region III the nonlinear interactions are developing and their transient behavior remains undescribed.

In Figure 14.2 the calculated values of the dissipation parameter $\alpha_R(\underline{k})$ are depicted for mean wind speeds of 10 m/sec and 5 m/sec and compared with the initial growth rates predicted for the Miles-Phillips mechanism. The rate at which the spectrum relaxes to its steady state level at late times is calculated to be fairly insensitive to the average wind speed in this frequency interval. The relaxation rate of perturbations of the short waves is one to two orders of magnitude faster than the initial growth rate. For example, a 6 mm wavelength wave has a growth rate of 5.6 sec^{-1} for a 10 m/sec wind. Whereas a perturbation of this wave near the steady state relaxes back to the steady state level, i.e., the perturbation vanishes, at a calculated rate of 24 sec^{-1} or in approximately one half of a cycle of the wave. The nonlinear interactions are thus very efficient in transferring energy out of the high frequency spectral interval once the waves have reached their near steady state levels.

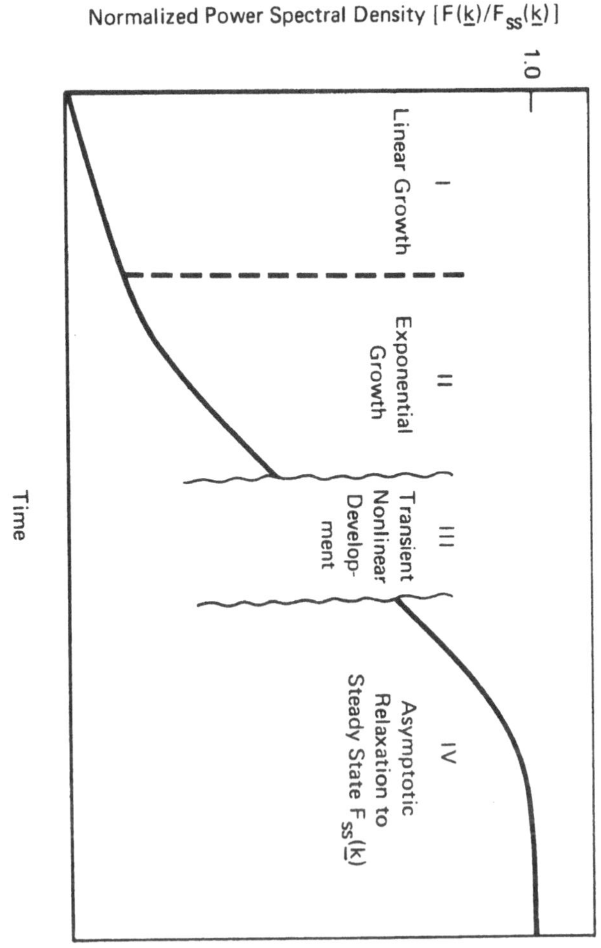

Figure 14.1: The growth as a function of time of the energy spectral density for the gravity-capillary waves normalized to its steady-state level is indicated schematically.

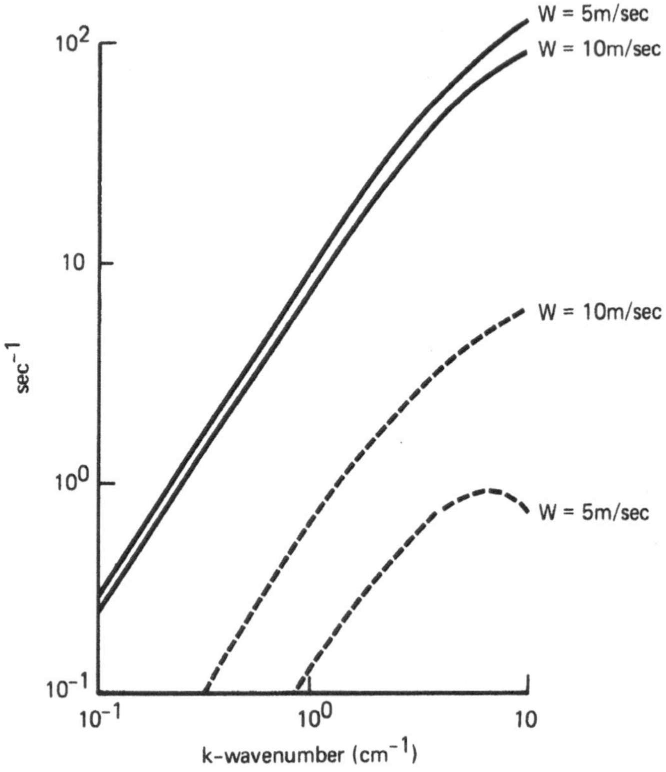

Figure 14.2: The magnitude of the initial growth rates of the gravity-capillary waves as a function of wavenumber (----) for wind speeds of 5m/sec and 10m/sec are contrasted with the relaxation rates in the asymptotic time domain (——).

15. Steady-State Linear Wave Field

As we have discussed, the evolution of the water wave field consists of two distinct types of motion. The deterministic movements of the free surface as governed by the nonlinear interaction terms in the Hamiltonian (3.34) and the fluctuations in the motion described by the stochastic terms in the dynamic equations (11.28, 12.23). Some of the difficulties associated with solving nonlinear stochastic differential equations directly were discussed in the preceeding lecture. To avoid many of these problems we adopt an alternative point of view in this and subsequent lectures. Instead of attempting to solve the dynamic equations directly we examine the evolution of the probability density for the water wave field in the phase space of the dynamic system. This was mentioned in conjunction with Brownian motion in lecture 13.

The phase space for the set of N dynamic variables $\{B_{\underline{k}}(t)\}$ consists of 2N orthogonal axis labeled by the continuous set of values $\{b_{\underline{k}}\}$ that these dynamic variables can assume. The probability density $P(\underline{b},t|\underline{b}_0)$ therefore determines the probability that the dynamic variable $\underline{B}(t)$ lies in the phase space interval $(\underline{b},\underline{b}+d\underline{b})$ at time t given an initial value $\underline{B}(t=0) \equiv \underline{b}_0$. We denote a point in phase space by $\Gamma_t \equiv \{b_{\underline{k}_1}, b_{\underline{k}_2}, \ldots b_{\underline{k}_n}; t\}$ which completely specifies the state of the N-component wave field at an instant of time t. The evolution of the field is determined by the unique trajectory Γ_t through the phase space where t is here a parameter specifying the position of the field along the trajectory. If the path Γ_t is predictable from a prescribed set of initial conditions, i.e., Γ_0, the evolution of the wave field is deterministic. If, however, a small change in initial conditions results in an arbitrarily large* change of all points in the *orbit* Γ_t the motion is said to be irregular or chaotic. Since a path in phase space describes the space-time evolution of the wave field, this small change results in two totally different asymptotic states of the field.

The KAM tori discussed in lecture 5 are examples of the type of trajectories one obtains in phase space for deterministic motion of the nonlinear water wave field. For chaotic motion, i.e., motion for which such trajectories are not stable, equilibrium statistical mechanics assumes that Γ_t wanders freely over an energy surface, asymptotically returning to the neighborhood of each point on this surface infinitely often.[81] Deterministic motion is confined to well defined closed orbits in phase space, whereas the irregular or chaotic motion of orbits uniformly fills phase space on the energy surface.[168] Both types of motion are thought by many to result from the self-same set of dynamic equations and not to be a consequence of the kind of reduced (coarse grained) description stressed in the previous lectures. The contention is that the resonances discussed in lecture 5 destroy the KAM tori describing the deterministic evolution of a mechanical system.[182] The state of the art of such analysis is not sufficiently advanced, however, that one can as yet make accurate statements about the evolution behavior of systems with many degrees of freedom, see eg. Tabor.[37] Therefore we here adopt the conventional viewpoint of statistical mechanics to discuss the statistical properties of the water wave field.[164]

---

*An arbitrarily large change is not precisely correct. The condition is that the initial separation of orbits must increase at least at an exponential rate for chaotic behavior to result, see eg. Arnold [181] for a discussion of the rigorous mathematical conditions.

Traditionally it is the very complication of the orbit Γ_t wandering through phase space that is used to simplify the description of a physical system. Consider an experiment in which we have prepared the wave field in some well defined state Γ_o. A measurement of the energy in the wave field, or any other phase function $H(\Gamma_t)$ for that matter, yields a number h. A sequence of such measurements determines a set of numbers $\{\underline{h}\}$. As time passes the state evolves from Γ_o to Γ_t with corresponding evolution of $\underline{h}(o)$ to $\underline{h}(t)$. The only information available to the experiments is the time dependence of the numerical values of $\underline{h}(t)$. A field of water waves is no exception. An experiment in the water wave system could, for example, be a time series of wave staff measurements yielding a time history of the vertical surface displacement at a point in space. An average can be defined from this record by breaking the trace into a large number of segments. If we have M segments of equal length, then each segment is a representation of the ocean surface. The M segments taken together constitute an ensemble, i.e., a number of copies of the surface which are statistically identical.[183] This empirical procedure is adequate for defining an ensemble as long as the time length of each segment is very much longer than any correlation time among the surface waves. The average surface displacement is then given by

$$<\zeta(\underline{x},t)> = \frac{1}{M}\sum_{j=1}^{M} \zeta^{(j)}(\underline{x},t) \qquad (1)$$

where $\zeta^{(j)}(\underline{x},t)$ is the value of the vertical displacement at position $\underline{x}$ and time t relative to the beginning of the record in the jth member of the ensemble. In the limit $M \to \infty$ it is assumed in equilibrium statistical mechanics that this average can be replaced by an average over an equilibrium ensemble distribution function having the statistical properties of the surface. We now consider a simple argument for constructing the equilibrium distribution for a field of linear gravity waves.

Consider the complex surface displacement $Z(\underline{x},t)$ generated by the linear superposition of M gravity waves with small amplitudes a_j and phases θ_j, i.e.,

$$Z(\theta_1,\theta_2, \ldots , \theta_M) = \sum_{j=1}^{M} a_j \, e^{i\theta_j} \quad . \tag{2}$$

The real and imaginary parts of (2) are indicated by

$$Z(\theta_1, \ldots , \theta_M) = Z_R(\theta_1, \ldots, \theta_M) + iZ_I(\theta_1, \ldots, \theta_M) \tag{3}$$

where the real vertical displacement of the surface is given by $Z_I(\theta_1, \ldots, \theta_M)$ and the real velocity potential at the surface is proportional to $Z_R(\theta_1, \ldots, \theta_M)$. The distribution of vertical displacements resulting from the superposition of M linear waves is that of $Z_I(\theta_1, \ldots, \theta_M)$ which results from allowing all θ_j's to take on values (with equal probability) in the interval $(0,2\pi)$. This assumption is often referred to as the random phase approximation. It is equivalent to a walker taking a number of steps on a two dimensional plane and after each step turning through an arbitrary angle θ. After M steps how far is the walker from the origin? The question, in this form, was first asked by Pierson[184] and the solution to the Pierson random walk was first given by Kluvyer.[185] A more recent solution to this problem is developed below.[186]

The first step in the determination of the probability distribution of surface height is to construct the joint probability density function of Z_R and Z_I, i.e., $P(Z_R,Z_I)$. In general, the probability density is the Fourier transformation of the characteristic function $p(\underline{k})$, i.e.,

$$p(\underline{k}) = \left\langle e^{i\underline{k}\cdot\underline{Z}}\right\rangle = \int_{-\infty}^{\infty} dZ_R \int_{-\infty}^{\infty} dZ_I\; e^{i\underline{k}\cdot\underline{Z}}\, P(Z_R,Z_I) \quad . \tag{4}$$

Now since

$$Z_R(\theta_1, \ldots, \theta_M) = \sum_{j=1}^{M} a_j \cos\theta_j \tag{5}$$

and

$$Z_I(\theta_1, \ldots, \theta_M) = \sum_{j=1}^{M} a_j \sin\theta_j \tag{6}$$

the required average is obtained by substituting (5) and (6) into the characteristic function expression (4) and integrating each θ_j over the interval $(0,2\pi)$

$$p(\underline{k}) = \frac{1}{(2\pi)^M} \int_0^{2\pi} d\theta_1 \cdots \int_0^{2\pi} d\theta_M \prod_{j=1}^{M} \exp\{ia_j[k_R\cos\theta_j + k_I\sin\theta_j]\}. \tag{7}$$

The integration over angle reduces each factor in the integrand of (7) to a zero order Bessel function $J_o(x)$ so that

$$p(\underline{k}) = \prod_{j=1}^{M} J_o[a_j(k_R^2 + k_I^2)^{1/2}] = p(k) \tag{8}$$

with $k^2 = k_R^2 + k_I^2$. Taking the inverse Fourier transform of (8) yields

$$P(Z_R, Z_I) = \frac{1}{(2\pi)^2} \int_{-\infty}^{\infty} dk_R \int_{-\infty}^{\infty} dk_I \; p(k) \exp[-i(k_R Z_R + k_I Z_I)]$$

$$= \frac{1}{(2\pi)^2} \int_{-\infty}^{\infty} dk_R \int_{-\infty}^{\infty} dk_I \prod_{j=1}^{M} J_o(ka_j) \exp[-i(k_R Z_R + k_I Z_I)] \quad (9)$$

Equation (9) can be simplified by using some properties of Bessel functions, see e.g., Watson[187]. The integrand of (9) is substantial for small ka_j and decreases for increasing ka_j because the oscillations in the J_o Bessel functions produce rapid interference. Therefore we use the small slope approximation, i.e., $ka_j \ll 1$, for $J_n(ka_j)$,

$$J_{n-1}(ka_j) \sim \frac{(\frac{1}{2}ka_j)^{n-1}}{\Gamma(n)} \exp[-k^2 a_j^2/4n] \quad (10)$$

and obtain from (9)

$$P(Z_R, Z_I) \cong \frac{1}{(2\pi)^2} \int_{-\infty}^{\infty} dk_R \int_{-\infty}^{\infty} dk_I \exp\left\{ - \frac{k^2 \Lambda^2}{4} - i(k_R Z_R + k_I Z_I) \right\} \quad (11)$$

with

$$\Lambda^2 = \sum_{j=1} a_j^2 \quad . \quad (12)$$

Integrating (11) yields

$$P(Z_R,Z_I) = \exp[-Z_R^2/\Lambda^2 - Z_I^2/\Lambda^2)]/(\pi\Lambda^2) \tag{13}$$

and the distribution for the surface displacement

$$\zeta = \frac{i}{2}(Z - Z^*) = -Z_I \tag{14}$$

is then given by

$$P(\zeta) = \int_{-\infty}^{\infty} P(Z_R,Z_I)dZ_R = \frac{1}{\sqrt{\pi}\,\Lambda}\exp[-\zeta^2/\Lambda^2] \quad . \tag{15}$$

From the above arguments the Gauss distribution for the vertical surface displacement is a consequence of superimposing a large number of linear waves with phases uniformly distributed in the interval $(0,2\pi)$. For this reason when the Gauss distribution is employed for a field variable, such as the height, it is often referred to as the random phase approximation. The distribution function (15) can now be used to replace the average defined by the discrete sum in (1) with the phase space ensemble average

$$<\zeta(\underline{x},t)> = \int_{-\infty}^{\infty} \zeta P(\zeta)d\zeta \tag{16}$$

for a linear, homogeneous, stationary wave field.

The above arguments can be generalized to include a distribution of wave amplitudes in the derivation of $P(\zeta)$. If we assume that the wave amplitudes a_j have a distribution function $P_0(a_j)$, then the mean square surface displacement Λ^2 defined by (12) is replaced with

$$\Lambda^2 = \sum_{j=1}^{M} \frac{1}{2} \int_0^\infty a_j^2 \, P_0(a_j) da_j = \frac{1}{2} \sum_{j=1}^{M} \langle a_j^2 \rangle \tag{17}$$

with the remainder of the argument remaining unchanged, see eg., Montroll and West[188].

Assumptions analogous to this simple model have been made to describe the interaction among waves on the ocean surface. The phase space for the nonlinear gravity wave field consists of the mode amplitudes for the individual waves with a wavevector $\underline{k}$ and frequency ω. Since the mode amplitudes $B_{\underline{k}}(t)$ are linear combinations of the Fourier transforms of the surface displacement $\zeta(\underline{x},t)$ and velocity potential $\phi_s(\underline{x},t)$, these quantities constitute a statistical wave field on the sea surface. The Gaussian approximation in this context is equivalent to assuming that the amplitudes of linear gravity waves have a Rayleigh distribution[189] and that the relative phases are uniformly distributed on the interval $(0,2\pi)$. The Rayleigh distribution of amplitudes can be constructed by noting that

$$R^2 \equiv Z_R^2 + Z_I^2 \tag{18}$$

and

$$dZ_R dZ_I = R dR d\theta \tag{19}$$

for a narrow band surface wave field so that (13) reduces to

$$P(R,\theta) = \frac{2R}{\Lambda^2} e^{-R^2/\Lambda^2} \frac{1}{2\pi} = P(R)P(\theta) \quad . \tag{20}$$

Therefore the distribution of the amplitude of a narrow band process is the Rayleigh distribution

$$P(R)dR = \frac{2RdR}{\Lambda^2} e^{-R^2/\Lambda^2} \tag{21}$$

as obtained by Longuet-Higgins[190] using a different argument, and the uniform distribution in angle

$$P(\theta)d\theta = \frac{d\theta}{2\pi} \quad . \tag{22}$$

We now refer back to the fourth lecture where a qualitative argument on the validity of expanding the exponential $e^{k\zeta}$ was considered in discussing the convergence of terms in the mode rate equations. We concluded that since the surface height $\zeta(\underline{x},t)$ contained a broad spectrum of waves that the expansion was only valid for low frequency waves, i.e, gravity waves, and could not be used when k referred to gravity-capillary waves. For a fully developed spectrum of sea waves we must be more circumspect in the analysis since the fluctuations in ζ make the condition $k\zeta < 1$ almost impossible to satisfy instantaneously. A more reasonable condition than $k\zeta < 1$, for the validity of the perturbation expansion on the fluctuating surface, is that the expansion be convergent *on the average*. To determine this average we use the equilibrium distribution (15) and write

$$\langle e^{k\zeta}\rangle = \int_{-\infty}^{\infty} e^{k\zeta}\, P(\zeta)d\zeta \quad . \tag{23}$$

Expanding the exponential in a Taylor series and recalling that $\Lambda^2 = \langle\zeta^2\rangle$ from (15) we have

$$\langle e^{k\zeta}\rangle = \sum_{n=0} \frac{(k\Lambda)^{2n}}{2^n\, n!} = e^{\frac{1}{2}(k\Lambda)^2} \tag{24}$$

which is an exact expression for a linear, homogenous, stationary wave field.

For the Taylor expansion truncated at second order to be a useful approximation to the exponential we require that $\frac{1}{2}(k\Lambda)^2 < 1$, where Λ^2 is the mean square height of the sea surface. We estimate this mean square height by assuming that a steady state spectrum of gravity waves $\Psi(\underline{k})$ is known from experiment so that

$$\Lambda^2 = \int \Psi(\underline{k})\; d^2k \quad . \tag{25}$$

To evaluate (25) we use the spectrum specified by Phillips[5] based on a scaling argument, i.e.,

$$\Psi(\underline{k}) = \frac{.0045}{2\pi\, k^4} \tag{26}$$

which is spatially isotropic, i.e., waves move with equal strength in all directions. Integrating (25) using (26) we obtain

$$\Lambda^2 = -\,2.25 \times 10^{-3}\left(\frac{1}{k_{max}^2} - \frac{1}{k_{min}^2}\right) \tag{27}$$

where k_{max} is the shortest and k_{min} the longest wavelength waves in the gravity wave spectrum. The minimum wavenumber in the spectrum is assumed to be determined by the longest wave generated by the wind of a given speed W. This wave is such that a given phase front travels with the wind, i.e., the phase velocity of the waves matches the average wind speed,

$$V_k = \frac{1}{2}\sqrt{\frac{g}{k_{min}}} = W \quad . \tag{28}$$

The convergence condition, neglecting the k_{max}^{-2} term in (27) and using (28) is then (in MKS units)

$$\frac{1}{2}\,(k\Lambda)^2 \;\cong\; 1.2 \times 10^{-5}\, k^2\, W^4 < 1 \tag{29}$$

which relates a test wavenumber k with a given wind speed. In Figure 15.1 the parabolic boundary between the convergent and divergent regions for the expansion as a function of k is given. The waves for which the expansion converges become longer with increasing wind speed. For a wind speed of 10m/sec, for example, one could not include waves shorter than ($\lambda = 2\pi/k$) 2.2 meters whereas for W = 5m/sec one could not include waves shorter than 54 centimeters. These waves would violate the convergence condition. Due to the uncertainty in the form of the spectral density, the wind speed, etc., the boundary between the convergence and divergence regions is made broad. There is still substantial uncertainty in which of the interactions in the mode rate equations are convergent and which are not.

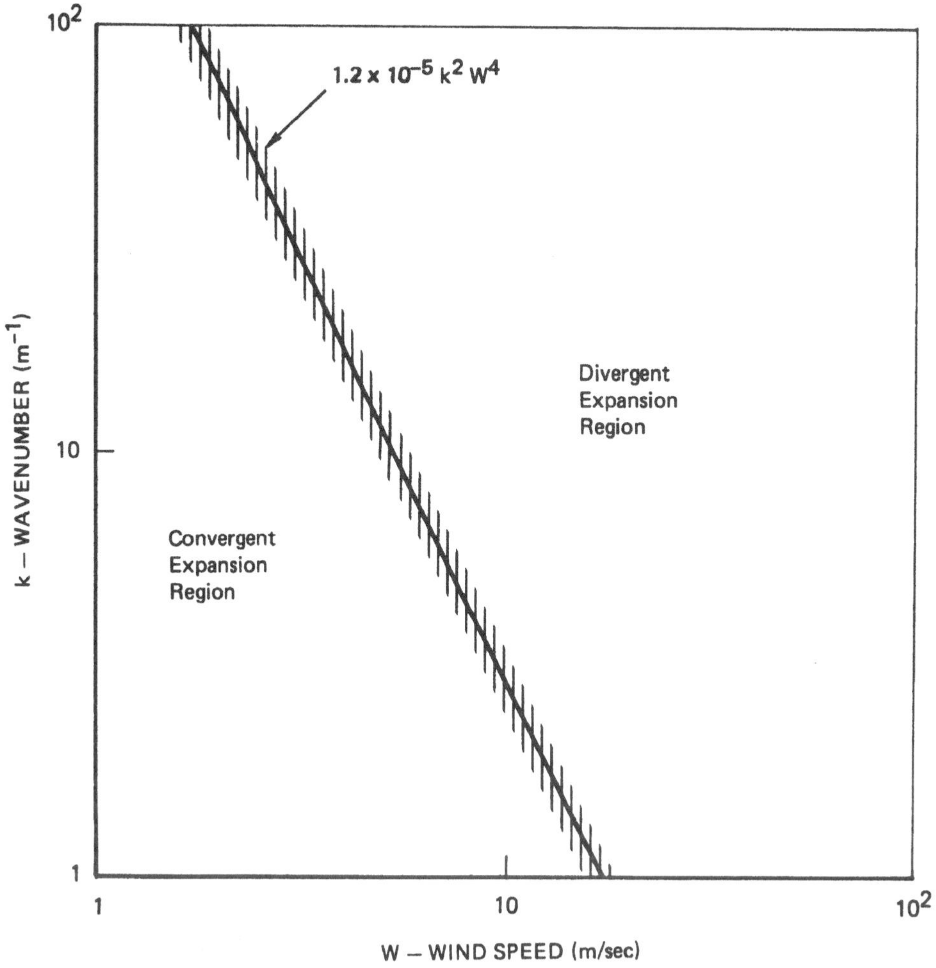

Figure 15.1: The boundary separating the regions of convergent and divergent series expansions of the exponential exp [k}] in the dynamic equations for the surface wave field is indicated.

16. Linear Stochastic Differential Equation (Cumulant Expansions)

In this lecture we consider a general technique for solving *linear* stochastic differential equations having additive fluxes of the Langevin form and also those with fluctuating parameters. There has been a great deal of study of both linear and quasi-linear stochastic equations of the Langevin type in the past seventy years. The most comprehensive review of the properties of physical systems in which these techniques can be important is given in the series of articles by M. Lax.[91,92,191] It is often not a simple matter to separate the mathematical formalism from the physical applications, however this has been done in a skillful way by van Kampen[94]. We draw heavily from the works of these two authors in this lecture, but in order to avoid a purely formal discussion of the technique we address the wave generation question again and develop the formalism in the context of solving a phenomenological model of this problem.

It would not be very interesting for us to reiterate the discussion of the preceeding lectures just to illustrate a mathematical technique. Let us consider then a physical mechanism not contained in the Miles-Phillips model. The mechanism is the coherent generation of water waves by the turbulent eddies in the wind field as first suggested by Stewart[192]. West and Seshadri[89] generalized the linear air-sea coupling models of Miles[87] and Phillips[85] to include this effect by considering the air-sea coupling parameter as a phenomenological fluctuating quantity. The model they considered was solved exactly, assuming these fluctuations to be delta correlated in time. The energy spectrum of the water wave field was shown by them to grow exponentially at rates higher than those predicted by the linear Miles-Phillips model. These initial growth rates are consistent with those observed in open ocean measurements.[88]

The original linear equations of motion for the gravity wave field is obtained from (12.17) by neglecting ν and γ to be

$$\frac{\partial}{\partial t}\begin{pmatrix}\phi_{\underline{k}}(t)\\ \zeta_{\underline{k}}(t)\end{pmatrix} + \begin{pmatrix}-2\mu_k\omega_k & g\\ -k & o\end{pmatrix}\begin{pmatrix}\phi_{\underline{k}}(t)\\ \zeta_{\underline{k}}(t)\end{pmatrix} = \begin{pmatrix}1\\ 0\end{pmatrix}\frac{p_{\underline{k}}(t)}{\rho_o} \quad . \tag{1}$$

West and Seshadri[89] assumed that the pressure fluctuations have the form

$$p_{\underline{k}}(t) = p_{\underline{k}}^{(o)}(t) + 2\rho_o\,\delta\mu_k(t)\omega_k\phi_{\underline{k}}(t) \tag{2}$$

where $p_{\underline{k}}^{(o)}(t)$ is the incoherent pressure fluctuation of Phillips and $\delta\mu_k(t)$ is the fluctuation in the air-sea coupling parameter due to turbulence in the air flow. An alternate to this expression could just as easily be postulated, i.e.

$$p_{\underline{k}}(t) = p_{\underline{k}}^{(o)}(t) + \rho_o\frac{\hat{\beta}_k(t)}{k}\zeta_{\underline{k}}(t) \tag{3}$$

where $\beta_k(t)/k$ is here a fluctuation in the frequency of the gravity wave (see the discussion by Phillips[5]). For the purpose of the present demonstration we prefer the latter model and using $\gamma_k(t) \equiv i\beta_k(t)/k$ rewrite (1) as

$$\frac{\partial}{\partial t}\begin{pmatrix}\phi_{\underline{k}}(t)\\ \zeta_{\underline{k}}(t)\end{pmatrix} + \begin{pmatrix}-2\mu_k\omega_k & g\\ -k & o\end{pmatrix}\begin{pmatrix}\phi_{\underline{k}}(t)\\ \zeta_{\underline{k}}(t)\end{pmatrix} = \gamma_k(t)\begin{pmatrix}0 & 1\\ 0 & 0\end{pmatrix}\begin{pmatrix}\phi_{\underline{k}}(t)\\ \zeta_{\underline{k}}(t)\end{pmatrix} + \frac{p_{\underline{k}}^{(o)}}{\rho_o}\begin{pmatrix}1\\ 0\end{pmatrix} \quad . \tag{4}$$

The similarity transformation $\underline{\underline{S}}$ in (12.21) still diagonalizes the deterministic part of (4) so that using

$$\underline{C}_{\underline{k}}(t) \equiv \begin{pmatrix}C_{\underline{k}}(t)\\ C_{\underline{k}}^{*}(t)\end{pmatrix} = \begin{pmatrix}\frac{1}{V_k} & -i\\ \frac{1}{V_k^{*}} & i\end{pmatrix}\begin{pmatrix}\phi_{\underline{k}}(t)\\ \zeta_{\underline{k}}(t)\end{pmatrix} \tag{5}$$

(4) can be written as

$$\dot{\underline{C}}_k(t) + \underline{\Lambda}_k \underline{C}_k(t) = \gamma_{\underline{k}}(t)\, \underline{B}\, \underline{C}_k(t) + \underline{S}\, \underline{f}_k(t) \qquad (6)$$

where

$$\underline{B} \equiv \frac{-i}{2V_k} \begin{pmatrix} -1 & e^{-i\theta_k} \\ -e^{i\theta_k} & 1 \end{pmatrix} \qquad (7)$$

and with the complex conjugate eigenvalues $\lambda_k = -\mu_k\omega_k + i\left(\omega_k^2 - \mu_k^2\omega_k^2\right)^{1/2} = -\mu_k\omega_k + i\omega_1(k)$

from (12.19) we have

$$e^{-i\theta_k} = \frac{V_k^*}{V_k} = \frac{\lambda_k^*}{\lambda_k} \quad . \qquad (8)$$

The harmonic oscillator system (6) has been solved exactly by West, Lindenberg and Seshadri[193] under relatively mild restrictions on the fluctuating fluxes $\gamma_{\underline{k}}(t)$ amd $f_{\underline{k}}(t)$. Note that the fluctuating frequency in (6) provides a term in the equation of evolution that is proportional to the mode amplitude so that fluctuations become dependent on the state of the wave field. A relatively weak fluctuation can therefore be amplified by a large amplitude mode, or suppressed by one with a small amplitude.

The solution to a linear stochastic differential equation such as (6) is meaningful only when the statistics of the fluctuations are specified. We assume the incoherent pressure fluctuations have a zero mean value so that denoting the average over an *ensemble* of realizations of $f_{\underline{k}}(t)$ by an overbar we have

$$\overline{f_{\underline{k}}(t)} = 0 \quad . \qquad (9)$$

This assumption presumes, as we did before, that the air-sea coupling parameter has adequately modeled the average interaction of the air and sea. We also assume that the incoherent fluctuations are stochastically stationary in time with a correlation function given by

$$\overline{f_{\underline{k}}(t)f^*_{\underline{k}'}(t-\tau)} = 2\Psi(\underline{k})\psi_{\underline{k}}(\tau)\delta_{\underline{k}-\underline{k}'}/\rho_0^2\ V_k^2 \tag{10}$$

where $\Psi(\underline{k})$ is the power spectral density of the pressure fluctuations and $\psi_{\underline{k}}(\tau)$ is the normalized correlation of the fluctuations over a time interval τ. The Kronecker delta $\delta_{\underline{k}-\underline{k}'} = 1$ if $\underline{k}=\underline{k}'$ and $= 0$ if $\underline{k}\neq\underline{k}'$ indicating that the fluctuations are homogeneous in space. Note that (10) is only slightly less general than (14.6). We denote the average over an ensemble of realizations of the coupling fluctuations (the coherent pressure fluctuations) by a bracket. We assume that these fluctuations are statistically independent of $f_{\underline{k}}(t)$,

$$\langle\gamma_{\underline{k}}(t)f_{\underline{k}}(t)\rangle = 0 \tag{11}$$

have a zero mean, i.e.

$$\langle\gamma_{\underline{k}}(t)\rangle = 0 \tag{12}$$

are statistically stationary in time with a correlation function given by

$$\langle\gamma_{\underline{k}}(t)\gamma_{\underline{k}'}(t-\tau)\rangle = 2D_2(\underline{k})\delta(\tau) \tag{13}$$

and the *nth* order semi-invariant (cumulant) is given by

$$\langle\langle\gamma_{\underline{k}}(t_1)\gamma_{\underline{k}}(t_2)\ldots.\gamma_{\underline{k}}(t_n)\rangle\rangle = 2^{n-1}D_n(\underline{k})\delta(t_1-t_2)\ldots.\delta(t_{n-1}-t_n). \tag{14}$$

Thus the coherent pressure fluctuations represented by $\gamma_k(t)$ are specified to be given by a zero-centered, delta correlated process of spectral strength $D_2(\underline{k})$ at wavevector $\underline{k}$ and higher cummulants given by $D_n(\underline{k})$, $n \geq 3$. We denote a cumulant average by the double bracket in (13) for $\gamma_{\underline{k}}$ - fluctuations and by a double overbar for $f_{\underline{k}}$ - fluctuations. Note that we have *not* assumed Gaussian statistics for $\gamma_{\underline{k}}(t)$.

We here follow the analysis of West, Lindenberg and Seshadri[193] quite closely in demonstrating the cumulant expansion technique for solving linear s.d.e.'s. To do this we transform (6) to the interaction representation and thus eliminate the systematic part of the evolution using

$$\underline{c}_{\underline{k}} \equiv \begin{pmatrix} c_{\underline{k}} \\ c^*_{\underline{k}} \end{pmatrix} = e^{-\underline{\underline{\Lambda}}_{\underline{k}} t} \underline{C}_{\underline{k}} \; . \tag{15}$$

In this new variable (6) becomes

$$\dot{\underline{c}}_{\underline{k}} = \gamma_{\underline{k}}(t)\, \tilde{\underline{\underline{B}}}(t)\, \underline{c}_{\underline{k}} + e^{-\underline{\underline{\Lambda}}_{\underline{k}} t} \underline{\underline{S}}\, \underline{f}_{\underline{k}}(t) \tag{16}$$

where the tilde over a matrix denotes the interaction representation, ie.

$$\tilde{\underline{\underline{B}}}(t) = e^{-\underline{\underline{\Lambda}}_{\underline{k}} t} \underline{\underline{B}}\, e^{\underline{\underline{\Lambda}}_{\underline{k}} t} = \frac{1}{2iV_k} \begin{pmatrix} -1 & e^{-i(2\omega_1(k)t+\theta_k)} \\ -e^{i(2\omega_1(k)t+\theta_k)} & 1 \end{pmatrix} \tag{17}$$

and

$$e^{-\underline{\Lambda}_{k} t} \underline{\underline{S}}\, \underline{f}_{\underline{k}}(t) = f_{\underline{k}}(t) \begin{pmatrix} e^{-\lambda_k t}/V_k \\ e^{-\lambda^*_k t}/V^*_k \end{pmatrix} = f_{\underline{k}}(t)\, \underline{g}_{\underline{k}}(t) \; . \tag{18}$$

The formal solution to (16) is

$$\underline{c}_k(t) = T\left\{\exp\left[-\int_0^t d\tau\gamma_{\underline{k}}(\tau)\ \tilde{\underline{\underline{B}}}_{\underline{k}}(\tau)\right]\right\} \underline{c}_{\underline{k}}(o)$$
$$+\int_0^t dt_1\ T\left\{\exp\left[-\int_0^t d\tau\gamma_{\underline{k}}(\tau)\ \tilde{\underline{\underline{B}}}_{\underline{k}}(\tau)\right]\right\}\ f_{\underline{k}}(t_1)\ \underline{g}_{\underline{k}}(t_1) \quad (19)$$

where T denotes a time ordering of the exponential factor. It is this time ordering of the formal solution which is one of the essential technical difficulties in obtaining useful solutions to linear stochastic systems.

The time ordered notation refers to the fact that the exponential term in (19) can be expressed as the solution of the differential equation for the evolution operator $\underline{\underline{U}}(t|t')$ as

$$\frac{\partial \underline{\underline{U}}(t|t')}{\partial t} = -\gamma_{\underline{k}}(t)\ \tilde{\underline{\underline{B}}}(t)\ \underline{\underline{U}}(t|t'),\ \text{with}\ \underline{\underline{U}}(t|t) = \underline{\underline{I}} \quad , \quad (20)$$

and $\underline{\underline{I}}$ is the 2 x 2 unit matrix. The solution to (20) is readily obtained by an iteration procedure to be

$$\underline{\underline{U}}(t|t') = 1 - \int_{t'}^{t} d\tau_1\gamma_{\underline{k}}(\tau_1)\ \tilde{\underline{\underline{B}}}(\tau_1) + \int_{t'}^{t} d\tau_1 \int_{t'}^{t} d\tau_2\gamma_{\underline{k}}(\tau_1)\gamma_{\underline{k}}(\tau_2)\ \tilde{\underline{\underline{B}}}(\tau_1)\ \tilde{\underline{\underline{B}}}(\tau_2) + \ldots$$

$$= \sum_{n=o}^{\infty} (-1)^n \int_{t'}^{t} d\tau_1 \int_{t'}^{\tau'} \ldots \int_{t'}^{\tau_{n-1}} d\tau_n\gamma_{\underline{k}}(\tau_1)\ldots\gamma_{\underline{k}}(\tau_n)\ \tilde{\underline{\underline{B}}}(\tau_1)\ldots\tilde{\underline{\underline{B}}}(\tau_n). \quad (21)$$

If the matrices $\tilde{\underline{\underline{B}}}(\tau)$ at different times commute, then all the integrals can be extended from $\tau_j = t'$ to $\tau_j = t$, $j = 1,2,\ldots n$, as long as a factor of $(n!)^{-1}$ is included to compensate for the additional area of integration introduced by this extension. When the matrices do not commute Dyson[194] noted that one could write (21) as

$$\underline{\underline{U}}(t|t') = \sum_{n=1}^{\infty} \frac{(-1)^n}{n!} T \left\{ \int_{t'}^{t} d\tau_1 \int_{t'}^{t} d\tau_2 .. \int_{t'}^{t} d\tau_n \gamma_{\underline{k}}(\tau_1) .. \gamma_{\underline{k}}(\tau_n) \, \underline{\underline{\tilde{B}}}(\tau_1) .. \underline{\underline{\tilde{B}}}(\tau_n) \right\} \tag{22}$$

where the time ordering symbol T indicates that the factors are to be rearranged in order of decreasing time prior to integrating. Note that (22) is merely the term by term rendering of the symbolic exponential factor in (19), ie, we can rewrite (19) in terms of the evolution operator (22) as

$$\underline{c}_{\underline{k}}(t) = \underline{\underline{U}}(t|0) \, \underline{c}_{\underline{k}}(0) + \int_0^t dt_1 \, \underline{\underline{U}}(t|t_1) \, \underline{g}_{\underline{k}}(t_1) \, f_{\underline{k}}(t_1) \quad . \tag{23}$$

The time ordering is necessary here because the matrices $\underline{\underline{\tilde{B}}}(\tau_1)$ and $\underline{\underline{\tilde{B}}}(\tau_2)$ do not commute if $\tau_1 \neq \tau_2$.

The average mode amplitudes can be calculated from (23) by sequentially taking the ensemble averages of $f_{\underline{k}}(t)$ and $\gamma_{\underline{k}}(t)$. Averaging over an ensemble of realizations of the incoherent pressure field fluctuations and using the zero mean property (9) we obtain

$$\overline{\underline{c}}_{\underline{k}}(t) = \underline{\underline{U}}(t|0) \, \underline{c}_{\underline{k}}(0) \quad . \tag{24}$$

If our interest is solely in the wave generation problem then initially there are no gravity waves present and $\underline{c}_{\underline{k}}(0) = \underline{0}$. Therefore the mean value of the wave amplitudes is zero regardless of the coherent fluctuations. Let us assume, however, that we initially have some waves on the sea surface, then the average of (24) over an ensemble of realizations of the coherent pressure fluctuations yields

$$\overline{\langle \underline{c}_{\underline{k}}(t) \rangle} = \langle \underline{\underline{U}}(t|0) \rangle \, \underline{c}_{\underline{k}}(0) \quad . \tag{25}$$

The average of the evolution operator over the $\gamma_{\underline{k}}(t)$ fluctuations must now be evaluated, but first we make note of a number of relations for an arbitrary statistical quantity, which are both needed here and will be useful later on.

For a stochastic variable X whose statistics are given by a probability density function P(x) we define the characteristic function p(k) by the relation

$$p(k) \equiv \langle e^{ikx} \rangle = \int_{-\infty}^{\infty} e^{ikx}\, p(x)\, dx \tag{26}$$

which is the Fourier transform of the probability density. We expand the exponential in the integrand and obtain the moment expansion

$$p(k) = \sum_{n=0}^{\infty} \frac{(ik)^n}{n!} \langle x^n \rangle \quad , \tag{27}$$

from which we conclude by Taylor expanding the left hand side of (27) and equating coefficients of equal powers of k that

$$\langle x^n \rangle = \frac{1}{i^n} \left[\frac{d^n}{dk^n} p(k) \right]_{k=0} \quad . \tag{28}$$

The characteristic function is thus the moment generating function for this process.[195] An equally valid relation exists for the semi-invariants or cumulants of the process indicated by a double bracket and is given by

$$\ln p(k) \equiv \sum_{n=1}^{\infty} \frac{(ik)^n}{n!} \langle\langle x^n \rangle\rangle \tag{29}$$

where the cummulant generating function is the logarithm of the characteristic function. Taylor expanding the left hand side of (29) and equating coefficients of equal powers of k yields

$$<<x^n>> = \frac{1}{i^n}\left[\frac{d^n}{dk^n}\ln p(k)\right]_{k=0} \quad . \tag{30}$$

A relation between cummulants and moments is obtained by equating the power series (27) and (29) in the form

$$\sum_{n=1}^{\infty} \frac{(ik)^n}{n!} <<x^n>> = \ln \sum_{n=0}^{\infty} \frac{(ik)^n}{n!} <x^n> \quad . \tag{31}$$

Expanding the logarithm around unity we obtain by equating coefficients of equal powers of (ik) the relations

$$\begin{aligned}
<x> &= <<x>> \\
<x^2> &= <<x^2>> + <<x>>^2 \\
<x^3> &= <<x^2>> + 3<<x>><<x^2>> + <<x>>^3 \\
<x^3> &= <<x^3>> + 3<<x^2>>^2 + 4<<x>><<x^3>> + 6<<x>>^2<<x^2>> \\
&\quad + <<x>>^4 \\
&\vdots
\end{aligned} \tag{32}$$

(see Gnedenko and Kolmogorov[196] for the formal proof of these equations).

If we now define X(t) as the integral $\int_0^t \gamma_{\underline{k}}(t_1)\, dt_1$, then we can rewrite (29), using the definition of the characteristic function (26), as

$$<e^{ik\int_0^t \gamma_{\underline{k}}(t_1)dt_1}> = \exp\left\{\sum_{n=1}^{\infty} \frac{(ik)^n}{n!} \int_0^t \cdots \int_0^t dt_1 \ldots dt_n <<\gamma_{\underline{k}}(t_1) \ldots \gamma_{\underline{k}}(t_n)>>\right\} . \tag{33}$$

We have introduced the cummulant expansion because although the moment expansion

$$< e^{ik\int_0^t \gamma_{\underline{k}}(t_1)dt_1}> = 1 + ik\int_0^t <\gamma_{\underline{k}}(t_1)>dt_1 - \frac{k^2}{2}\int_0^t dt_1 \int_0^t dt_2 <\gamma_{\underline{k}}(t_1)\, \gamma_{\underline{k}}(t_2)>$$
$$+ \ldots$$

can be evaluated term by term, it is well known that any finite number of terms in this expression is a poor representation of the exponential (see van Kampen[191] for a further discussion of the inadequacy of this representation). In the same way, although we can write $\underline{\underline{U}}(t|0)$ from (22) as

$$\langle\underline{\underline{U}}(t|0)\rangle = \sum_{n=0}^{\infty} \frac{(-1)^n}{n!} T \left\{ \int_0^t d\tau_1 \int_0^t d\tau_2 \ldots \int_0^t d\tau_n \langle\gamma_{\underline{k}}(\tau_1)\ldots\gamma_{\underline{k}}(\tau_n)\rangle \tilde{\underline{\underline{B}}}(\tau_1)\ldots\tilde{\underline{\underline{B}}}(\tau_n) \right\} \quad (34)$$

we suspect that it also will not yield a valid representation of the evolution operator. However, Kubo[197] has generalized the arguments sketched between (26) and (32), so that one can write the cummulant expansion

$$\langle\underline{\underline{U}}(t|0)\rangle = \exp\left\{ \sum_{n=1}^{\infty} (-1)^n \int_0^t d\tau_n \ldots \int_0^{\tau_2} d\tau_1 \langle\langle\gamma_{\underline{k}}(\tau_1)\ldots\gamma_{\underline{k}}(\tau_1)\rangle\rangle \tilde{\underline{\underline{B}}}(\tau_n)\ldots\tilde{\underline{\underline{B}}}(\tau_1) \right\} \quad (35)$$

which is the natural extension of the cummulant expansion (33) to non-commuting matrices[195].

We have assumed the $\gamma_{\underline{k}}(t)$ fluctuations to be delta correlated in time, so the cummulants in (35) vanish except when the time arguments are all equal, $\tau_1 = \tau_2 = \ldots = \tau_n$. Thus using (14) we can write

$$\langle\underline{\underline{U}}(t|0)\rangle = \exp\left\{ \sum_{n=1}^{\infty} (-1)^n \int_0^t D_n \left[\tilde{\underline{\underline{B}}}(\tau_1)\right]^n d\tau_1 \right\} \quad (36)$$

where each delta function contributes a factor of 1/2 since the limit of each integral matches one of the arguments in the delta functions. On the other hand, it can be seen from (17) that the matrix $\tilde{\underline{\underline{B}}}(\tau)$ is nilpotent, ie.

$$\tilde{\underline{\underline{B}}}^2(\tau) = 0 \quad (37)$$

which is a consequence of the nilpotency of the original interaction matrix $\underline{\underline{B}}$ in (6). It is of course this property which was known from the recent work of West, Lindenberg and Seshadri[193] that motivated our choice of air-sea coupling in the dynamic equations (4). With the property (37) we see that all the $n \geq 2$ terms in the exponent of (36) vanish because of this nilpotency, while the n=1 term vanishes because the average of $\gamma_{\underline{k}}(t)$ is zero. The average mode amplitude $\overline{<\underline{c}_{\underline{k}}(t)>}$ is therefore time independent.

Inverting (14) we therefore obtain for the evolution of the mean mode amplitudes

$$\overline{<\underline{C}_{\underline{k}}(t)>} = e^{\underline{\underline{\Lambda}}_{\underline{k}} t} \, \underline{C}_{\underline{k}}(0) \tag{38}$$

which is the same as the average motion in the absence of frequency fluctuations. In particular, by inverting the transformations (5) and (14) we obtain for the mean surface elevation and mean velocity potential due to the wave at wave vector $\underline{k}$

$$\overline{<\phi_{\underline{k}}(t)>} = e^{\mu_k \omega_k t} \left\{ \phi_{\underline{k}}(0) \left[\cos\omega_1 t - \mu_k \sin\omega_1 t\right] + \frac{\zeta_k(0)}{\omega_1} \sin\omega_1 t \right\} \tag{39}$$

$$\overline{<\zeta_{\underline{k}}(t)>} = e^{\mu_k \omega_k t} \left\{ -\phi_{\underline{k}}(0) \, \frac{\omega_k^2}{\omega_1} \sin\omega_1 t + \zeta_{\underline{k}}(0) \left[\cos\omega_1 t + \mu_k \sin\omega_1 t\right] \right\} .$$

The instability in the average surface properties is the same as that in the Miles-Phillips model and is due to the direct coupling to the average wind field. There is no additional instability generated by the coherent pressure fluctuations so that $\gamma_{\underline{k}}(t)$ does not affect the average dynamics of the surface.

However, we find that the second order moment properties (spectral density) are altered by these fluctuations and we examine these alterations in the next lecture.

One should not be too concerned about the failure to establish a new result for the mean field quantities using this elaborate formalism for two reasons. Firstly, because it is the second moment quantities which are of interest and these do have a new instability as we see in the next lecture. Secondly, the cumulant expansion technique is central to the understanding of the phase space evolution of the probability density and the calculation of the first order statistics was a relatively painless way of presenting the main ideas.

17. Gravity Wave Instability

In the preceeding lecture we specified that the coherent pressure fluctuations are described by a process which is delta correlated in time but that the incoherent pressure fluctuations have time correlations of arbitrary length. We have assumed the latter since one normally expects that fluctuations in the air flow near the sea surface will have some non-vasishing correlation time. However the delta correlation assumptions (16.13) and (16.14) enables us to obtain *simple exact* solutions to the first order statistics of the air-sea coupling model. In a similar way we could continue the analysis and use the formal solution (16.19) to evaluate the power spectral density for the surface properties. The general technique has been applied in detail by West, Lindenberg and Seshadri[193] but the algebraic analysis is lengthy. We therefore present an alternate discussion of the second order moments and construct a closed set of transport equations for these moments.[198,199]

Here we are concerned with the *initial* growth rates of the energy spectral density of the surface wave field. Since asymptotically the nonlinear terms become important and invalidate this model we restrict our investigation to equal time second-order statistics of the surface field quantities $\phi_{\underline{k}}(t)$ and $\zeta_{\underline{k}}(t)$. We evaluate the second order surface quantities $\overline{\langle\zeta_{\underline{k}}^{2}(t)\rangle}$, $\overline{\langle\phi_{\underline{k}}^{2}(t)\rangle}$ and $\overline{\langle\zeta_{\underline{k}}(t)\phi_{\underline{k}}(t)\rangle}$ to determine the modifications in the initial growth rates of the surface waves produced by the $\gamma_{\underline{k}}(t)$- fluctuations. We show here, following West and Seshadri[89], that the inclusion of these fluctuations in the linear wave field dynamics leads to an *enhanced growth rate* of the spectral energy density in general. We view this *soluable* model as a first order description of the effect of coherent pressure fluctuations on the evolution of the water wave field which is a modest extension of the lowest order description of the evolution given by the Miles-Phillips model[87].

We construct a column vector $\underline{X}_k(t)$ of the second order quantities of interest

$$\underline{X}_k(t) = \begin{pmatrix} \frac{1}{2}\zeta_{\underline{k}}^2(t) \\ \frac{1}{2}\phi_{\underline{k}}^2(t) \\ \zeta_{\underline{k}}(t)\phi_{\underline{k}}(t) \end{pmatrix} . \qquad (1)$$

The equation of motion for $\underline{X}_k(t)$ is constructed from (16.2) by multiplying the individual expressions by either $\phi_{\underline{k}}(t)$ or $\zeta_{\underline{k}}(t)$ to obtain the set of linear equations

$$\frac{d}{dt}\,\underline{X}_k(t) + \left[\underline{\underline{M}}_o + \gamma_{\underline{k}}(t)\,\underline{\underline{M}}_1\right]\underline{X}_k(t) = \underline{Q}_k(t) \qquad (2)$$

with the resulting coupling matrices, using the model (16.2) rather than (16.3),[89]

$$\underline{\underline{M}}_o \equiv \begin{pmatrix} 0 & 0 & -k \\ 0 & -4\mu_k\omega_k & g \\ 2g & -2k & -2\mu_k\omega_k \end{pmatrix},$$

$$\underline{\underline{M}}_1 \equiv \begin{pmatrix} 0 & 0 & 0 \\ 0 & 2 & 0 \\ 0 & 0 & 1 \end{pmatrix} ; \qquad \underline{Q}_k(t) = \begin{pmatrix} 0 \\ \phi_{\underline{k}}(t)\ f_{\underline{k}}(t) \\ \zeta_{\underline{k}}(t)\ f_{\underline{k}}(t) \end{pmatrix}. \qquad (3)$$

We first average (2) over an ensemble of realizations of the incoherent pressure fluctuations $f_{\underline{k}}(t)$ and defining $\underline{Z}_k(t) \equiv \overline{\underline{X}}_{\underline{k}}(t)$ and $\underline{R}_{\underline{k}} \equiv \overline{\underline{Q}_k(t)}$ we obtain the set of transport equations

$$\frac{d}{dt}\,\underline{Z}_k(t) + \left[\underline{\underline{M}}_o + \gamma_{\underline{k}}(t)\,\underline{\underline{M}}_1\right]\underline{Z}_k(t) = \underline{R}_k \quad . \qquad (4)$$

Note that we have smoothed over the $f_{\underline{k}}$ fluctuations in (4) but the $\gamma_{\underline{k}}$ - fluctuations are still explicit in the dynamic equations.

We evaluate the vector $\underline{R}_{\underline{k}}$ by calculating the correlated quantities $\overline{f_{\underline{k}}(t)\phi_{\underline{k}}(t)}$ and $\overline{f_{\underline{k}}(t)\zeta_{\underline{k}}(t)}$. We do this using the solution to the dynamic equations given by (16.19) with the initial conditions $\phi_{\underline{k}}(0) = 0$ and $\zeta_{\underline{k}}(0) = 0$ so that $\underline{C}_{\underline{k}}(0) = \underline{0}$. Multiplying (16.19) by $f_{\underline{k}}(t)$ and bar averaging we obtain

$$\overline{f_{\underline{k}}(t)\underline{C}_{\underline{k}}(t)} = \int_0^t dt_1 \; T \left\{ \exp \left[- \int_0^t d\tau \gamma_{\underline{k}}(t) \underline{\tilde{B}}_{\underline{k}}(t) \right] \right\} \overline{f_{\underline{k}}(t) f_{\underline{k}}(t_1) \underline{g}_{\underline{k}}(t_1)} \; . \tag{5}$$

Recalling the definitions (14.6) and (16.10) we obtain

$$\overline{f_{\underline{k}}(t)\underline{C}_{\underline{k}}(t)} = - \frac{2\Psi(\underline{k})}{\rho_0^2 \; V_k^2} \int_0^t \psi_{\underline{k}}(t-t_1) \; T \left\{ \exp \left[- \int_{t_1}^t d\tau \gamma_{\underline{k}}(\tau) \underline{\underline{B}}_{\underline{k}}(\tau) \right] \right\} \underline{g}_{\underline{k}}(t_1) \tag{6}$$

where if the correlation time in $\psi_{\underline{k}}(t-t_1)$ is sufficiently short we may replace the time ordered integral by unity and applying the inverse transforms (16.5) and (16.15) to (6) we obtain

$$\begin{pmatrix} \overline{f_{\underline{k}}(t)\phi_{\underline{k}}(t)} \\ \\ \overline{f_{\underline{k}}(t)\zeta_{\underline{k}}(t)} \end{pmatrix} = \frac{-2\Psi(\underline{k})}{\rho_0^2 \; V_k^2} \int_0^t \psi_{\underline{k}}(t-t_1) \underline{\underline{S}}^{-1} e^{\underline{\Lambda}_k (t-t_1)} \underline{\underline{S}} \begin{pmatrix} 1 \\ 0 \end{pmatrix} dt_1 \; .$$

$$= \frac{-2\Psi(\underline{k})}{\rho_0^2 \; V_k^2} \int_0^t \psi_{\underline{k}}(t-t_1) \; e^{\mu_k \omega_k (t-t_1)} \, dt_1 \begin{pmatrix} \cos \omega_1 (t-t_1) - \mu_k \sin \omega_1 (t-t_1) \\ - \dfrac{\omega_k^2}{\omega_1} \sin \omega_1 (t-t_1) \end{pmatrix} .$$

In the approximation that the correlation time τ_c of the incoherent pressure fluctuations is short compared to t_1 i.e. $t \gg \tau_c$, we may replace the upper limit of the integral in (7) by ∞ without significant loss of accuracy. Introducing the notation

$$D_c(\underline{k}) \equiv \frac{2}{\rho_0^2\, V_k^2} \int_0^\infty \Psi_{\underline{k}}(t')\, e^{\mu_k \omega_k t'} \cos \omega_1 t'\, dt' \quad , \tag{8a}$$

$$D_s(\underline{k}) \equiv \frac{2}{\rho_0^2\, V_k^2} \int_0^\infty \Psi_{\underline{k}}(t')\, e^{\mu_k \omega_k t'} \sin \omega_1 t'\, dt' \quad , \tag{8b}$$

we can write the inhomogeneous vector in the transport equation (4) as

$$\underline{R_k} = \Psi(\underline{k}) \begin{pmatrix} 0 \\ -D_c(\underline{k}) + \mu_k \dot{D}_s(\underline{k}) \\ \dfrac{\omega_k^2}{\omega_1} D_s(\underline{k}) \end{pmatrix} . \tag{9}$$

Note that both in evaluating the time ordered term in the integrand of (6) and in replacing the upper limit of the integral in (7) by infinity we have essentially expanded quantities in orders of the correlation time τ_c. In the limit $\tau_c \to 0$ which obtains for delta correlated incoherent pressure fluctuations these results become *exact*. Therefore the results below may be considered exact for delta correlated $f_{\underline{k}}$-fluctuations or approximate when the fluctuations have a finite but short correlation time τ_c.

To solve (4) we again go to the interaction representation using

$$\underline{\tilde{Z}_k}(t) \equiv e^{\underline{\underline{M}}_0 t}\, \underline{Z_k}(t) \tag{10}$$

$$\frac{d}{dt}\underline{\tilde{Z}}_k(t) = -\gamma_{\underline{k}}(t)\ \underline{\underline{\tilde{M}}}_1(t)\underline{\tilde{Z}}_{\underline{k}}(t) + \underline{\tilde{R}}_{\underline{k}}(t) \tag{11}$$

where

$$\underline{\underline{\tilde{M}}}_1(t) \equiv e^{\underline{\underline{M}}_o t}\ \underline{\underline{M}}_1 e^{-\underline{\underline{M}}_o t} \quad ; \ \underline{\tilde{R}}_{\underline{k}}(t) \equiv e^{\underline{\underline{M}}_o t}\ \underline{R}_{\underline{k}} \quad . \tag{12}$$

For the initial value problem $\overline{\zeta^2_{\underline{k}}(o)} = 0$, $\overline{\phi^2_{\underline{k}}(o)} = 0$ and $\overline{\zeta_{\underline{k}}(0)\phi_{\underline{k}}(o)} = 0$, ie. an initially smooth water surface, the formal solution to (11) is

$$\underline{\tilde{Z}}_k(t) = \int_0^t dt_1\ T\left\{\exp\left[-\int_0^t \gamma_{\underline{k}}(\tau)\ \underline{\underline{\tilde{M}}}_1(\tau)\ d\tau\right]\right\} e^{\underline{\underline{M}}_o t_1}\underline{R}_{\underline{k}} \quad . \tag{13}$$

The solution (13) can now be averaged over an ensemble of realizations of the coherent fluctuations and the cummulant expansion (16.35) applied.

Averaging (13) and inverting the transformation (10) yields

$$<\underline{Z}_{\underline{k}}(t)> = e^{-\underline{\underline{M}}_o t}\int_0^t dt_1\ \exp\left[\sum_{n=1}^{\infty} (-1)^n \int_{t_1}^{t} d\tau_1 \ldots \int_{t_1}^{t_2} d\tau_n \ <<\gamma_{\underline{k}}(\tau_n)\ldots\gamma_{\underline{k}}(\tau_1)>> \right.$$

$$\left. \cdot\ \underline{\underline{\tilde{M}}}_1(\tau_1)\ \ldots\ldots \underline{\underline{\tilde{M}}}_1(\tau_n)\right] e^{\underline{\underline{M}}_o t_1}\ \underline{R}_{\underline{k}} \tag{14}$$

where we have replaced the time ordered exponential by the infinite sum over cummulants using (16.35). Since the $\gamma_{\underline{k}}(t)$ - fluctuations are assumed to be delta correlated in time, using the expression for the *nth* cummulant (16.13), we reduce (14) to

$$<\underline{Z}_{\underline{k}}(t)> = e^{-\underline{\underline{M}}_o t}\int_0^t dt_1\ \exp\left[\sum_{n=2}^{\infty} D_n(\underline{k})\int_{t_1}^{t} \underline{\underline{\tilde{M}}}_1^n\ (t')\ dt'\right] e^{\underline{\underline{M}}_o t_1}\ \underline{R}_{\underline{k}} \quad . \tag{15}$$

Taking the time derivative of (15) we obtain the *full exact* transport equation

$$\frac{d}{dt} \langle \underline{Z}_k(t) \rangle + \left[\underline{\underline{M}}_0 - \sum_{n=1}^{\infty} D_n \underline{\underline{M}}_1^n \right] \langle \underline{Z}_{\underline{k}}(t) \rangle = \underline{R}_{\underline{k}} \tag{16}$$

with constant coefficients. The growth rates of $\overline{\langle \phi_{\underline{k}}^2(t) \rangle}$, $\overline{\langle \zeta_{\underline{k}}^2(t) \rangle}$ and $\overline{\langle \zeta_{\underline{k}}(t)\phi_{\underline{k}}(t) \rangle}$ are thus determined by the eigenvalues of (16).

We recall from lecture 16 that the earlier analysis was dramatically simplified because the interaction matrix $\underline{\underline{B}}_{\underline{k}}$ was nilpotent. A related simplification results here from the properties of $\underline{\underline{M}}_1$, ie. we find that $\underline{\underline{M}}_1^n$ is a diagonal matrix. Thus we assume a solution to (16) of the form $e^{\varepsilon_{\underline{k}} t}$ and reduce (16) to the secular equation

$$\det \left[\sum_{n=1}^{\infty} D_n \underline{\underline{M}}_1^n - \underline{\underline{M}}_0 - \varepsilon_{\underline{k}} \underline{\underline{I}} \right] = 0 \tag{17}$$

where $\varepsilon_{\underline{k}}$ is the eigenvalue and $\underline{\underline{I}}$ is a 3 x 3 unit matrix. The diagonal property of $\underline{\underline{M}}_1^n$ simplifies the contribution to (17), i.e.

$$\sum_{n=1}^{\infty} D_n M_1^n = \begin{pmatrix} 0 & 0 & 0 \\ 0 & \sum_{n=1}^{\infty} 2^n D_n & 0 \\ 0 & 0 & \sum_{n=1}^{\infty} D_n \end{pmatrix} \tag{18}$$

so that as long as the two sums in (18) are positive, a physically reasonable restriction, there is at least one root of (17) which yields an enhanced rate of growth.

The remarkable fact is that we have not had to specify the statistics of the coherent pressure fluctuations to determine its effect on the stability of the second moments of the linear water wave field (spectral densities).

In particular we have *not* assumed that the turbulent fluctuations in the air flow are represented by a Gaussian random process. To solve the cubic equations for the roots of (17) using data for the cummulants D_n, however, West and Seshadri[89] restricted the sums to n=2 since the experimental data is only available for power spectral densities. This restriction is in fact equivalent to assuming Gaussian statistics for the $\gamma_{\underline{k}}(t)$ - fluctuations in the determination of the eigenvalues, since it sets all the higher semi-invariants to be equal to zero.

We do not carry out the analysis of the roots of (17) here but instead refer the student to the literature[89,193,198] for a more complete discussion. We do note however that the effect of the coherent pressure fluctuations is to enhance the instability of the water wave field beyond that provided by the direct energy influx from the wind, i.e., the exponential growth rate of the waves is increased. In Figure (17.1) we show the real part of the largest positive root of (17) for a phenomenological power spectral density of the coherent pressure fluctuations. The dashed curves are growth rates from the Miles instability mechanism alone. The solid curves indicate the effect of the coherent fluctuations of the air-sea coupling parameter. The increase in growth rate for decreasing wavenumber is dramatic. A systematic study of this mechanism remains to be done, since as yet there is no fundamental calculation of $\gamma_{\underline{k}}(t)$, only the spectrum of $\gamma_{\underline{k}}(t)$ was used in evaluating the eigenvalues of (17).

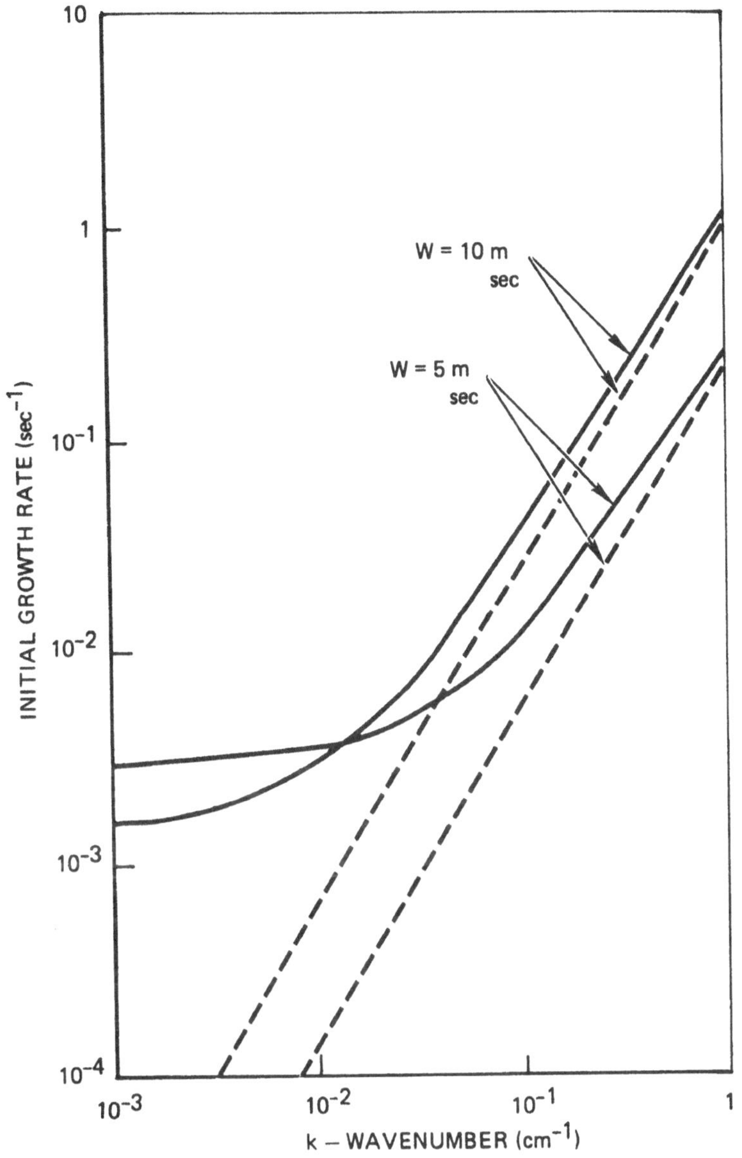

Figure 17.1: The positive and eigenvalue of Eq. (17.17) (growth rate of gravity wave energy spectral density) is indicated (——) for wind speeds of 5m/sec and 10m/sec and compared with the Miles growth rates (---) for the same wind speeds.

18. Stochastic Nonlinear Gravity Waves

In order to describe the cumulant transport technique in the preceeding lecture we assumed that the gravity wave field was driven by both coherent and incoherent fluctuations of the turbulent wind field. In discussing these two mechanisms we neglected the presence of the high frequency waves on the ocean surface. We now remedy that oversight and consider the effect of the high frequency waves on the evolution of the gravity waves. Since the gravity-capillary waves are in a steady state with the influx of energy from the fluctuating air flow, they themselves have fluctuating amplitudes and phases. Here we make the following assumptions about the generation and dissipation of high and low frequency waves: 1) the high frequency waves are generated primarily by the direct influx of energy from the fluctuating air flow[84,151,144]; 2) the low frequency waves are generated primarily, at least initially, by the nonlinear coupling between high and low frequency waves[152, 154-157]; 3) the water waves are damped by kinematic viscosity with a linear coefficient of the form νk^2 which is essentially negligible for the long waves, and 4) the short time-scales for the generation and the strong viscous damping of the high frequency waves as well as the energy transfer to lower frequency waves causes them to be essentially in a steady state at all points in space on the time-scale of the evolution of the gravity waves.

In this lecture we follow the arguments of Seshadri and West[150] and assume that the motion of the ocean surface is partitioned into waves which are resolved, with wavelengths longer than $2\pi/K_0$ say, and waves which are unresolved with wavelengths shorter than $2\pi/K_0$. The length scales are partitioned if a field measurement is done by means of a wave staff, for example. The mode rate equations (12.23) completely specify the dynamics of both the resolved and unresolved waves of our model water wave system. For a sufficiently large surface areas Σ_0 the short waves are essentially continuous in wavenumber; whereas the low frequency, longer wavelength gravity waves remain discrete. Explicitly separating the scales between these two spectral regions we *assume* that the complex surface displacement $Z(\underline{x},t)$ using the similarity transform (12.21) is given by

$$\begin{pmatrix} Z(\underline{x},t) \\ \\ Z^*(\underline{x},t) \end{pmatrix} = \begin{pmatrix} \frac{1}{V_K} & -i \\ \\ \frac{1}{V_K^*} & i \end{pmatrix} \begin{pmatrix} \phi_s(\underline{x},t) \\ \\ \zeta(\underline{x},t) \end{pmatrix} , \tag{1}$$

and can be segmented into resolved and unresolved wave parts, i.e.,

$$Z(\underline{x},t) = \sum_{|\underline{K}|<|\underline{K}_0|} B_{\underline{K}}(t)\, e^{iK\cdot x} + \int_{|\underline{k}|>>|\underline{K}_0|} d^2k\, C_{\underline{k}}(t)\, e^{i\underline{k}\cdot\underline{x}} . \tag{2}$$

We indicate the discrete set of long wavelength resolved waves with mode amplitudes $B_{\underline{K}}(t)$ by the capital Roman letters $\underline{K}$, $\underline{L}$, etc. The high frequency unresolved waves form a continuum and are indicated by lower case Roman letters $\underline{k}$, $\underline{\ell}$, etc., with mode amplitudes $C_{\underline{k}}(t)$. A similar separation of amplitudes was made by Yerkmakov and Pelinovshiy[200] for a related problem.

When we use the partitioning indicated in (2) in the equations of motion (12.23) we obtain a coupled system of rate equations involving both the B and C waves. In previous lectures we have restricted our attention to either the short fetch situations in which the C-waves have had sufficient time to reach a steady state, but the B-waves could not develop a significant amplitude, or the long fetch situation but neglecting the effect of the C-waves. To proceed to longer fetches over which the gravity waves will attain their asymptotic level we must re-examine the steady state of the unresolved (gravity-capillary) waves because now they are no longer isolated and will be modulated by the evolving resolved (gravity) waves. Here we model the gravity wave response to the steady state, modulated spectrum of unresolved waves using first order perturbation theory.

We use the properties of the unresolved waves determined in earlier lectures to eliminate these degrees of freedom from the dynamic description of the resolved surface wave field. This can be done because the unresolved waves oscillate through many periods before there is appreciable change in the resolved waves. Therefore, it is possible to solve the equations of motion for the C-waves assuming the B-waves are constants. This procedure is known as adiabatic elimination of the fast variables as discussed in lecture 13. The equation of evolution for the resolved B-waves is then given by

$$\dot{B}_{\underline{K}}(t) + \lambda_K B_{\underline{K}}(t) + i\sum_{\underline{L}} \Omega_{\underline{KL}}(t)\, B_{\underline{L}}(t) = g_{\underline{K}}(t) + T_{\underline{K}}(\underline{B};t) \quad . \tag{3}$$

The diagonal linear coefficient λ_K consists of the Miles-Phillips air-sea coupling parameter $\mu_K\omega_K$ and the unperturbed frequency of a linear gravity wave, i.e.,

$$\lambda_K = -\mu_K\omega_K + i\omega_K \quad . \tag{4}$$

The function $T_{\underline{K}}(\underline{B};t)$ represents the quadraic and cubic interactions among the gravity waves themselves. The explicit forms of these terms are unchanged from those given earlier except that the viscosity (νK^2) and surface tension (γ) are now negligible so the dynamic equations are given by (11.28) rather than (12.23).

The remaining terms $g_{\underline{K}}(t)$ and $\Omega_{\underline{KL}}(t)$ are the lowest order non-vanishing modulation of the B-waves by the unresolved C-waves. The function $g_{\underline{K}}(t)$ is obtained by substituting the partitioned surface displacement (2) into the definition of the Hamiltonian H_g in lecture 3 and is given by the product of two C-waves in order to satisfy the wave vector matching restriction. This triad interaction of two high frequency and one low frequency wave gives[150]

$$g_{\underline{K}}(t) \equiv \int d^2m\, d^2\ell\, \Gamma_{\underline{\ell}}^{Km}\, \delta(\underline{\ell}-\underline{m}-\underline{K})\, C_{\underline{\ell}}(t)\, C_{\underline{m}}^{*}(t) \quad . \tag{5}$$

Similarly, the time dependent frequency matrix is determined by the wave vector matching condition in the cubic interaction term. This is the interaction of two high and two low frequency waves yielding

$$\Omega_{\underline{KL}}(t) \equiv \frac{1}{2}\int d^2m\, d^2\ell\, \bar{\Gamma}_{\underline{L\ell}}^{Km}\, \delta(\underline{m}+\underline{K}-\underline{\ell}-\underline{L})\, C_{\underline{\ell}}(t)\, C_{\underline{m}}^{*}(t) \tag{6}$$

in which we can also include the diagonal coherent pressure fluctuation, if desi[red]. The coupling coefficients are given by

$$\bar{\Gamma}_{\underline{\ell}}^{Km} = \Gamma_{\underline{\ell}}^{Km} + \Gamma_{\underline{m}}^{K\ell} \quad , \tag{7a}$$

$$\bar{\Gamma}_{\underline{L\ell}}^{Km} = \Gamma_{\underline{L\ell}}^{Km} + \Gamma_{\underline{\ell L}}^{Km} \quad . \tag{7b}$$

There are two effects induced in the evolution of the resolved waves by the function $g_{\underline{K}}(t)$. First, since the unresolved waves have: 1) a higher natural frequency; 2) are strongly damped by viscosity and 3) are rapidly driven by the turbulent air flow over the surface, they act as fluctuating fluxes in the evolution of the B-waves. Second, although C-waves are rapidly fluctuating, they can, through their nonlinear coupling to the B-waves, provide an average or systematic modulation of the B-waves. These two effects can be separated by writing $g_{\underline{K}}(t)$ as

$$g_{\underline{K}}(t) = \langle g_{\underline{K}}(t) \rangle_{ss} + \hat{g}_{\underline{K}}(t) \quad , \tag{8}$$

where $\hat{g}_{\underline{K}}(t)$ is the deviation of the flux from its steady state value

$$\hat{g}_{\underline{K}}(t) \equiv g_{\underline{K}}(t) - \langle g_{\underline{K}}(t) \rangle_{ss} \quad , \tag{9}$$

and the brackets with a ss subscript denote an average over a steady state distribution of unresolved waves. The average denotes the systematic modulation of the resolved waves by the unresolved waves and the deviation (9) is a zero centered fluctuating driver in (3).

The separation of $g_{\underline{K}}$ into a mean and a fluctuating part in (8) makes explicit use of the steady state properties of the high frequency waves. The C-waves are in a steady state on the time scale of the evolution of the B-waves so that averaging over the spectral distribution of the C-waves

is in fact a well defined operation. The existence of an asymptotic steady state value for the correlation of the C-wave amplitudes is well established experimentally and has the theoretical basis discussed in lecture 14. There we consider the effects of the turbulent wind driving the short waves, the nonlinear transfer of energy and visious dissipation, but not the reverse modulation which provides the non-zero average value of $g_{\underline{K}}$, i.e., $\langle g_{\underline{K}}\rangle_{ss} \neq 0$. This reverse modulation mechanism was also discussed by Longuet-Higgins[152] as well as Garrett and Smith[156] and is a crucial ingredient of the present argument.

The procedure of decoupling the modulation of the resolved waves by the unresolved waves using a time scale separation argument has its origin in the statistical mechanical theory of Brownian motion discussed in lecture 13. The evolution of the long wavelength B-waves is analogous to the evolution of the massive Brownian particles, and the short wavelength C-waves evolve in a manner corresponding to the bath particles. The non-zero average value of $g_{\underline{K}}(t)$ characterizes the net effect of the C-waves on the B-waves. There are important differences between the two systems, however, just as there was in the analysis of the C-waves in isolation. The kinematic wave vector matching requirements do not allow the B-waves to interact with a full spectrum of C-waves. Also a minimum of three waves are needed to satisfy the wave vector requirements for the mutual interactions of gravity and gravity-capillary waves. Thus, the effect of the systematic (average) term $\langle g_{\underline{K}}(t)\rangle_{ss}$ on the evolution of the gravity waves can either enhance the growth *or* the dissipation of these waves. This is in contrast to traditional Brownian motion where the heat bath always induces a dissipation in the dynamics of the Brownian particle. This damping is a consequence of the lowest order coupling between the bath and Brownian particle, i.e., the interaction energy, being bi-linear in the bath and particle mode amplitudes. The interactions do not have a kinematic constraint in Brownian motion, so the heavy particle can interact with a continuous spectrum of perturbations.

Thus, the unresolved waves can either absorb energy from the resolved waves as in Brownian motion [see e.g., Hasselmann[153]] or giveup energy to the resolved waves as argued by Longuet-Higgins[152]. Valenzuela and Wright[145] point out that, when the correlation between the energy input from the air flow and the large scale surface flow (these are assumed to be uncorrelated by Hasselmann) are included in the energy balance equation, the modulated, short-wave, surface-stress supplies the energy influx to maintain the growth of the large scale gravity waves, see also, Keller and Wright[201] and Garrett and Smith[156] take this effect into account perturbatively.

The equations of motion for the C-waves when the B-waves have non-negligible amplitudes is given by the coupling of the high frequency waves to the orbital currents of the low frequency waves. To first order in this coupling strength we have

$$\frac{d}{dt} C_{\underline{k}}(t) + \alpha(\underline{k})\, C_{\underline{k}}(t) = f_{\underline{k}}(t) + \sum_{\underline{K}} \int d^2\ell\, \delta(\underline{k}-\underline{\ell}-\underline{K})\, C_{\underline{\ell}}(t)$$

$$\times \left\{ 2\Gamma^{\underline{k}}_{\underline{\ell}\underline{K}}\, B_{\underline{K}}(t) + \Gamma^{\underline{k},-\underline{K}}_{\underline{\ell}}\, B^{*}_{-\underline{K}}(t) \right\} \quad , \tag{10}$$

where

$$\alpha(\underline{k}) = \nu k^2 - \mu_k \omega_k + i\sqrt{\omega_k^2 - (\nu k^2 - \mu_k \omega_k)} - h_{\underline{k}} \quad . \tag{11}$$

Here $f_{\underline{k}}(t)$ is the incoherent pressure fluctuation and $h_{\underline{k}}$ is the parameter from statistical linearization. The solution to the linear equation (10) is given by the integral equation

$$C_{\underline{k}}(t) = \hat{C}_{\underline{k}}(t) + \sum_{\underline{K}} \int d^2\ell\, \delta(\underline{k}-\underline{\ell}-\underline{K}) \int_0^t dt'\; e^{-\alpha(\underline{k})(t-t')} C_{\underline{\ell}}(t')$$

$$\times \left\{ 2\Gamma_{\underline{\ell}\underline{K}}^{\underline{k}}\, B_{\underline{K}}(t') + \Gamma_{\underline{\ell}}^{\underline{k},-\underline{K}}\, B_{-\underline{K}}^{*}(t') \right\} \quad . \tag{12}$$

We recall from (14.26) that the unperturbed solution to (10) is

$$\hat{C}_{\underline{k}}(t) = \int_0^t e^{-\alpha(\underline{k})(t-t')} f_{\underline{k}}(t')\, dt' \tag{13}$$

when the ocean is assumed to be initially quiescent. The lowest order perturbative solution to (12) is given by replacing $C_{\underline{\ell}}(t')$ by $\hat{C}_{\underline{\ell}}(t')$ inside the integral. This approximation should be adequate for our purposes since the coupling between B-waves and C-waves is weak away from the region of wave breaking.

The average of the term $g_{\underline{K}}(t)$ in (3) is obtained by substituting the approximate solution to (12) into (5) and averaging over the steady state distribution of C's. It can be shown, in the limit of short correlation times, that the correlator $\langle \hat{C}_{\underline{r}}(t)\hat{C}_{\underline{s}}(t')\rangle_{ss}$ is given by [150]

$$\frac{1}{2}\langle \hat{C}_{\underline{r}}(t)\hat{C}^{*}_{\underline{s}}(t')\rangle_{ss} \cong \delta(\underline{r}-\underline{s})\ F(\underline{r})\ e^{-\alpha(\underline{r})\ (t-t')} \quad ;\ t > t'\ . \tag{14}$$

The quantity F(r) is the steady state spectrum of the gravity-capillary wave and was found in lecture 14 to be

$$F(\underline{r}) = \frac{1}{2}\langle |\hat{C}_{\underline{r}}|^2\rangle_{ss}$$

$$= \Phi(\underline{r})/\ 4\rho_0^2 V_r^2 \alpha_R(\underline{r}) \tag{15}$$

where $\tilde{\Phi}(\underline{r})$ is the power spectral density of the pressure fluctuations at wavevector $\underline{r}$ present on the surface and $\alpha(\underline{r})=\alpha_R(\underline{r})+i\alpha_I(\underline{r})$ with both α_R and α_I real. Thus in (5) we have

$$\langle g_{\underline{K}}(t)\rangle = \int d^2\ell d^2m\, \delta(\underline{\ell}-\underline{m}-\underline{K})\ \bar{\Gamma}^{\underline{K}m}_{\ell} \sum_{\underline{L}}\int d^2\ell' \int_0^t dt'\ e^{-\alpha(\underline{\ell})(t-t')}$$

$$\times \Big\{ \delta(\underline{\ell}-\underline{\ell}'-\underline{L})\ \delta(\underline{\ell}'-\underline{m})\ F(\underline{m})\ e^{-\alpha^*(\underline{m})(t-t')} \left[2\bar{\Gamma}^{\ell}_{\underline{\ell}';\underline{L}}\ B_{\underline{L}}(t') + \bar{\Gamma}^{\underline{\ell}',-\underline{L}}_{\underline{\ell}'}\ B^{*}_{-\underline{L}}(t')\right]$$

$$+\ \delta(\underline{m}-\underline{\ell}'-\underline{L})\ \delta(\underline{\ell}-\underline{\ell}')\ F(\underline{\ell})e^{-\alpha(\underline{\ell})(t-t')} \left[2\left(\bar{\Gamma}^{\underline{m}}_{\underline{\ell}';\underline{L}}\right)^{*} B^{*}_{\underline{L}}(t') + \left(\bar{\Gamma}^{\underline{m},-\underline{L}}_{\underline{\ell}}\right)^{*} B_{-\underline{L}}(t')\right]\ . \tag{16}$$

and we use the separation in scale between the long and short water waves to replace $F(\underline{\ell}-\underline{K})$ by $F(\underline{\ell})$; the coupling coefficient $\overline{\Gamma}^{\underline{\ell}-\underline{K}}_{\underline{\ell},\underline{K}}$ by $\overline{\Gamma}^{\underline{\ell}}_{\underline{\ell},\underline{K}}$ and $\overline{\Gamma}^{\underline{\ell},-\underline{K}}_{\underline{\ell}-\underline{K}}$ by $\overline{\Gamma}^{\underline{\ell},-\underline{K}}_{\underline{\ell}}$; then by ignoring the imaginary part of the coupling coefficients introduced by the wind and viscosity in the interaction between the long and short waves we obtain

$$\langle g_{\underline{K}}(t)\rangle \cong \int d^2\ell \; 4|\overline{\Gamma}^{\underline{\ell}}_{\underline{\ell}\,\underline{K}}|^2 F(\underline{\ell}) \int e^{-2\alpha_R(\underline{\ell})(t-t')} dt' [B_{\underline{K}}(t') - B^*_{-\underline{K}}(t')]. \tag{17}$$

We used the symmetry properties $\overline{\Gamma}^{\underline{\ell},-\underline{K}}_{\underline{\ell}} = -\overline{\Gamma}^{\underline{\ell}\underline{K}}_{\underline{\ell}}$ and $\overline{\Gamma}^{\underline{\ell}\underline{K}}_{\underline{\ell}} = 2\overline{\Gamma}^{\underline{\ell}}_{\underline{\ell}\underline{K}}$ in (17). Finally, if the unresolved waves relax sufficiently fast on the time scales of the B-waves, the $B_{\underline{K}}$ coefficients may be taken outside the integral in (17) and the upper limit of the integral extended to infinity. We therefore define the coefficient $\beta_{\underline{K}}$ by

$$\beta_{\underline{K}} \equiv \int d^2\ell \;\; |\overline{\Gamma}^{\underline{\ell}}_{\underline{\ell}\underline{K}}|^2 \, \frac{F(\ell)}{\alpha_R(\underline{\ell})} \tag{18}$$

and rewrite (17) as

$$\langle g_{\underline{K}}(t)\rangle = \beta_{\underline{K}}\left[B_{\underline{K}}(t) - B^*_{-\underline{K}}(t)\right] . \tag{19}$$

The fluctuating coefficient term in (3) given by (6) is a rapidly varying real quantity and can be regarded as a frequency fluctuation of the B-waves. Since this fluctuating quantity multiplies the variable B itself, it introduces a multiplicative fluctuation in the evolution process as opposed to the purely additive fluctuations introduced by the inhomogeneous term $\hat{g}_{\underline{K}}(t)$. West and Seshadri[89] determined that such a process could arise in the coupling between the sea surface and the turbulent air flow when the air-sea coupling parameter was assumed to have such fluctuations. The effect of these fluctuations on the predicted initial growth rates of the gravity wave field was found to be substantial. Here, however, we do not expect the wave induced frequency fluctuation to significantly effect the evolution of a broad spectrum of gravity waves since the imaginary part of the coupling coefficient $\Gamma^{-\underline{K}\ \underline{m}}_{\ \underline{L}\ \underline{\ell}}$ is so small. We therefore replace $\Omega_{\underline{KL}}(t)$ by its average value in the equation of motion, i.e.,

$$\langle\Omega_{\underline{KL}}(t)\rangle_{ss} = \frac{1}{2}\int \Gamma^{-\underline{K}\ \underline{\ell}}_{\ \underline{K}\ \underline{\ell}}\ \delta_{\underline{KL}}\langle|B_{\underline{\ell}}(t)|^2\rangle_{ss}\, d^2\ell$$

$$\equiv -\Delta\omega_{\underline{K}}\ \delta_{\underline{KL}} \tag{20}$$

to lowest order and neglect the frequency fluctuations. Indicating the total frequency by $\Omega_{\underline{K}} = \omega_K + \Delta\omega_{\underline{K}}$ we now write the mode rate equation as

$$\frac{d}{dt} B_{\underline{K}}(t) + i(\Omega_{\underline{K}} + i\mu_{\underline{K}}\omega_{\underline{K}})\ B_{\underline{K}}(t) = \beta_{\underline{K}}\left[B_{\underline{K}}(t) - B^{*}_{-\underline{K}}(t)\right] + \hat{g}_{\underline{K}}(t) + T_{\underline{K}}(\underline{B};t)\ . \tag{21}$$

In (21) the long wavelength waves grow even when only a gentle breeze is blowing due to the $\beta_{\underline{K}}$ term. This form for the coupling was also obtained by Garrett and Smith[156].

19. Steady-State Nonlinear Wave Field

The picture we have constructed of the physical mechanisms operative in the generation and evolution of the nonlinear field of water waves is now fairly extensive. The turbulent wind blows over the water surface and the incoherent fluctuations in the pressure induce a high frequency surface response.[87] The air flow efficiently couples to this surface "roughness" and the generated water waves modulate the air flow extracting energy from the mean wind.[85] The combined effects of the nonlinear hydrodynamic interactions and viscous dissipation act to balance the energy influx from the air flow producing an asymptotic steady state in a relatively short fetch.[154] At longer fetches, longer surface waves are stimulated by the incoherent pressure fluctuations. These linearly growing waves also couple to the mean wind field resulting in an instability with a subsequent exponential growth of the waves with fetch. The rate of growth of these waves is enhanced by two mechanisms: (i) the coherent pressure fluctuations and (ii) the modulation of the steady state spectrum of the high frequency waves by the growing gravity waves.[150] The short scale waves augment both the additive flux of energy to the long waves and the multiplicative coherent fluctuations.

As the energy is accumulated in the gravity wave region of the water wave spectrum, their amplitudes increase and the nonlinear interactions become important. Each of the above mechanisms is included in the *nonlinear stochastic differential equation* in Lecture 18 describing the dynamics of the gravity wave field. How one can use the information contained in that equation is a long story and it must be told with care. Therefore we do not concern ourselves in this lecture with the dynamics of the wave field, but rather we *assume* the existence of an asymptotic steady state for the gravity waves and examine its statistical properties. Later we shall address the question of how this steady state is reached.

Earlier on we discussed the properties of a linear stochastic wave field by exploiting its similarity to Brownian motion. This similarity enabled us to study the transient stages of growth as well as the relaxation to the asymptotic steady state. We were even able to determine the effects of a new physical mechanism on the growth rate of the waves because of the assumed linear nature of the wave field. To develop an understanding of how the nonlinear interactions modify the wave dynamics we initiate this part of our study with a review of the steady state properties of the nonlinear water wave field. As with much of the other work in this area the papers of Longuet-Higgins[202,203] and Phillips[204] are found at the foundation of what is presently understood.

Here we generalize the argument presented in lecture 15 for constructing the steady state distribution of a linear wave field to include the nonlinear interactions among the gravity waves. Consider the gravity wave field at an infinite fetch, ie. in its asymptotic steady state where the nonlinear interactions have reached a balance with the multitude of source terms just mentioned. We assume the initial state of the gravity wave field to be given by shuting off the wind and waiting a short time for the high frequency waves to relax to zero due to viscous damping. The mode rate equations (4.16) describing the Hamiltonian wave dynamics are then the deterministic set of equations

$$\dot{B}_{\underline{K}}(t) + i\omega_K \; B_{\underline{K}}(t) = T_{\underline{K}}(\underline{B};t) \tag{1}$$

where in the absence of wind and high frequency waves we have only the initial value problem to solve.

We eliminate the rapid phase variation in (1) by introducing the slowly changing mode amplitude $b_{\underline{K}}(t)$ by

$$B_{\underline{K}}(t) = e^{-i\omega_K t} \; b_{\underline{K}}(t) \tag{2}$$

and obtain

$$\dot{b}_{\underline{K}}(t) = \sum_{\underline{L},\underline{M}} \delta_{\underline{L}+\underline{M}-\underline{K}} \Big\{ \Gamma_{\underline{L}\underline{M}}^{\underline{K}} b_{\underline{L}} b_{\underline{M}} e^{i(\omega_K - \omega_L - \omega_M)t}$$

$$+ \Gamma_{\underline{L}}^{\underline{K},-\underline{M}} b_{\underline{L}} b^*_{-\underline{M}} e^{i(\omega_K + \omega_M - \omega_L)t} + \Gamma^{\underline{K},-\underline{L},-\underline{M}} b^*_{-\underline{L}} b^*_{-\underline{M}} e^{i(\omega_K + \omega_L + \omega_M)t} \Big\}$$

$$+ O(b^3) \qquad (3)$$

where we are not interested in the third order terms for this argument. We assume a smallness parameter ε and expand $b_{\underline{K}}(t)$ in the perturbation series

$$b_{\underline{K}}(t) = \varepsilon b_{\underline{K}}^{(0)}(t) + \varepsilon^2 b_{\underline{K}}^{(1)}(t) + \varepsilon^3 b_{\underline{K}}^{(2)}(t) + \ldots \qquad (4)$$

Substituting (4) into (3) and equating equal powers of the ε coefficients yields the hierarchy of equations

$$\dot{b}_{\underline{K}}^{(0)}(t) = 0 \qquad (5a)$$

$$\dot{b}_{\underline{K}}^{(1)}(t) = \sum_{\underline{L},\underline{M}} \delta_{\underline{L}+\underline{M}-\underline{K}} \Big\{ \Gamma_{\underline{L}\,\underline{M}}^{\underline{K}} b_{\underline{K}}^{(0)} b_{\underline{M}}^{(0)} e^{i(\omega_K - \omega_L - \omega_M)t}$$

$$+ \Gamma_{\underline{L}}^{\underline{K},-\underline{M}} b_{\underline{L}}^{(0)} b_{-\underline{M}}^{(0)*} e^{i(\omega_K + \omega_M - \omega_L)t}$$

$$+ \Gamma^{\underline{K},-\underline{L},-\underline{M}} b_{-\underline{L}}^{(0)*} b_{-\underline{M}}^{(0)*} e^{i(\omega_K + \omega_M + \omega_L)t} \Big\} \qquad (5b)$$

We see from (5a) that $b_{\underline{K}}^{(0)}$ is a constant in time. Thus only the phases in (5b) vary in time and we can integrate these equations to obtain

$$B_{\underline{K}}^{(0)}(t) = e^{-i\omega_K t} b_{\underline{K}}^{(0)}(0) \qquad (6a)$$

$$B_{\underline{K}}^{(1)}(t) = \sum_{\underline{L},\underline{M}} \delta_{\underline{L}+\underline{M}-\underline{K}} \left\{ \frac{\Gamma_{\underline{L}\ \underline{M}}^{\underline{K}}}{i(\omega_K-\omega_L-\omega_M)} B_{\underline{L}}^{(0)}(t)\, B_{\underline{M}}^{(0)}(t) \right.$$

$$\left. + \frac{\Gamma_{\underline{L}}^{\underline{K},-\underline{M}}}{i(\omega_K+\omega_M-\omega_L)} B_{\underline{L}}^{(0)}(t)\, B_{\underline{M}}^{(0)}(t) + \frac{\Gamma^{\underline{K},-\underline{L},-\underline{M}}}{i(\omega_K+\omega_M+\omega_L)} B_{-\underline{L}}^{(0)*}(t)\, B_{-\underline{M}}^{(0)*}(t) \right\} \tag{6b}$$

The mode amplitudes $b_{\underline{K}}(0)$ are set equal to their steady state values, i.e., the initial values for the wave field.

We use the solutions given by (6) to write the surface deflection (4.7b) near the steady state as

$$\zeta(\underline{x},t) = \sum_{\underline{K}} r_{\underline{K}} \sin \chi_{\underline{K}} + \sum_{\underline{L},\underline{M}} r_{\underline{L}} r_{\underline{M}} \left\{ \bar{\Gamma}_{\underline{L}\underline{M}}^{\underline{K}} \cos \chi_{\underline{L}} \cos \chi_{\underline{M}} + \bar{\Gamma}'^{\underline{K}}_{\underline{L}\ \underline{M}} \sin \chi_{\underline{L}} \sin \chi_{\underline{M}} \right\} \tag{7}$$

where the total phase is given by

$$\chi_{\underline{K}} \equiv \underline{K}\cdot\underline{x} - \omega_K t + \theta_{\underline{K}} \tag{8}$$

and the mode amplitude $b_{\underline{K}}$ is expressed in terms of a real amplitude $r_{\underline{K}}$ and phase $\theta_{\underline{K}}$ as

$$b_{\underline{K}}^{(0)} \equiv r_{\underline{K}}\, e^{i\theta_{\underline{K}}} . \tag{9}$$

Here we take $\theta_{\underline{K}}$ as the random initial phase and $r_{\underline{K}}$ as the real initial mode amplitude and assume $r_{\underline{K}} \cos \theta_{\underline{K}}$ and $r_{\underline{K}} \sin \theta_{\underline{K}}$ have joint Gaussian distributions. The coupling coefficients in (7) are given by

$$\bar{\Gamma}_{\underline{L}\ \underline{M}}^{\underline{K}} \equiv \left(\frac{V_L V_M}{2V_K}\right)^{\frac{1}{2}} \left\{ \frac{\Gamma_{\underline{L}\ \underline{M}}^{\underline{k}}}{\omega_K-\omega_L-\omega_M} \delta_{\underline{L}+\underline{M}-\underline{K}} + \frac{\Gamma_{\underline{L}}^{\underline{K}\ \underline{M}}}{\omega_K+\omega_M-\omega_L} \delta_{\underline{L}-\underline{M}-\underline{K}} - \frac{\Gamma^{\underline{K}\ \underline{L}\ \underline{M}}}{\omega_K+\omega_L+\omega_M} \delta_{\underline{L}+\underline{M}+\underline{K}} \right\} . \tag{10a}$$

$$\overline{\Gamma}'^{\underline{K}}_{\underline{L}\,\underline{M}} \equiv \left(\frac{V_L V_M}{2V_K}\right)^{\frac{1}{2}} \left\{ - \frac{\Gamma^{\underline{K}}_{\underline{L}\,\underline{M}}}{\omega_{\underline{K}}-\omega_{\underline{L}}-\omega_{\underline{M}}}\,\delta_{\underline{L}+\underline{M}-\underline{K}} + \frac{\Gamma^{\underline{K}\,\underline{M}}_{\underline{L}}}{\omega_K+\omega_M-\omega_L}\,\delta_{\underline{L}-\underline{M}-\underline{K}} - \frac{\Gamma^{\underline{K}\,\underline{L}\,\underline{M}}}{\omega_K+\omega_L+\omega_M}\,\delta_{\underline{L}+\underline{M}+\underline{K}} \right\}. \tag{10b}$$

The surface deflection defined by (7) is therefore a nonlinear superposition of fluctuating linear waves. We assume the statistical properties to be homogeneous and stationary so that evaluating the characteristics of ζ at $x = \underline{0}$ and $t = 0$ is sufficient to determine the steady state statistical properties of the gravity wave field at all space-time points.

Indicating an average over an ensemble of realizations of the independent variables $r_{\underline{K}} \cos \theta_{\underline{K}}$ and $r_{\underline{K}} \sin \theta_{\underline{K}}$ by a bracket, we define the discrete energy spectral density $F(\underline{K})$ as

$$F(\underline{K}) \equiv \frac{1}{2} \langle r^2_{\underline{K}} \rangle = \langle |B^{(0)}_{\underline{K}}|^2 \rangle \, V_K^{-1} \tag{11}$$

The weighting V_K^{-1} arises from the definition of the canonical variables in lecture (4) so that the units of $F(\underline{K})$ are the fourth power of distance. We will also have occasion to use the continuum energy spectral density $\Psi(\underline{K})$ which we define as

$$\Psi(\underline{K}) = \frac{\Sigma_0}{(2\pi)^2} F(\underline{K}) \tag{12}$$

where Σ_0 is the two dimensional area of the ocean surface overwhich we have imposed periodic boundary conditions. The discrete sums over $\underline{K}$ can be replaced by integrals in the continuum limit using

$$\sum_{\underline{K}} \rightarrow \frac{\Sigma_0}{(2\pi)^2} \int d^2K \tag{13}$$

so that

$$\sum_{\underline{K}} F(\underline{K}) \to \int d^2K \, \Psi(\underline{K}) = \langle\zeta^2\rangle_{ss} \; . \qquad (14)$$

We have made the Gaussian assumption for the linear waves based on observational data.[205]

To evaluate the probability distribution for the nonlinear field variable ζ given by (7) at $\underline{x} = 0$ and $t = 0$, i.e. $P(\zeta)$, we use the characteristic function introduced in lecture 15. The characteristic function $p(\alpha)$ is given by

$$p(\alpha) \equiv \int_{-\infty}^{\infty} P(\zeta) e^{i\alpha\zeta} d\zeta \qquad (15)$$

and can be expressed in terms of an infinite series of cumulants of the field variable as

$$p(\alpha) = \exp\left\{\sum_{n=1} \frac{(i\alpha)^n}{n!} \langle\langle \zeta^n \rangle\rangle\right\} \; . \qquad (16)$$

In the context of diagrammatic expansions the Thiele semi-invariants or cumulants are the linked cluster moments used extensively in the study of many body systems by Kubo[206,207] and others.[195,208] Since $p(\alpha)$ can also be expressed as an infinite series of moments of ζ, i.e.

$$p(\alpha) = \sum_{n=0}^{\infty} \frac{(i\alpha)^n}{n!} \langle\zeta^n\rangle \qquad (17)$$

the two sets of coefficients can be expressed in terms of each other. The first few cumulants are related to the moments, as shown by inverting the relations in (16.32), to be

$$<<\zeta>> = <\zeta> \tag{18a}$$

$$<<\zeta^2>> = <\zeta^2> - <\zeta>^2 \tag{18b}$$

$$<<\zeta^3>> = <(\zeta-<\zeta>)^3 = <\zeta^3>-3<\zeta><\zeta^2>+2<\zeta>^3 \tag{18c}$$

$$<<\zeta^4>> = <(\zeta-<\zeta>)^4>-3[<(\zeta-<\zeta>)^2>]^2 \tag{18d}$$

The term 'linked cluster' arises from the factored form of the terms which are subtracted in (18d), for example. Only the unfactorable, i.e. completely linked, terms remain in the definition of a cumulant so that they represent intrinsic correlations up to a given order. The cumulant $<<\zeta^n>>$ does not contain correlations from lower order moments $<\zeta^m>, m<n$. Thus if $<<\zeta^n>> = 0$ then the *nth* moment of ζ can be expressed solely in terms of lower order moments.[191,207]

The first order cumulant is obtained by direct substitution of (7) into (18a) to be

$$<\zeta> = \sum_{\underline{K}} <r_{\underline{K}} \sin\theta_{\underline{K}}> + \sum_{\underline{L},\underline{M}} \bar{\Gamma}^{\underline{K}}_{\underline{L},\underline{M}} <r_{\underline{L}} r_{\underline{M}} \cos\theta_{\underline{L}} \cos\theta_{\underline{M}}>$$
$$+ \sum_{\underline{L},\underline{M}} \bar{\Gamma}'^{\underline{K}}_{\underline{L}\,\underline{M}} <r_{\underline{L}} r_{\underline{M}} \sin\theta_{\underline{L}} \sin\theta_{\underline{M}}> + O(<r^3>) \tag{19}$$

The Gaussian property of the fluctuations requires that all the odd moments in (19) vanish immediately, yielding

$$<\zeta> = \frac{1}{2} \sum_{\underline{L}} \left(\bar{\Gamma}^{\underline{K}}_{\underline{LL}} + \bar{\Gamma}'^{\underline{K}}_{\underline{LL}}\right) <r^2_{\underline{L}}> + O(<r^4>) \,. \tag{20}$$

However from (10)

$$\overline{\Gamma}^{\underline{K}}_{\underline{L}\,\underline{M}} + \overline{\Gamma}'^{\underline{K}}_{\underline{L}\,\underline{M}} = \frac{2\Gamma^{\underline{K}\,\underline{M}}_{\underline{L}}}{\omega_K+\omega_M-\omega_L}\,\delta_{\underline{L}-\underline{M}-\underline{K}}\left(\frac{V_L V_M}{2V_K}\right)^{1/2} \tag{21}$$

so that since $\underline{K} \neq 0$, when $\underline{L} = \underline{M}$ the second order contribution to (20) vanishes and the mean surface displacement vanishes to fourth order in the mode amplitude, i.e.

$$\langle\zeta\rangle = 0 + O(\langle r^4\rangle) \quad . \tag{22}$$

The second cumulant can be evaluated by direct substitution to be

$$\langle\langle\zeta^2\rangle\rangle = \sum_{\underline{K},\underline{K}'} \langle r_{\underline{K}} \sin\theta_{\underline{K}}\, r_{\underline{K}'} \sin\theta_{\underline{K}'}\rangle + O(\overline{\Gamma\Gamma}\langle r^4\rangle) - O(\langle r^4\rangle) \tag{23}$$

where we have already set the average of the term cubic in the mode amplitude to zero and schematically indicated the fourth order cross term. To leading order the second cumulant becomes

$$\langle\langle\zeta^2\rangle\rangle = \sum \tfrac{1}{2}\langle r^2_{\underline{K}}\rangle = \sum_{\underline{K}} F(\underline{K}) \tag{24}$$

because of the high degree of cancellation between the two fourth order terms.

The third cumulant is evaluated in the same way. The lowest order contribution arises from the three cross products consisting of two terms lin in $r_{\underline{K}}$ and one quadratic in $r_{\underline{K}}$, i.e.,

$$\langle\langle\zeta^3\rangle\rangle \cong 3 \sum_{\underline{LMK}} (\overline{\Gamma}^{\underline{K}}_{\underline{LM}} + \overline{\Gamma}'^{\underline{K}}_{\underline{LM}})\, \tfrac{1}{4}\langle r^2_{\underline{L}}\rangle\langle r^2_{\underline{M}}\rangle \quad . \tag{25}$$

Using the expression for the coupling coefficients (21) we then have

$$<<\zeta^3>> \cong 6 \sum_{\underline{L},\underline{M},} \left(\frac{V_L V_M}{2V_K}\right)^{1/2} \frac{\Gamma_{\underline{L}}^{\underline{L}-\underline{M},\underline{M}}}{\omega_{|\underline{L}-\underline{M}|} + \omega_M - \omega_L} F(\underline{L})F(\underline{M}) \ . \qquad (26)$$

In the continuum limit we have for the first three cumulants of the field variable[202]

$$<<\zeta>> = 0 \qquad (27a)$$

$$<<\zeta^2>> = \int d^2K \ \Psi(\underline{K}) \qquad (27b)$$

$$<<\zeta^3>> = 6 \int d^2K d^2K' \ \Gamma(\underline{K},\underline{K}') \ \Psi(\underline{K}) \ \Psi(\underline{K}') \qquad (27c)$$

where using (4.20) we have for the coupling coefficient

$$\Gamma(\underline{K},\underline{K}') = \sqrt{2}\, V_{|\underline{K}-\underline{K}'|}[\omega_{|\underline{K}-\underline{K}'|} + \omega_{K'} - \omega_K] \Big\{ V_K V_{|\underline{K}-\underline{K}'|}[|\underline{K}-\underline{K}'|K + \underline{K}\cdot\underline{K}' - K^2]$$

$$+ V_K V_{K'}(KK' - \underline{K}\cdot\underline{K}') - V_{K'} V_{|\underline{K}-\underline{K}'|}[|\underline{K}-\underline{K}'|K' + \underline{K}\cdot\underline{K}' - K'^2] \Big\} \qquad (28)$$

It is clear that since the third moments of the surface deflection is non-zero that $P(\zeta)$ cannot be a Gaussian distribution. Thus, as is well known, a nonlinear function of a Gaussian variable is non-Gaussian. However, it remains to determine what the steady state probability density for ζ is. If we assume that a probability density is uniquely determined by all its cumulants then by inverting the expression for the characteristic function (15) we have

$$P(\zeta) = \frac{1}{2\pi} \int_{-\infty}^{\infty} d\alpha \ e^{-i\alpha\zeta} \ p(\alpha) \ . \qquad (29)$$

In terms of cumulants (29) becomes

$$P(\zeta) = \frac{1}{2\pi} \int_{-\infty}^{\infty} d\alpha \, \exp\left\{ -i\zeta\alpha + \sum_{n=1}^{\infty} \frac{(i\alpha)^n}{n!} <<\zeta^n>> \right\} \qquad (30)$$

so that introducing the normalized quantities λ_n, by

$$\lambda_n \equiv \frac{<<\zeta^n>>}{<<\zeta^2>>^{n/2}} \qquad (31)$$

and following Longuet-Higgins[202] we introduce the scaled variables

$$\beta \equiv <<\zeta^2>>^{\frac{1}{2}}\alpha \quad ; \; q \equiv \zeta/<<\zeta^2>>^{\frac{1}{2}} \qquad (32)$$

and write (30) as

$$P(\zeta) = \frac{1}{2\pi} \int_{-\infty}^{\infty} \frac{d\beta}{<<\zeta^2>>^{\frac{1}{2}}} \exp\left\{ -\frac{1}{2}(\beta^2 + 2i\beta q) + \sum_{n=3}^{\infty} \frac{(i\beta)^n}{n!} \lambda_n \right\}. \qquad (33)$$

If we assume that the skewness λ_3 and kurtosis λ_4 as well as the higher measures of the deviation from a Gaussian distribution[205] are small, we can expand the exponential in (33) to obtain

$$P(\zeta) = \frac{1}{2\pi<<\zeta^2>>^{\frac{1}{2}}} \int_{-\infty}^{\infty} d\beta \, e^{-1/2(\beta^2+2iq\beta)} \left\{ 1 + \sum_{n=3}^{\infty} \frac{(i\beta)^n}{n!} \lambda_n \right.$$

$$\left. + \frac{1}{2!} \left[\sum_{n=3}^{\infty} \frac{(i\beta)^n}{n!} \lambda_n \right]^2 + \dots \right\} . \qquad (34)$$

To simplify (34) we make use of the identity

$$\frac{1}{\sqrt{2\pi}}\int_{-\infty}^{\infty} e^{-1/2(\beta^2+2iq\beta)}\,(i\beta)^n d\beta = \frac{(-1)^n}{\sqrt{2\pi}}\frac{d^n}{dq^n}\int_{-\infty}^{\infty} e^{-1/2(\beta^2+2iq\beta)}\,d\beta$$

$$= (-1)^n \frac{d^n}{dq^n}\, e^{-q^2/2}\ . \qquad (35)$$

The right hand side of (35) is the generating function for the Hermite polynomial of degree n, H_n, i.e.,

$$H_n(q) \equiv (-1)^n\, e^{q^2/2}\, \frac{d^n}{dq^n}\left(e^{-q^2/2}\right) \qquad (36)$$

Thus in terms of the zero-centered Gaussian distribution

$$P_G(\zeta) = \frac{1}{\sqrt{2\pi\langle\langle\zeta^2\rangle\rangle}}\, e^{-\zeta^2/2\langle\langle\zeta^2\rangle\rangle}\ , \qquad (37)$$

we may write (34) as

$$P(\zeta) = P_G(\zeta)\left\{1 + \sum_{n=3}^{\infty} C_n\, H_n(\zeta/\langle\langle\zeta^2\rangle\rangle^{1/2})\right\} \qquad (38)$$

where the C_n's are constants determined from the expansion in (34). The first few terms in the expansion with their associated Hermite polynomials are:

$$C_3 = \lambda_3/3! \quad ; \; H_3 = \zeta^3/\langle\langle\zeta^2\rangle\rangle^{3/2} - 3\zeta/\langle\langle\zeta\rangle\rangle^{1/2}$$

$$C_4 = \lambda_4/4! \quad ; \; H_4 = \zeta^4/\langle\langle\zeta^2\rangle\rangle^{2} - 6\zeta^2/\langle\langle\zeta^2\rangle\rangle + 3 \tag{39}$$

$$C_5 = \lambda_5/5! \quad ; \; H_5 = \zeta^5/\langle\langle\zeta^2\rangle\rangle^{5/2} - 10\zeta^3/\langle\langle\zeta^2\rangle\rangle^{3/2} + 15\zeta/\langle\langle\zeta^2\rangle\rangle^{1/2} .$$

In terms of the first correction to the Gaussian distribution we have

$$P(\zeta) \cong \frac{e^{-\zeta^2/2\langle\zeta^2\rangle}}{\sqrt{2\pi\langle\zeta^2\rangle}} \left\{ 1 + \frac{1}{6}\lambda_3 \left[\frac{\zeta^3}{\langle\zeta^2\rangle^{3/2}} - 3\frac{\zeta}{\langle\zeta^2\rangle^{1/2}} \right] \right\} \tag{40}$$

where, given a steady state spectrum $\Psi(\underline{K})$, we can calculate $\langle\zeta^2\rangle$ and λ_3 using (27). A careful determination of the distribution for ζ from wind generated wave data was made by Kinsman[205] and is shown in Figure (19.1). The skewness of the distribution toward larger values of ζ than predicted by a Gaussian with the same variance is evident. It was determined empirically that the Gram-Charlier distribution [cf., Eq. (38)] gives a significantly better representation of the wind wave data than did $P_G(\zeta)$ above[205]. Phillips[5] comments that the suppression of negative values of ζ and the enhancement of positive values appears to be a consequence of the tendency of the waves to form sharp crests and shallow troughs.

We close this lecture with a note of caution and a question. The question is what happened to the effect of the resonances at third order? Longuet-Higgins[203] points out that the calculations given here break down if carried beyond the second order terms, because of the existence of resonances in the four wave interactions. The resonance gives rise to time dependent coefficients of the third order terms but he states that the method is consistent to the order calculated. The agreement with experiment suggests that the resonant terms do not significantly influence the asymptotic steady state distribution. How and why this should be, is unclear.

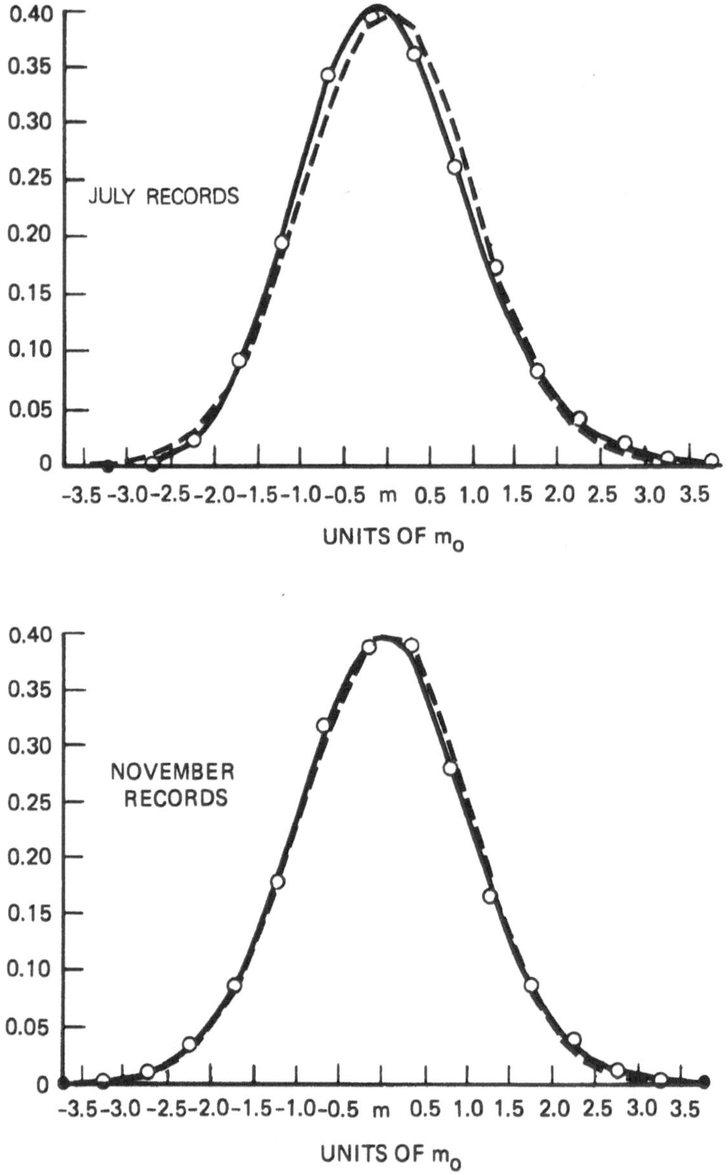

Figure 19.1 The measured frequency distribution water surface displacement (0) is fit with the best Gram-Charlier series distribution (——) and contrasted with the optimum Gaussian distribution (---). (From Kinsman, pg. 275)

20. Phase Space Equation of Evolution

The evolution of the stochastically driven field of gravity waves is determined by (18.3) which describes the development of the set of dynamic variables $\underline{B}(t) \equiv \{B_{\underline{K}}(t)\}$ for a particular realization of the wind field and gravity-capillary wave field, i.e., a particular realization of the set of fluctuations $\underline{F}(t) \equiv \{\hat{g}_{\underline{K}}(t), \Omega_{\underline{K}\,\underline{L}}(t), f_{\underline{K}}(t)\}$. In lectures 16 and 17 the difficulties associated with solving even a linear system of s.d.e.'s became clear. The presence of the mode mixing term $T_{\underline{K}}(\underline{B};t)$ in (18.3) invalidates the techniques developed in those lectures as applied directly to the non-linear dynamic equations. The cumulant expansion method can still be used in the present context, however, if we shift our attention from the dynamic equations to the phase space probability density. This is because the equation of evolution for the probability density is always linear so that the cumulant expansion technique can be applied directly. In this lecture we construct the equation of evolution for the probability density using the cumulant expansion technique of lectures 16 and 17.

We define a phase space for the dynamic system $\Gamma(\underline{b})$ by the values $\underline{b}$ that the dynamic vector $\underline{B}(t)$ can assume. For each realization of the set of fluctuations $\underline{F}(t)$ there corresponds a unique trajectory in this phase space which describes the complete evolution of the gravity wave system. Note that here the time t is a parameter specifying the position of the wave field along such an n-dimensional trajectory. A large number of realizations of $\underline{F}(t)$ defines a corresponding ensemble of trajectories in phase space. This ensemble of gravity wave fields is distributed in phase space and the density of systems, i.e., trajectories, in the interval $(\underline{b},\underline{b}+d\underline{b})$ at time t is given by the distribution function $\rho_F(\underline{b},t)$. The time rate of change of $\rho_F(\underline{b},t)$ then determines how the ensemble of systems redistributes itself in phase space as a function of time. A related concept is the conditional probability density $P(\underline{b},t|\underline{b}_0)$, where $P(\underline{b},t|\underline{b}_0)d\Gamma(\underline{b})$ is the probability that $\underline{B}(t)$ has a value in the interval $(\underline{b},\underline{b}+d\underline{b})$ at time t given an initial value $\underline{b}_0$ and $d\Gamma(\underline{b})$ is a differential volume element of the phase space. Part of the discussion here is to resolve some of the confusion in the literature on the distinction between $\rho_F(\underline{b},t)$ and $P(\underline{b},t|\underline{b}_0)$ and to show how to use the equation of evolution for $\rho_F(\underline{b},t)$ to determine the equation of evolution for $P(\underline{b},t|\underline{b}_0)$.

The time rate of change of the phase space distribution function $\rho_F(\underline{b},t)$ is determined by the rate equation (18.3). Introducing the function

$$\underline{Q}[\underline{B}(t);\underline{F}(t)] \equiv \hat{\underline{g}}(t) + \underline{T}\,(\underline{B};t) + \underline{f}(t) - \underline{\lambda}\,\underline{B}(t) - \underline{\underline{\Omega}}(t)\,\underline{B}(t) \qquad (1)$$

we rewrite (18.3) in the more convenient form

$$\frac{d}{dt}\,\underline{B}(t) = \underline{Q}\,[\underline{B}(t);\underline{F}(t)]. \qquad (2)$$

Note that the dependence of $\underline{Q}$ on the fluctuating quantities $\hat{\underline{g}}(t)$, $\underline{f}(t)$ and $\underline{\underline{\Omega}}(t)$ has been made explicit. Similarly, we explicitly denote the solution to (2) for a particular realization of $\underline{F}(t)$ as $\underline{B}_F(t)$. The phase space distribution function is then given by

$$\rho_F(\underline{b},t) \equiv \delta(\underline{b} - \underline{B}_F(t)) \qquad (3)$$

indicating that the phase space vector $\underline{b}$ has non-zero values only along the trajectory corresponding to the solution of (2) for a particular choice of $\underline{F}(t)$. The initial value of ρ_F is $\delta(\underline{b}-\underline{B}_F(0))$ and we will assume that all systems we consider start from the same initial state, i.e., $\underline{b}_0 \equiv \underline{B}_F(0)$. The solution $\underline{B}_F(t)$ to the s.d.e. (2) is a fluctuating quantity, therefore the phase space distribution (3) is also a fluctuating quantity and *cannot* be interpreted as a probability density. The appropriate probability density is obtained by averaging (3) over all realizations of the fluctuations $\underline{F}(t)$ denoted by a bracket with an F subscript so that[91,92,208]

$$P(\underline{b},t|\underline{b}_0) \equiv \langle\rho_F(\underline{b},t)\rangle_F \quad . \tag{4}$$

This smoothing procedure indicates that the probability of being in the phase space interval $(\underline{b},\underline{b}+d\underline{b})$ at time t given in initial state $\underline{b}_0$ is given by the distribution of trajectories at time t averaged over an ensemble of realizations of the complete set of fluctuations $\underline{F}(t)$. The relation between the probability density and phase space distribution functions in (4) is therefore used to determine the evolution of $P(\underline{b},t|\underline{b}_0)$ from the mean evolution of $\rho_F(\underline{b},t)$.

To determine the evolution of the phase space distribution function consider the time rate of change of the arbitrary phase space function $G(\underline{b})$ evaluated for the particular solution $\underline{B}_F(t)$, i.e.,

$$\frac{dG}{dt}[\underline{B}_F(t)] = \frac{\partial G}{\partial \underline{B}_F} \cdot \frac{d}{dt}\underline{B}_F(t) \quad . \tag{5}$$

Equation (5) can be written as a phase space equation by using

$$G[\underline{B}_F(t)] = \int G(\underline{b})\ \delta(\underline{b}-\underline{B}_F(t))\ d\Gamma(\underline{b}) \tag{6}$$

which from (3) we can identify the delta function with the phase space distribution function to obtain

$$G[\underline{B}_F(t)] = \int G(\underline{b})\ \rho_F(\underline{b},t)\ d\Gamma(\underline{b}). \tag{7}$$

The time rate of change of (7) is then

$$\frac{d}{dt} G[\underline{B}_F(t)] = \int G(\underline{b})\ \frac{\partial \rho_F(\underline{b},t)}{\partial t}\ d\Gamma(\underline{b}), \tag{8}$$

the right hand side of which is a phase space equation. The time derivative only operates on ρ_F under the integral because the phase space variables $\underline{b}$ are time independent, i.e., they are not the dynamic variables $\underline{B}(t)$.

In a similar manner we can express the right hand side of (5) as the phase space integral

$$\frac{\partial G}{\partial \underline{B}_F} \cdot \frac{d}{dt} \underline{B}_F(t) = \int \delta(\underline{b} - \underline{B}_F(t))\ \frac{\partial G(\underline{b})}{\partial \underline{b}} \cdot \underline{Q}[\underline{b},\underline{F}(t)]\ d\Gamma(\underline{b})\ . \tag{9}$$

Note that the time derivative of $\underline{B}_F(t)$ is replaced by the time dependent phase space function $\underline{Q}[\underline{b},\underline{F}(t)]$. Again using (3) and integrating (9) by parts yields

$$\frac{\partial G}{\partial \underline{B}_F} \cdot \frac{d}{dt} \underline{B}_F(t) = - \int G(\underline{b}) \frac{\partial}{\partial \underline{b}} \cdot \left[\underline{Q}[\underline{b},\underline{F}(t)]\ \rho_F(\underline{b},t) \right] d\Gamma(\underline{b}). \quad (10)$$

Subtracting (10) from (3) we now obtain

$$\frac{d}{dt} G[\underline{B}_F(t)] - \frac{\partial G}{\partial \underline{B}_F} \cdot \frac{d}{dt} \underline{B}_F(t) = \int G(\underline{b}) \left\{ \frac{\partial \rho_F}{\partial t} + \frac{\partial}{\partial \underline{b}} \cdot (\underline{Q}\rho_F) \right\} d\Gamma(\underline{b}) \quad (11)$$

the left hand side of which is zero by construction. Since the phase space function $G(\underline{b})$ is arbitrary, the expression in brackets under the integral must also vanish yielding

$$\frac{\partial}{\partial t} \rho_F (\underline{b},t) + \frac{\partial}{\partial \underline{b}} \cdot \left[\underline{Q}[\underline{b},\underline{F}(t)]\ \rho_F(\underline{b},t) \right] = 0\ . \quad (12)$$

Equation (12) is a phase space equation for a particular realization of the fluctuating fluxes $\underline{F}(t)$, but since it is linear, a superposition of an ensemble of similar systems obeys the same equation. Lax[91] and others[208-211] refer to (12) as the Liouville equation describing the evolution of the phase space distribution function.

Although (12) has the form of the Liouville equation we stress that there are a number of fundamental differences. The first difference is that the phase space function $\underline{Q}$ replaces the time derivatives of the canonical variables present in the microscopic Liouville equation for a Hamiltonian system. These "velocities" are deterministic in a Hamiltonian system so that the solutions to the analogue of (12) can be interpreted as probability densities[212].

In (12) however the vector $\underline{Q}[\underline{b},\underline{F}(t)]$, which replaces the time derivatives of the canonical variables, depends explicitly on $\underline{F}(t)$ and is therefore a fluctuating quantity in phase space. These fluctuations induce rapid variations in $\rho_F(\underline{b},t)$ which must be smoothed over an ensemble of realizations of the fluctuations to define a probability density $P(\underline{b},t|\underline{b}_0)$ as indicated in (4). Also for the traditional Liouville equation the flow of probability in the phase space is incompressible and the order of the "divergence" $\frac{\partial}{\partial \underline{b}}\cdot$ and "velocity" $\underline{Q}$ can be interchanged. The resulting expression then defines trajectories of constant probability in the phase space. This interchange cannot be done in the coarse grained description (12) so the flow of ρ_F is not incompressible. We can however exploit the formal similarity of (12) to the Liouville equation to construct the equation of evolution for the probability density (4). For convenience we delete the dependence on the initial state $\underline{b}_0$ in constructing the equation of motion for $P(\underline{b},t|\underline{b}_0)$, since if $P(\underline{b}_0)$ is a distribution of initial states we can always write the average over an ensemble of initial conditions as

$$P(\underline{b},t) = \int P(\underline{b},t|\underline{b}_0)\, P(\underline{b}_0)\, d\Gamma(\underline{b}_0) \tag{13}$$

so that $P(\underline{b},t)$ and $P(\underline{b},t|\underline{b}_0)$ obey the same equation of evolution.

Consider the two phase space operators L_0 and L_F defined to operate on the arbitrary phase space function $G(\underline{b})$ by

$$L_0\, G(\underline{b}) \equiv \sum_{\underline{K}} \frac{\partial}{\partial b_{\underline{K}}} \left\{ \left[\gamma_{\underline{K}}\, b_{\underline{K}} - T_{\underline{K}}(\underline{b}) \right] G(\underline{b}) \right\} \quad +cc \tag{14}$$

and

$$\varepsilon L_F(t)G(\underline{b}) \equiv -\sum_{\underline{K}} \frac{\partial}{\partial b_{\underline{K}}} \left\{[\hat{g}_{\underline{K}}(t) + f_{\underline{K}}(t) - \sum_{\underline{L}} \Omega_{\underline{K}\,\underline{L}}(t) b_{\underline{L}}] G(\underline{b})\right\} + cc \quad (15)$$

where cc indicates complex conjugate. We identify these operators as the phase space analogs of the dynamic equations for the water wave field (18.21). Here we restrict our discussion to the case of gravity waves travelling in the $\underline{K}$ direction and neglect the B^*_{-K} term in (18.21). The linear coupling coefficient in (14) is therefore given by

$$\gamma_{\underline{K}} = \mu_K \omega_K + \beta_{\underline{K}} - i\,\Omega_{\underline{K}} .$$

We could equally well have diagonalized the linear part of (18.21) by means of the transformation

$$\underline{S} = \frac{1}{1 + \frac{\beta_{\underline{K}}^2}{\Lambda_{\underline{K}}^2}} \begin{pmatrix} 1 & -\frac{\beta_K}{\Lambda_{\underline{K}}} \\ \frac{\beta_K}{\Lambda_{\underline{K}}} & 1 \end{pmatrix}$$

with

$$\Lambda_{\underline{K}} \equiv \sqrt{(\mu_K \omega_K + \beta_{\underline{K}})^2 + \beta_{\underline{K}}^{\,2}} - i\,\Omega_{\underline{K}}$$

but this would only introduce unnecessary algebraic complications. A new sytem of variables

$$\begin{pmatrix} q_{\underline{K}}^{(1)} \\ q_{\underline{K}}^{(2)} \end{pmatrix} = \underline{S} \begin{pmatrix} B_{\underline{K}} \\ B^*_{-K} \end{pmatrix}$$

would then be used to describe the evolution of the wave field. Using (14) and (15) we can rewrite the Liouville equation (12) as

$$\frac{\partial}{\partial t} \rho_F(\underline{b},t) = L_0\, \rho_F(\underline{b},t) + \varepsilon\, L_F(t)\, \rho_F(\underline{b},t). \tag{16}$$

The behavior of (16) has been conveniently separated into two parts. The first is associated with the average evolution of the gravity wave field, i.e., L_0 is a deterministic operator. The second is associated with the stochastic evolution of the gravity waves, i.e., $L_F(t)$ is an explicitly time dependent operator due to the fluctuations $\underline{F}(t)$. We have introduced an ordering parameter ε suggesting that we may treat these fluctuations as perturbations at some later stage of the analysis.

There are a number of procedures for deriving an equation of evolution for $P(\underline{b},t)$ starting from the Liouville equation (16).[91,94,213,214] The critical physical parameter that determines the form of this equation is the correlation time τ_c for the fluctuations. Since we have defined all the fluctuating terms in (15) by $\underline{F}(t)$, i.e.,

$$\underline{F}(t) \equiv \hat{\underline{g}}(t) + \underline{f}(t) - \underline{\underline{\Omega}}(t)\underline{b} \tag{17}$$

the correlation time is defined by the condition that the correlation function $\langle F_{\underline{K}}(t)F^*_{\underline{K}'}(t')\rangle_F$ have an appreciable value for $|t-t'| \ll \tau_c$ but is arbitrarily small for $|t-t'| \gg \tau_c$. Here we take τ_c to be the longest correlation time of any of the constituent processes contributing to $\underline{F}(t)$. The mathematical convention of assuming some fluctuations to be delta correlated in time [eg., cf. equ. (16.12)], an approximation we sometimes make, must be implemented only after the effects of a finite correlation time have been included in the equation of evolution[210]. We consider the correlation time τ_c to be finite and use the cumulant expansion procedure from lecture 16 in conjunction with (16) to construct the equations of evolution for $P(\underline{b},t|\underline{b}_0)$.

Recall that in the analysis of the linear system we first transformed the dependent variable to a representation in which the deterministic evolution of the system had been removed. Thus we introduce $\hat{\rho}_F$ by the transformation

$$\hat{\rho}_F(\underline{b},t) \equiv e^{-L_0 t} \rho_F(\underline{b},t) \tag{18}$$

noting that L_0 is a time independent operator in phase space. Direct substitution of (18), in the form $\rho_F(\underline{b},t) = e^{L_0 t} \hat{\rho}_F(\underline{b},t)$, into the Liouville equation (16) yields

$$\frac{\partial \hat{\rho}_F(\underline{b},t)}{\partial t} = e^{-L_0 t} \varepsilon L_F(t) e^{L_0 t} \hat{\rho}_F(\underline{b},t)$$

which in the interaction representation is

$$\frac{\partial \hat{\rho}_F(\underline{b},t)}{\partial t} = \varepsilon \tilde{L}_F(t) \hat{\rho}_F(\underline{b},t) \tag{19}$$

The formal solution to (19) is given in terms of the time evolution operator introduced in lecture 16

$$\hat{\mathcal{U}}(t|t_1) \equiv T\left\{\exp\left[\int_{t_1}^{t} \varepsilon\tilde{L}_F(t')\,dt'\right]\right\} \tag{20}$$

as

$$\hat{\rho}_F(\underline{b},t) = \hat{\mathcal{U}}(t|0)\,\delta(\underline{b}-\underline{b}_0). \tag{21}$$

The average of (21) over an ensemble of realizations of $\underline{F}(t)$ yields

$$\langle\hat{\rho}_F(\underline{b},t)\rangle_F = \langle\hat{\mathcal{U}}(t|0)\rangle_F\,\delta(\underline{b}-\underline{b}_0) \tag{22}$$

and the average evolution operator can be expressed as an expansion in the time order cumulant operators [cf. lecture 16]

$$\langle\hat{\mathcal{U}}(t|o)\rangle_F = T\exp\left[\sum_{n=1}^{\infty} \varepsilon^n K_n(t)\right] \tag{23}$$

where the n*th* order cumulant *operator* is defined as

$$K_n(t) \equiv \int_0^t d\tau_1 \int_0^{\tau_1} d\tau_2 \cdots \int_0^{\tau_{n-1}} d\tau_n \,\langle\langle\tilde{L}_F(\tau_1)\tilde{L}_F(\tau_2)\cdots\tilde{L}_F(\tau_n)\rangle\rangle_F\,. \tag{24}$$

We can now take the time derivative of (22) and recalling that L_0 is a deterministic operator, so that it commutes with the averaging procedure, we obtain

$$\frac{\partial}{\partial t}\langle\rho_F(\underline{b},t)\rangle_F = L_0\langle\rho_F(\underline{b},t)\rangle_F + \sum_{n=1}^{\infty} \varepsilon^n\, e^{L_0 t}\,\frac{dK_n(t)}{dt}\,e^{-L_0 t}\,\langle\rho_F(\underline{b},t)\rangle_F\,. \tag{25}$$

Substituting from (4) we obtain the *exact* equation of evolution for the conditional probability density $P(\underline{b},t|\underline{b}_0)$:

$$\frac{\partial}{\partial t} P(\underline{b},t|\underline{b}_0) = L_0 P(\underline{b},t|\underline{b}_0) + \sum_{n=1}^{\infty} \varepsilon^n \; \overbrace{\frac{d}{dt} K_n(t)} \; P(\underline{b},t|\underline{b}_0) \tag{26}$$

where the time derivative of the cumulants in the interaction representation is indicated.

Equation (26) takes on a particularly simple form if all the fluctuation processes in $\underline{F}(t)$ are generated by Gaussian processes. In this case, which is the one we examine in some detail, the cumulants $K_n(t)$ vanish identically for $n > 2$ for the *partially ordered prescription* (POP) given by Mukamel, et al.[213] In this case (26) reduces to the *still exact* expression

$$\frac{\partial}{\partial t} P(\underline{b},t|\underline{b}_0) = L_0 P(\underline{b},t|\underline{b}_0) + \varepsilon^2 \int_0^t dt \, \langle L_F(t) e^{L_0 \tau} \, L_F(t-\tau) e^{-L_0 \tau} \rangle_F \times P(\underline{b},t|\underline{b}_0) \tag{27}$$

which was also obtained by Lax[92] using a perturbation theory argument.

A less symbolic expression for (27) is obtained by specifying the correlation function of the function $\underline{F}(t)$. Since we have made the assumption of zero-centered Gaussian statistics on all the contributions to $\underline{F}(t)$ we make the following further assumptions on the constituent processes;

$$\langle f_{\underline{K}}(t) f^{*}_{\underline{K}'}(t-\tau)\rangle_f = \delta_{\underline{K}-\underline{K}'} \frac{2\Psi(\underline{K})k^2}{\rho_o^2|\lambda_K|^2} \quad \psi_{\underline{K}}(\tau) \tag{28a}$$

$$\langle f_{\underline{K}}(t) f_{\underline{K}'}(t-\tau)\rangle_f = \delta_{\underline{K}+\underline{K}'} \frac{2\Psi(\underline{K})k^2}{\rho_o^2\lambda_K^2} \quad \psi_{\underline{K}}(\tau) \tag{28b}$$

$$\langle \hat{g}_{\underline{K}}(t)\hat{g}^{*}_{\underline{K}'}(t-\tau)\rangle_{\hat{g}} = -\langle \hat{g}_{\underline{K}}(t)\hat{g}_{\underline{K}'}(t-\tau)\rangle_{\hat{g}} = \delta_{\underline{K}-\underline{K}'}\, 2\tilde{D}_2(\underline{K})\delta(\tau) \tag{28c}$$

$$\langle \gamma_{\underline{K}}(t)\gamma_{\underline{K}'}(t-\tau)\rangle_\gamma = \delta_{\underline{K}-\underline{K}'}\, 2\, D_2(\underline{K})\, \delta(\tau) \tag{28d}$$

where $\Psi(\underline{K})$ is the spectrum of incoherent pressure fluctuations with correlation function $\psi_{\underline{K}}(\tau)$; $\tilde{D}_2(\underline{K})$ is the spectral strength of the additive flux of energy from the gravity-capillary wave and $D_2(\underline{K})$ is the spectral strength for the coherent fluctuations of the pressure field. In this discussion we neglect the frequency fluctuations of the gravity waves produced by the gravity-capillary waves and assume the three processes to be statistically independent. Using the statistical properties listed in (28) we write (27) in the form

$$\begin{aligned}
\frac{\partial P(\underline{b},t|\underline{b}_o)}{\partial t} &= L_o P(\underline{b},t|\underline{b}_o) + \sum_{\underline{K}} 2\, D_2(\underline{K}) \frac{\partial^2}{\partial b_{\underline{K}} \partial b^{*}_{\underline{K}}} \left[|b_{\underline{K}}|^2\, P(\underline{b},t|\underline{b}_o)\right] \\
&- \sum_{\underline{K}} \tilde{D}_2(\underline{K}) \left(\frac{\partial}{\partial b_{\underline{K}}} - \frac{\partial}{\partial b^{*}_{\underline{K}}}\right)^2 P(\underline{b},t|\underline{b}_o) \\
&+ \sum_{\underline{K}} \frac{\Psi(\underline{K})k^2}{\rho_o^2} \int_0^t \psi_{\underline{K}}(\tau)d\tau \left(\frac{\partial}{\partial \lambda_K b_{\underline{K}}} + \frac{\partial}{\partial \lambda^{*}_K b^{*}_{\underline{K}}}\right) e^{L_o\tau} \left(\frac{\partial}{\partial \lambda_K b_{\underline{K}}} + \frac{\partial}{\partial \lambda^{*}_K b^{*}_{\underline{K}}}\right) e^{-L_o\tau} \\
&\times P(\underline{b},t|\underline{b}_o)
\end{aligned} \tag{29}$$

in which we see that the contribution of the three fluctuation terms are quite different. Note that the Gaussian assumption has eliminated all the higher order cumulants of $\gamma_{\underline{K}}(t)$ that we studied in lecture 16. We have set ε to unity in (29).

To further simplify (29) we introduce the correlation time

$$\tau_c(\underline{K}) \equiv \int_0^\infty \psi_{\underline{K}}(\tau)d\tau \tag{30}$$

which qualitatively is the length of time a pressure fluctuation of wavelength scale $2\pi/K$ remains correlated. If the fluctuations in the pressure field are sufficiently short that we can replace $e^{L_0 \tau_c}$ by unity, then we can replace the upper limit of integration in (29) by ∞ and write

$$\begin{aligned}\frac{\partial}{\partial t} P(\underline{b},t|\underline{b}_0) \cong & \sum_{\underline{K}} \frac{\partial}{\partial b_{\underline{K}}} \left\{ \gamma_{\underline{K}} b_{\underline{K}} - \sum_{\underline{L},\underline{P},\underline{N}} \delta_{\underline{K}+\underline{N}-\underline{L}-\underline{P}} \, i\Gamma^{\underline{K}\,\underline{N}}_{\underline{L}\,\underline{P}} \, b_{\underline{L}} b_{\underline{P}} b^*_{\underline{N}} \right\} P(\underline{b},t|\underline{b}_0) \\ & + cc + \sum_{K} \left\{ [D_{\underline{K}} - \mathcal{D}_2(\underline{K})] \frac{\partial^2}{\partial b_{\underline{K}}^2} + [D^*_{\underline{K}} - \mathcal{D}_2(\underline{K})] \frac{\partial}{\partial b^{*2}_{\underline{K}}} \right. \\ & \left. + 2 \frac{\partial^2}{\partial b_{\underline{K}} \partial b^*_{\underline{K}}} [|D_{\underline{K}}| + \mathcal{D}_2(\underline{K}) + D_2(\underline{K})|b_{\underline{K}}|^2] \right\} P(\underline{b},t|\underline{b}_0) \end{aligned} \tag{31}$$

which clearly has the form of a Fokker-Planck equation with a state dependent diffusion coefficient and $D_{\underline{K}} \equiv \tau_c(\underline{K}) \, \Psi(\underline{K}) K^2/\rho_0^{\,2} \lambda_{\underline{K}}^{\,2}$.

Equations (29) and (31) are the phase space equivalents of the dynamic equation (18.31) given the specified properties of the stochastic driving function $\underline{F}(t)$. Both (29) and (31) require the statistics of the non-resonant interactions to be Gaussian and the latter has the additional requirement that all the correlation times of the constitutive processes are very short on the time scale of the gravity waves, i.e., they are all essentially delta correlated. The Fokker-Planck equation (31) therefore describes a Markoff process, whereas (29) has a memory effect of the pressure fluctuations and is thus not Markoffian.

Equation (31), albeit a simplified model of the original dynamic wave system still defies exact solution. The properties of this expression can, however, be used to give insight into the general structure of the evolving gravity wave field by using a combination of physical arguments and perturbation theory. The value of applying perturbation arguments to (31) is that it is a linear equation and the response of such equations to perturbations is well known. On the other hand, the original dynamic system (18.3) is nonlinear and its response to linear perturbations is uncertain at best.

21. Entropy and the Fokker-Planck Approximation

The applicability of statistical mechanical concepts to non-equilibrium systems with mesoscopic fluctuations has not been established in any definitive manner although there are a large number of extended discussions of this problem, see e.g., DeGroot and Mazur[55]; or Glansdorff and Prigogine[215] or Lavenda[54]. In this section we *assume* a form for the entropy of the gravity wave field. The form is that due to Gibbs[216] from his discussion of the equilibrium properties of physical systems with many degrees of freedom. We here apply his postulate to non-equilibrium processes and use the results of the preceding lecture to determine whether this notion is of value in describing the model evolution of such geophsyical systems. To emphasize the distinction between the entropy introduced here and that employed in the description of a microscopic process, we again stress the difference between microscopic and mesoscopic fluctuations in a system variable.

In a quiescent simple fluid the individual water molecules are moving erratically due to thermal excitation. The equilibrium thermodynamic properties of such a fluid are determined by Liouville's equation and an equilibrium probability distribution function. The concept of such a distribution function requires that any part of space be statistically equilvalent to any other part. The canonical distributions of statistical mechanics, i.e., the micro-canonical, canonical, grand canonical, etc., are expressed in terms of the extensive system variables (energy, momentum, number of particles, etc.) without reference to spatial or temporal behavior. The measurable properties of the fluid are related to the thermal fluctuations by statistical mechanical arguments associating the thermodynamic parameters such as viscosity, specific heat, etc. to the mean square fluctuations of the appropriate extensive variable. The entropy for a simple fluid system such as this is a well-defined equilibrium concept.

The equilibrium distribution function for the linear surface wave field discussed in lecture 15 is unlike the canonical distribution of statistical mechanics. The distribution for the linear field is Gaussian in the surface displacement $\zeta(\underline{x},t)$ and the statistics of the surface are unchanged as different locations in space-time are examined. The canonical distributions of statistical mechanics would not be expressed in terms of the field variable directly, but rather in terms of the conserved quantities in the field, e.g., the total energy. This situation persists when we examine the equilibrium distribution function for the nonlinear wave field discussed in lecture 18. The corrections to the Gaussian distribution of the linear wave field arising from the nonlinear interaction among the waves are related to the form of the canonical distribution functions when a more general point of view is adopted. In this lecture we make our first contact between the concepts of statistical thermodynamics, i.e., the stochastic formulation of thermodynamics[54,215], and the stochastic wave field on the ocean surface. This contact will be made using the concept of entropy as a point of commonality.

Einstein[217] was the first to extend the concept of entropy beyond its definition in an absolute equilibrium state of a system. He developed a macroscopic theory of thermodynamic fluctuations using the generic concept of entropy and hypothesized an inverse relation between the equilibrium distribution function for the system and the entropy. This hypothesis was proven by Greene and Callen[218] for the response of a system's extensive parameters to an applied static force. They have given an elegant formulation of this theory in terms of $W(\delta\zeta)$, the probability of a macroscopic fluctuation $\delta\zeta$ occurring in the system variable ζ [here taken to be the surface displacement] and the entropy of the system $S(\zeta)$, i.e.,

$$W(\delta\zeta) = \exp\left\{\frac{1}{k_B}\left[S(\delta\zeta) - S_0 - X * \delta\zeta\right]\right\} \tag{1}$$

where X is the thermodynamic force conjugate to ζ, i.e.,

$$X \equiv \frac{\delta S}{\delta\zeta} \tag{2}$$

and the asterisk (*) in (1) indicates a spatial convolution of the functions X and $\delta\zeta$. Here the entropy is taken as the basic thermodynamic potential and X is the force tending to restore $\delta\zeta$ back to the reference state S_0. The reference level of the surface is taken to be the horizontal plane $z = 0$ so that $\delta\zeta$ is the total surface deflection away from $z = 0$. The term $(X * \delta\zeta)$ is the work required to restore the perturbed system back to its steady state. The argument of Greene and Callen[218] requires that the interactions allow the system to achieve a statistical equilibrium, i.e., the reference state S_0 in (1) must be an absolute equilibrium state of the system, and is *independent of all special assumptions regarding the laws which may govern the elementary processes.* The argument was extended by Nicolis and Babloyantz[219] in such a way that the assumption of an absolute equilibrium was replaced by a statistical steady state in (1), i.e., the reference state S_0 need only be a steady state of the system.

The entropy in (1) is a functional of the two-dimensional scalar surface displacement $\delta\zeta(\underline{x},t)$. In the neighborhood of the statistical steady state we expand the entropy about its steady state value to obtain

$$S(\delta\zeta) = S_{ss} + \left(\frac{\delta S}{\delta\zeta}\right)_{ss} * \delta\zeta + \frac{1}{2}\,\delta\zeta * \left(\frac{\delta^2 S}{\delta\zeta^2}\right)_{ss} * \delta\zeta + \ldots \tag{3}$$

where the two asterisks denote spatial convolutions over both the spatial coordinates $\underline{x}$ and $\underline{x}'$. Introducing the matrix

$$S^{(2)} \equiv -\frac{1}{k_B}\left(\frac{\delta^2 S}{\delta\zeta^2}\right)_{ss} \tag{4}$$

and using the definition of the thermodynamic force (2) to remove the first order term, we obtain the distribution function from (1)

$$W(\delta\zeta) \cong C \exp\left\{-\frac{1}{2}k_B\ \delta\zeta * S^{(2)} * \delta\zeta\right\} \tag{5}$$

Equation (5) is the approximate probability density for the surface wave field fluctuations with C the normalization constant in the above approximation. Greene and Callen[218] prove that the second moments calculated using the approximate distribution function (5) are *exact* when the reference state is an absolute equilibrium; Nicolis and Babloyantz[219] demonstrate this also for the steady state where (5) is a local equilibrium distribution function. Although (5) *suggests* that the gravity wave fluctuations are a Gaussian random process, if we restrict our calculations to second-order terms, there is no way to distinguish (5) from the exact distribution (1). Therefore, using (5) to calculate second-order properties of the fluctuating gravity wave field does not restrict us to Gaussian processes.

Note that (5) appears to be different from the distribution function we obtained in lecture 15, because of the integration over the intermediate spatial locations within the convolution. Let us examine this apparent difference by making the assumption that the second variation in the entropy is solely a function of the relative distance between fluctuations in the wave field so that the convolution in (5) becomes

$$\delta\zeta * S^{(2)} * \delta\zeta = \int d^2x_1 d^2x_2 \ \delta\zeta(\underline{x}-\underline{x}_1)\ S^{(2)}(\underline{x}_1-\underline{x}_2)\ \delta\zeta(\underline{x}-\underline{x}_2) \quad . \tag{6}$$

Now introducing the Fourier transform of $S^{(2)}$ as

$$S^{(2)}(\underline{x}_1-\underline{x}_2) \equiv \sum_{\underline{K}} R_{\underline{K}}\, e^{-i\underline{K}\cdot(\underline{x}_1-\underline{x}_2)} \tag{7}$$

and applying Parsavel's theorem to (6) we obtain

$$\delta\zeta * S^{(2)} * \delta\zeta = \sum_{\underline{K}} \frac{1}{2V_K} (B_{\underline{K}} - B^*_{-\underline{K}}) \, R_{\underline{K}} (B_{\underline{K}} - B^*_{-\underline{K}})^* \tag{8}$$

or in terms of the displacement mode amplitudes $\zeta_{\underline{K}}$

$$= \sum_{\underline{K}} \zeta_{\underline{K}} \, R_{\underline{K}} \, \zeta^*_{\underline{K}} \quad . \tag{9}$$

If we now associate $R_{\underline{K}}$ with the mean square mode amplitude, i.e., $k_B R_{\underline{K}} = \langle |\zeta_{\underline{K}}|^2 \rangle^{-1}$, then the probability density (5) can be written as

$$W(\delta\zeta) = C \exp\left\{- \frac{1}{2} \sum_{\underline{K}} |\zeta_{\underline{K}}|^2 / \langle |\zeta_{\underline{K}}|^2 \rangle \right\} \quad . \tag{10}$$

The distribution (10) is clearly that of a linear superposition of a large number (spectrum) of surface deflections produced by a linear field of water waves as discussed in lecture (15). With the above associations for the entropy of the linear wave field the two descriptions become equivalent.

The hypothesis of Einstein concerns the behavior of systems near an absolute equilibrium and although later investigations have extended the argument to near steady state configurations, none have definitely shown the approach to such a steady state using the dynamic equations for a system with mesoscopic fluctuations. Here we *postulate* a form for the entropy of the water wave field using the conditional probability $P(\underline{b},t|\underline{b}_0)$ and obtain an H-theorem for the wave field. The H-theorem is shown to be a consequence of the detailed interactions among the water waves.

We assume the entropy for the fluctuating water wave field is given by the *Gibbs entropy postulate*[216], i.e.,

$$S(t) = -k_B \int d\Gamma(\underline{b})P(\underline{b},t|\underline{b}_0)\ln\left[\frac{P(\underline{b},t|\underline{b}_0)}{\Lambda P_0(\underline{b})}\right] \tag{11}$$

where $P_0(\underline{b})$ is an asymptotic reference distribution for the wave field, either a steady state or an absolute equilibrium, and Λ is a constant[54]. From its definition

$$P_0(b) \equiv \lim_{t\to\infty} P(\underline{b},t|\underline{b}_0) \tag{12}$$

so that we obtain from (11)

$$\lim_{t\to\infty} S(t) = k_B \ln\Lambda \tag{13}$$

and Λ determines the entropy of the fluctuating water wave field in the asymptotic reference state. It is useful to introduce the relative entropy $\Delta S(t)$, i.e., the deviation of the entropy from its reference state, by the equation

$$\Delta S(t) = -k_B \int d\Gamma(\underline{b}) P(\underline{b},t|\underline{b}_o) \ln \left[\frac{P(\underline{b},t|\underline{b}_o)}{P_o(\underline{b})} \right] \tag{14}$$

which by construction vanishes as $t \to \infty$. The stability properties of the non-equilibrium water wave field are determined by the time rate of change of the relative entropy of the surface motion.

The time rate of change of the relative entropy of the gravity wave field is given by

$$\frac{d}{dt} \Delta S(t) = -k_B \int d\Gamma(\underline{b}) \left[1 + \ln \frac{P(\underline{b},t|\underline{b}_o)}{P_o(\underline{b})} \right] \frac{\partial}{\partial t} P(\underline{b},t|\underline{b}_o) \quad . \tag{15}$$

For the gravity wave field to monotonically approach an asymptotic steady state we must have $d\Delta S(t)/dt \geq 0$. Prigogine has expressed the consequence of this requirement in a general thermodynamic system along with the inequality $\frac{d^2}{dt^2} \Delta S(t) < 0$ indicating that $-\Delta S(t)$ is a concave function of time[215,220]. These properties have been examined in a microscopic context in the recent work of Levine and co-workers[221,223] and Moreau[224] from an information theory point of view starting with the early work of Jaynes[225]. The examination of the fundamental problems in a geophysical context remain to be explored. As a preliminary step in this exploration we examine (15) for consistency with the phase space equation of evolution for the probability density (20.29) or (20.31).

We restrict the discussion here to the Fokker-Planck equation (20.31) and examine the rate of production of entropy in the gravity wave field. In this approximation we recall that the fluctuations driving the gravity wave field, i.e., $\underline{F}(t)$, are independent from one (physically) infinitesimal time increment to the next. The correlation functions are then modeled as delta functions in time and the equation of evolution for $P(\underline{b},t|\underline{b}_0)$ is given by

$$\frac{\partial}{\partial t} P(\underline{b},t|\underline{b}_0) = \sum_{\underline{K}} \frac{\partial}{\partial b_{\underline{K}}} \left[\gamma_{\underline{K}} b_{\underline{K}} - \sum_{\underline{L}\,\underline{P}\,\underline{N}} \delta_{\underline{K}+\underline{N}-\underline{L}-\underline{P}}\; i\Gamma_{\underline{L}\,\underline{P}}^{\underline{K}\,\underline{N}}\; b_{\underline{L}} b_{\underline{P}} b_{\underline{N}}^{*} \right] P(\underline{b},t|\underline{b}_0) + cc$$

$$+ \sum_{\underline{K}} \Bigg\{ \left[D_{\underline{K}} - \tilde{D}_2(\underline{K}) \right] \frac{\partial^2}{\partial b_{\underline{K}}^2} + \left[D_{\underline{K}}^{*} - \tilde{D}_2(\underline{K}) \right] \frac{\partial^2}{\partial b_{\underline{K}}^{*2}} + 2 \frac{\partial^2}{\partial b_{\underline{K}} \partial b_{\underline{K}}^{*}} \Big[|D_{\underline{K}}|$$

$$+ \tilde{D}_2(\underline{K}) + D_2(\underline{K}) |b_{\underline{K}}|^2 \Big] \Bigg\} P(\underline{b},t|b_0) \qquad (16)$$

We assume that $P_0(\underline{b}_0)$ describes the ensemble of initial states of the gravity wave. If we integrate $P(\underline{b},t|\underline{b}_0)$ over all these possible initial states we obtain

$$\int P_0(\underline{b}_0)\, P(\underline{b},t|\underline{b}_0)\, d\underline{b}_0 = P_0(\underline{b}) \qquad (17)$$

so that multiplying (16) on the left by $P_0(\underline{b}_0)$ and integrating over the

phase space volume of the initial states $\Gamma(\underline{b}_o)$, then since $\frac{\partial P_o(\underline{b})}{\partial t} = 0$,
we obtain

$$\left\{\sum_{\underline{K}} \frac{\partial}{\partial b_{\underline{K}}}\left[\gamma_{\underline{K}} b_{\underline{K}} - \sum_{\underline{L}\,P\,\underline{N}} \delta_{\underline{K}+\underline{N}-\underline{L}-\underline{P}}\; i\Gamma\frac{\underline{K}\,\underline{N}}{\underline{L}\underline{P}}\; b_{\underline{L}} b_{\underline{P}} b_{\underline{N}}^*\right] + cc + \left[D_{\underline{K}} - \tilde{D}_2(\underline{K})\right]\frac{\partial^2}{\partial b_{\underline{K}}^2}\right.$$

$$\left. + \left[D_{\underline{K}}^* - \tilde{D}_2(\underline{K})\right]\frac{\partial^2}{\partial b_{\underline{K}}^{*2}} + 2\,\frac{\partial^2}{\partial b_{\underline{K}}\partial b_{\underline{K}}^*}\left[|D_{\underline{K}}| + \tilde{D}_2(\underline{K}) + D_2(\underline{K})|b_{\underline{K}}|^2\right]\right\} P_o(\underline{b}) = 0 \tag{18}$$

The distribution $P_o(\underline{b})$ is therefore a stationary solution of the Fokker-Planck equation (16), i.e., $P_o(\underline{b}) = P_{ss}(\underline{b})$. Further, since $\frac{\partial P_{ss}}{\partial t} = 0$ we also have that $d\Delta S(t)/dt = 0$ from (15) implying that the asymptotic steady state for the fluctuating gravity wave field is also a stationary solution to the Fokker-Planck equation.

A condition *sufficient* to ensure that the time variation of $P(\underline{b},t|\underline{b}_o)$ vanish asymptotically, i.e.,

$$\lim_{t\to\infty} \frac{\partial P(\underline{b},t|\underline{b}_o)}{\partial t} = 0 \quad \Rightarrow \quad \frac{\partial P_{ss}(\underline{b})}{\partial t} = 0 \tag{19}$$

is

$$\sum_{\underline{K}}\left\{\frac{\partial}{\partial b_{\underline{K}}}\left[T_{\underline{K}}(\underline{b})\, P_{ss}(\underline{b})\right] + \frac{\partial}{\partial b_{\underline{K}}^*}\left[T_{\underline{K}}^*(\underline{b})\, P_{ss}(\underline{b})\right]\right\} = 0 \quad . \tag{20}$$

Therefore, a stationary solution to the Fokker-Planck equation (18), $P_{ss}(\underline{b})$, exists under the condition that the mode mixing terms cancel against each other. Thus, if we could integrate the resulting equation directly we would have the steady state solution for n gravity wave modes. To integrate (18) under the constraint of no mode mixing (20) a number of possible approximations could be made since we cannot solve (18) even with the simplifying constraint. We assume that the coherent pressure fluctuations are negligible near the steady state even though they can be important initially, i.e., $D_2(\underline{K}) = 0$. Further, since the average energy flux from the wind is positive, $\mu_K > 0$, as is the energy flux from the high frequency waves, $\beta_{\underline{K}} > 0$, the Fokker-Planck equation as it stands will not yield any *straightforward* steady state solution. Therefore, before we impose the condition (20) we apply the method of statistical linearization to the gravity wave field, just as we did to the gravity-capillary waves in lecture 14.

We segment the contributions of the nonlinear interactions $T_{\underline{K}}(\underline{b})$ into those that are in resonance, i.e., $\underline{K} + \underline{N} = \underline{L} + \underline{P}$ and $\omega_K + \omega_N = \omega_L + \omega_P$, and those that are not in resonance. The resonant terms become part of the subsidiary condition (20) and the non-resonant terms enter into the linearization parameter $\gamma(\underline{K})$, i.e.,

$$\gamma(\underline{K}) \equiv -\mu_K\omega_K - \beta_K + i\Omega_{\underline{K}} - \frac{\left\langle b_{\underline{K}}^{*} T_{\underline{K}}^{\overline{R}}(\underline{b})\right\rangle_{ss}}{\left\langle |b_{\underline{K}}|^2\right\rangle_{ss}} \quad . \tag{21}$$

The superscript $\bar{R}$ refers to non-resonant interactions and the average is over an ensemble of realizations of the statistical steady state. The stationary form of the Fokker-Planck equation subject to the above restrictions is

$$\sum_{\underline{K}}\left\{\frac{\partial}{\partial b_{\underline{K}}}\left(\gamma(\underline{K})b_{\underline{K}}\right)+\frac{\partial}{\partial b^*_{\underline{K}}}\left(\gamma^*(\underline{K})b^*_{\underline{K}}\right)+\left[D_{\underline{K}}-\tilde{D}_2(\underline{K})\right]\frac{\partial^2}{\partial b_{\underline{K}}^2}\right.$$

$$\left.+\left[D^*_{\underline{K}}-\tilde{D}_2(\underline{K})\right]\frac{\partial^2}{\partial b^{*2}_{\underline{K}}}+2\left[|D_{\underline{K}}|+\tilde{D}_2(\underline{K})\right]\frac{\partial^2}{\partial b_{\underline{K}}\partial b^*_{\underline{K}}}\right\}P_{ss}(\underline{b})=0\;. \qquad (22)$$

To solve (22) in simple form we assume that $\tilde{D}_2(\underline{K}) \cong |D_{\underline{K}}|$ so that making an error of the order of the ratio of the wave frequency to the wave growth rate in neglecting the complex coefficient of the second derivatives, (22) reduces to

$$\sum_{\underline{K}}\left\{\frac{\partial}{\partial b_{\underline{K}}}\left(\gamma(\underline{K})b_{\underline{K}}\right)+\frac{\partial}{\partial b^*_{\underline{K}}}\left(\gamma^*(\underline{K})b^*_{\underline{K}}\right)+4\,\tilde{D}_2(\underline{K})\frac{\partial^2}{\partial b_{\underline{K}}\partial b^*_{\underline{K}}}\right\}P_{ss}(\underline{b})=0\;. \qquad (23)$$

Equation (23) can be integrated exactly to obtain the steady state solution

$$P_{ss}(\underline{b}) = N\exp\left\{-\sum_{\underline{K}}\frac{\gamma_R(\underline{K})}{2\tilde{D}_2(\underline{K})}|b_{\underline{K}}|^2\right\} \qquad (24)$$

where $\gamma(\underline{K}) = \gamma_R(\underline{K}) + i\ \gamma_I(\underline{K})$ and N is the normalization coefficient determined by

$$\int P_{ss}(\underline{b})\ d\Gamma(\underline{b}) = 1 \quad . \tag{25}$$

It is convenient to introduce the nxn correlation matrix $\underline{\underline{Q}}(t)$ whose elements are defined by the non-equilibrium ensemble average

$$[\underline{\underline{Q}}(t)]_{\underline{k}\underline{k}'} \equiv \langle b_k b^*_{k'};t\rangle \quad . \tag{26}$$

In the statistical steady state $\underline{\underline{Q}}(t)$ and its inverse $\underline{\underline{Q}}^{-1}(t)$ are diagonal with matrix elements determined by

$$(\underline{\underline{Q}}_{ss})^{-1}_{\underline{K}\ \underline{K}'} = \lim_{t\to\infty} \quad Q^{-1}_{\underline{K}\ \underline{K}'}(t) = \frac{\gamma_R(\underline{K})}{2\tilde{D}_2(\underline{K})}\ \delta_{\underline{K}-\underline{K}'} \quad . \tag{27}$$

Comparing (27) with (26) we obtain a fluctuation-dissipation relation of the first kind

$$\langle |b_{\underline{K}}|^2\rangle_{ss} = \frac{2\tilde{D}_2(\underline{K})}{\gamma_R(\underline{K})} \tag{28}$$

and thus obtain the exact steady state probability density,

$$P_{ss}(\underline{b}) = \frac{1}{(2\pi)^{n/2}||\underline{\underline{Q}}_{ss}||^{1/2}} \exp\left\{-\frac{1}{2}\ \underline{b}^+ \underline{\underline{Q}}^{-1}_{ss}\ \underline{b}\right\} \tag{29}$$

where $||\underline{\underline{Q}}(t)||$ is the nxn determinant of the correlation matrix (26). The term exact is used here to refer to the solution to (23) not to the gravity wave field problem, i.e., we have an exact solution to an approximate model.

Comparing (29) with the distribution based on Einstein's hypothesis (9) we see that if the function obtained from the Fourier transform of the second derivative of the entropy, $R_{\underline{K}}$, is interpreted as the inverse of the diagonal elements of the correlation matrix $\underline{\underline{Q}}_{ss}^{-1}$, then $W(\delta\zeta)$ is equivalent to the stationary solution to the Fokker-Planck equation (23), i.e., $P_{ss}(\underline{b})$. Also recall the many approximations required to obtain (23) from (16).

22. State of Minimum Entropy Production

The steady state soluton to the Fokker-Planck equation obtained in the previous lecture is

$$P_{ss}(\underline{b}) = N \exp\left\{ -\sum_{\underline{K}} \frac{\gamma_R(\underline{K})}{2\tilde{D}_2(\underline{K})} |b_{\underline{K}}|^2 \right\} \tag{1}$$

where N is the normalization chosen such that

$$\int P_{ss}(\underline{b})\, d\Gamma(\underline{b}) = 1 \tag{2}$$

If we insert (1) into the expression for the time rate of change the relative entropy (21.15) and recall the condition on the resonant nonlinear interactions (21.20) we obtain

$$\left.\frac{d\Delta S}{dt}\right|_{ss} = 0 \tag{3}$$

Therefore the asymptotic steady state for the gravity wave field is a state of minimum entropy production. This is a global property of the spectrum so there is no restriction on the *local* entropy production properties of the system. We will in fact find later that the nonlinear interacton behaves in such a way as to decrease the entropy on the large scale by increasing the entropy on the small scale[226].

We know, of course, from our analysis of the steady state distribution for the nonlinear surface deflection in lecture 19 that the Gaussian distribution (1) is only a first order approximation of the more accurate Gram-Chalier distribution for the nonlinear function ζ. It is therefore necessary at some future date to extend the present analysis to a non-Gaussian solution to the Fokker-Planck equation. However, the present analysis is still of interest since it produces a number of previously derived results and thereby highlights their limitations.

Consider the condition on the nonlinear interactions in the Fokker-Planck equation (21.20) using the steady state solution (1), i.e.,

$$\sum_{\underline{K}} \left\{ \frac{\gamma_R(\underline{K})}{2\tilde{D}_2(\underline{K})} \left[b^*_{\underline{K}}\, T^R_{\underline{K}}(\underline{b}) + b_{\underline{K}}\, T^{R*}_{\underline{K}}(\underline{b}) \right] - \left[\frac{\partial T^R_{\underline{K}}(\underline{b})}{\partial b_{\underline{K}}} + \frac{\partial T^{R*}_{\underline{K}}(\underline{b})}{\partial b^*_{\underline{K}}} \right] \right\} = 0 \quad . \tag{4}$$

The equations of motion for an *isolated* system of n gravity wave modes (n large) are given by Hamilton's equations (4.15). The resonant nonlinear interactions $T^R_{\underline{K}}(\underline{b})$ are derivable from the perturbation Hamiltonian containing only the resonant nonlinear interactions (H_R) as discussed in the preceeding lecture, i.e.,

$$T^R_{\underline{K}}(\underline{b}) = - i \frac{\partial H_R}{\partial b^*_{\underline{K}}} \tag{5}$$

Therefore in (4) the divergence of the nonlinear interactions among the gravity waves vanishes, i.e.,

$$\frac{\partial T^R_{\underline{K}}(\underline{b})}{\partial b_{\underline{K}}} + \frac{\partial T^{R*}_{\underline{K}}(\underline{b})}{\partial b^*_{\underline{K}}} = 0 \quad . \tag{6}$$

Using (5) in the remaining terms in (4) then yields

$$\sum_{\underline{k}} i \frac{\gamma_R(\underline{K})}{2\tilde{D}_2(\underline{K})} \left[b_{\underline{K}}^* \frac{\partial H_R}{\partial b_{\underline{K}}^*} - b_{\underline{K}} \frac{\partial H_R}{\partial b_{\underline{K}}} \right] = 0 \quad , \tag{7}$$

which is just the Poisson bracket of $|b_{\underline{K}}|^2$ and the Hamiltonian H_R, i.e.,

$$\sum_{\underline{K}} \frac{\gamma_R(\underline{K})}{2\tilde{D}_2(\underline{K})} \left\{ |b_{\underline{K}}|^2, H_R \right\}_{PB} = 0 \quad . \tag{8}$$

The total time derivative of any function of the canonical variables $\underline{b}$, say $G(\underline{b})$, can be expressed in terms of the Poisson bracket of that function and the system Hamiltonian[110,113], i.e.,

$$\frac{dG(\underline{b})}{dt} = \left\{ G(\underline{b}), H \right\}_{PB} + \frac{\partial G(\underline{b})}{\partial t} \tag{9}$$

The partial derivative term in (9) vanishes if $G(\underline{b})$ is not an explicit function of time. Since the Poisson bracket of $|b_{\underline{K}}|^2$ with the linear part of the Hamiltonian H_2 vanishes, i.e.,

$$\left\{ |b_{\underline{K}}|^2; H_2 \right\}_{PB} = \sum_{\underline{L},\underline{M}} \left[\frac{\partial}{\partial b_{\underline{L}}} |b_{\underline{K}}|^2 \frac{\partial}{\partial b_{\underline{L}}^*} \omega_M |b_{\underline{M}}|^2 - \frac{\partial}{\partial b_{\underline{L}}^*} |b_{\underline{K}}|^2 \frac{\partial}{\partial b_{\underline{L}}} \omega_M |b_{\underline{M}}|^2 \right]$$

$$= \sum_{\underline{L},\underline{M}} \delta_{\underline{L}-\underline{K}} \, \delta_{\underline{M}-\underline{K}} \, \omega_M \left[b_{\underline{K}}^* b_{\underline{M}} - b_{\underline{K}} b_{\underline{M}}^* \right] = 0 \quad ,$$

we can replace $G(\underline{b})$ by the series in (8) and obtain the time derivative

$$\frac{d}{dt} \sum_{\underline{K}} \frac{\gamma_R(\underline{K})}{2\tilde{D}_2(\underline{K})} |b_{\underline{K}}|^2 = \sum_{\underline{K}} \frac{\gamma_R(\underline{K})}{2\tilde{D}_2(\underline{K})} \left\{ |b_{\underline{K}}|^2, H \right\}_{PB} = 0 \tag{10}$$

which implies that

$$\sum_{\underline{K}} \frac{\gamma_R(\underline{K})}{2\tilde{D}_2(\underline{K})} |b_{\underline{K}}|^2 = \text{constant} . \tag{11}$$

Equation (10) and (11) clearly indicate one or more conservation laws for the gravity wave interactions depending on the functional form of $\gamma_R(\underline{K})/\tilde{D}_2(\underline{K})$

The conservation laws imbedded in these equations can be made explicit by introducing the form of $T^R_{\underline{K}}$ into (7) to obtain

$$\sum^{R}_{\underline{K},\underline{L},\underline{M},\underline{P}} \Gamma^{\underline{KP}}_{\underline{LM}} \left[b_{\underline{L}} b_{\underline{M}} b^*_{\underline{P}} b^*_{\underline{K}} - b^*_{\underline{L}} b^*_{\underline{M}} b_{\underline{P}} b_{\underline{K}} \right] \frac{\gamma_R(\underline{K})}{2\tilde{D}_2(\underline{K})} = 0 , \tag{12}$$

where the R superscript indicates that only those terms which are in resonance are included in (12). The terms in the series (12) may be re-ordered if we recall the symmetry properties for the interaction of gravity waves, i.e., $\Gamma^{\underline{KP}}_{\underline{LM}} = \Gamma^{\underline{PK}}_{\underline{LM}} = \Gamma^{\underline{KP}}_{\underline{ML}} = \Gamma^{\underline{LM}}_{\underline{KP}}$. This re-ordering yields

$$\sum^{R}_{\underline{K},\underline{L},\underline{M},\underline{P}} \Gamma^{\underline{KP}}_{\underline{LM}} b_{\underline{L}} b_{\underline{M}} b^*_{\underline{P}} b^*_{\underline{K}} \left[\frac{\gamma_R(\underline{K})}{\tilde{D}_2(\underline{K})} + \frac{\gamma_R(\underline{P})}{\tilde{D}_2(\underline{P})} - \frac{\gamma_R(\underline{L})}{\tilde{D}_2(\underline{L})} - \frac{\gamma_R(\underline{M})}{\tilde{D}_2(\underline{M})} \right] = 0 \tag{13}$$

which is a restriction on the weighting coefficients in the steady state distribution, i.e.,

$$\frac{\gamma_R(\underline{K})}{\tilde{D}_2(\underline{K})} + \frac{\gamma_R(\underline{P})}{\tilde{D}_2(\underline{P})} - \frac{\gamma_R(\underline{L})}{\tilde{D}_2(\underline{L})} - \frac{\gamma_R(\underline{M})}{\tilde{D}_2(\underline{M})} = 0 \quad . \tag{14}$$

In (14) if the coefficients are all equal to a wavevector independent constant, α say, then relation (14) is satisfied. This restriction corresponds to the conservation of wave action during a resonant interaction among four gravity waves [cf. Eq. (11)]. If the coefficients are proportional to the frequency, $\beta\omega_K$ say, then the relation (14) reduces to the frequency resonance condition $(\omega_K + \omega_P - \omega_L - \omega_M) = 0$ corresponding to the conservation of energy during the four wave interaction. If the coefficients are equal to the projection of the wavevector onto a constant vector, say $\underline{\gamma}\cdot\underline{K}$, then (14) reduces to the wavevector matching condition $(\underline{K}+\underline{P}-\underline{L}-\underline{M}) = 0$, corresponding to the conservation of momentum during a four wave interaction. Finally, (for completeness), if the coefficients are proportional to the scalar product of some power of the wavevector onto a constant tensor, say $\underline{\phi}_\mu\cdot\underline{k}^\mu$ where $\mu > 1$, then (14) represents the conservation of the corresponding physical quantity during an interaction. The most general algebraic solution to (14) is then given by

$$\frac{\gamma(\underline{K})}{\tilde{D}_2(\underline{K})} = \alpha + \beta\omega_K + \underline{\gamma}\cdot\underline{K} + \underline{\phi}_\mu\cdot\underline{k}^\mu \quad . \tag{15}$$

The mean square values of the mode amplitudes in the steady state are related to the coefficients in the steady state distribution by (21.28). We choose the convenient scale factor $[\Sigma_0/(2\pi)^2]$ and using (21.28) in (15) write

$$\frac{1}{2}\left\langle |b_{\underline{K}}|^2 \right\rangle_{ss} \frac{\Sigma_0}{(2\pi)^2} = \alpha + \beta\omega_K + \underline{\gamma}\cdot\underline{K} + \underline{\phi}_\mu \cdot \underline{k}^\mu \tag{16}$$

where Σ_0 is the area of the two dimensional ocean surface. In the present context (16) is a fluctuation-dissipation relation of the first kind, i.e., it relates the spontaneous fluctuations of the steady state gravity wave field to the dissipation in the wave field (induced by the non-resonant interactions) when the surface is linearly displaced from its steady state.

In many discussions of the properties of (1) it is convenient to change from discrete to continuum normalization by replacing the sum over discrete wavenumbers by an integral through the substitution (7.4). The mean action in the wavenumber interval $k_0 \leq k \leq k_{max}$ for positive traveling gravity waves is given in the continuum limit, i.e., limit of densely packed modes, by

$$J = \int_{-\pi/2}^{\pi/2} \int_{k_0}^{k_{max}} \frac{k dk d\theta_k}{\alpha + \beta\,\omega_k + \underline{\gamma}\cdot\underline{k} + \underline{\phi}_\mu \cdot \underline{k}^\mu} \tag{17a}$$

the mean energy by

$$E = \int_{-\pi/2}^{\pi/2} \int_{k_0}^{k_{max}} \frac{\omega_k k dk d\theta_k}{\alpha + \beta\,\omega_k + \underline{\gamma}\cdot\underline{k} + \underline{\phi}_\mu \cdot \underline{k}^\mu} \tag{17b}$$

the mean momentum by

$$\underline{M} = \int_{-\pi/2}^{\pi/2} \int_{k_o}^{k_{max}} \frac{\underline{k}\ kdkd\theta_k}{\alpha + \beta\ \omega_k + \underline{\gamma}\cdot\underline{k} + \underline{\phi}_\mu\cdot\underline{k}^\mu} \tag{17c}$$

and the remaining average integral constraints by

$$\underline{I}^\mu = \int_{-\pi/2}^{\pi/2} \int_{k_o}^{k_{max}} \frac{\underline{k}^\mu k\ dkk\theta_k}{\alpha + \beta\ \omega_k + \underline{\gamma}\cdot\underline{k} + \underline{\phi}_\mu\cdot\underline{k}^\mu} \quad . \tag{17d}$$

In the appropriate scaling limits the steady state distribution (1) then becomes

$$P_{ss}\ (J,E,\underline{M},\underline{I}_\mu) = \bar{N}\ \exp\left\{-\alpha J-\beta E-\underline{\gamma}\cdot\underline{M}-\underline{\phi}_\mu\cdot\underline{I}^\mu\right\} \tag{18}$$

where $\bar{N}$ is the continuum normalization such that

$$\int P_{ss}\ (J,E,\underline{M},\underline{I}^\mu)\ dJdEd\underline{M}d\underline{I}^\mu = 1. \tag{19}$$

The steady state distribution (18) has a form analogous to the canonical distributions of classical statistical mechanics.[212] The parameters α, β, $\underline{\gamma}$ and $\underline{\phi}_\mu$ are analogous to the thermodynamic potentials required to maintain the integral constraints of the system. Such a distribution also arises in the study of two dimensional turbulence where in addition to the total energy, the total enstrophy, i.e., the integrated vortex intensity, is conserved as argued by Onsager[227] for point vortices and later by Kraichnan[228]. Pursuing this association we can interpret $\overline{N}$ as the inverse of the partition function for the wave field and β as the analogue of the inverse temperature of the sea waves. The canonical distribution ($\underline{\phi}_\mu = \underline{\gamma} = \alpha = 0$) is traditionally obtained in statistical physics by segmenting a closed system into a number of equilvalent cells and considering the energy for a single cell when the remaining cells can be characterized by a single scalar temperature, i.e., β^{-1}. This argument was constructed by Gibbs[216] to circumvent the paradox of describing a closed conservative system by a distribution function. This resolution of the paradox yields a distribution which is not only rigorous in the subsystem, but can be used with success for any closed system. Conceptually one places the closed system in contact with an imaginary "heat bath" to obtain the distribution. As pointed out by Landau and Lifshitz[50] the difference between the open and closed system is only of consequence when one considers the fluctuations in the total energy which has a meaning only when the system is open. In the present context the "temperature" may be interpreted as the level of excitation of the ambient wave field.

To associate the total action in the distribution (18) ($\alpha \neq 0$, $\beta \neq 0$ but $\underline{\gamma} = \underline{\phi}_\mu = \underline{0}$) with the number of waves in a macroscopic system we use an analogy between the number of particles in a specified volume and the number of waves in a given specified area. The number of gravity waves in this area is then allowed to fluctuate by placing it in contact with the ambient "heat bath" of waves with which they non-resonantly interact. This is the "grand canonical" distribution for the gravity wave field. This analogy between the action density of water waves and the number density of quantum mechanics was made quite strong by Hasselman[11] where he noted that $J_{\underline{K}} \equiv B_{\underline{K}} B^{*}_{\underline{K}}$ is the number density operator in quantum mechanics.

To interpret the momentum term in (18) we use an association often made in statistical mechanics for hydrodynamic systems in local equilibrium, see e.g. Piccirelli[229]. We associate the vector parameter $\underline{\gamma}$ with the mean velocity of the fluid $\underline{V}$, i.e., $\underline{\gamma} \equiv -\beta\underline{V}$. Recalling the definitions of the total energy (17a) and total momentum (17c) we observe that with this association these two terms combine in the exponential (18) to yield (with $\underline{\phi}_{\mu} = \underline{0}$)

$$P_{ss}(J,E') = \overline{N}\, e^{-\alpha J - \beta E'} \quad . \tag{20}$$

The total energy E' is defined by (17a) with the frequency replaced by the Doppler shifted frequency; $\omega_k \to \omega_k' = \omega_k + \underline{k}\cdot\underline{V}$. Equation (20) is the steady state distribution of the gravity wave field observed in a coordinate system in which the fluid surface is at rest on the average. The interpretation of the parameter $\underline{\phi}_{\mu}$ when it exists, remains open.

An alternative to the assumed form of the solution (16) is to construct a scaling argument to determine the ratio $\gamma_R(\underline{K})/2\tilde{D}_2(\underline{K})$. One can argue as does Phillips[5] that the homogenous wavenumber spectrum is solely a function of the acceleration of gravity g and the wind friction speed u_*, so that since the continuum gravity wave spectrum $\Psi(\underline{K})$ has the dimensions L^4 we can write

$$\Psi(\underline{K}) = K^{-4} G(\theta_K, Ku_*^2/g) \quad (K_0 << K << k_\gamma) \tag{21}$$

where $G(\theta_K, Ku_*^2/g)$ is an undetermined function; θ_K is the angle specifying the direction of the wavenumber vector $\underline{K}$; K_0 is the maximum spectral component generaged by the wind and k_γ is $(g/\gamma)^{\frac{1}{2}}$ where γ is the kinematic surface tension. If the effects of surface drift are not important, then (21) reduces to

$$\Psi(\underline{K}) = K^{-4} G(\theta_K) \tag{22}$$

which neglecting the angular dependence of the wave field is of the form (16) with $\phi_4 k^4 >> \alpha + \beta\omega_k + \underline{\gamma}\cdot\underline{k}$ indicating that $\mu \cong 4$ phenomenologically. Experimental observations of steady state power spectral densities such as reviewed by Kitaigorodskii[7], indicate that $\Psi(\underline{K})$ has a dominant behavior given by $3 \leq \mu \leq 4$ in the gravity wave regime. In the present context we would attribute this rapid decrease of the observed spectrum with wavenumber with integral constraints on the gravity field dynamics beyond those of action, energy and momentum.

23. Entropy Generation in the Approach to the Steady-State

The solution to the phase space equation of motion (20.29) or (20.31) contains the same information as the complete solution to the dynamic equations (18.31). If one calculates $P(\underline{b},t|\underline{b}_0)$ from (20.29) then the evolution of the relative entropy (21.14) is unambiguous since the problem is solved. What is more germane to the discussion here is whether there exists an H-theorem for the relative entropy $\Delta S(t)$ defined by (21.14) and if so, how can we use it without solving for the full dynamic equations of the gravity wave field?

It is often assumed that the regression of fluctuations obey the same laws as the corresponding macroscopic irreversible process, such as was first assumed by Onsager[230]. This assumption is reasonable for linear systems and one can go quite far in the statistical analysis of linear mechanical systems using the Langevin equation as a phenomenological model, see eg., lectures 13-17 and in a more general context de Groot and Mazur[55]. Some progress has also been made in the analysis of nonlinear mechanical systems with mesoscopic additive fluctuations using both linear response theory, see eg., the review by Fox[53], and nonlinear response theory, see eg., Weare and Oppenheim[231]. A linear response analysis yields a probability density having the Gaussian form for the steady state solution to the Fokker-Planck equation (20.31) with a time dependent correlation matrix $\underline{\underline{Q}}(t)$. We therefore *assume* that the non-equilibrium evolution of a fluctuating field of gravity waves in the neighborhood of its steady state has a probability density of the form

$$P_M(\underline{b},t|\underline{b}_o) = \frac{1}{(2\pi)^{n/2}||\underline{\underline{Q}}(t)||^{\frac{1}{2}}} \exp\left\{ - \tfrac{1}{2}(\underline{b}-e^{-\underline{\underline{M}}t}\underline{b}_o)^+ \underline{\underline{Q}}^{-1}(t)\, (\underline{b}-e^{-\underline{\underline{M}}t}\underline{b}_o)\right\} \quad (1)$$

where the matrix $\underline{\underline{M}}$ describes the linear evolution of the test wave field from the initial state $\underline{b}_o$. The matrix $\underline{\underline{M}}$ includes a linearization of the nonlinear interactions in addition to the dissipation parameter (see previous lecture).

To make this last assumption more palatable we consider the linearized form of the equation of evolution (18.21) neglecting the coherent pressure fluctuations, i.e.,

$$\frac{d}{dt}\underline{B}(t) + \underline{\underline{M}}\underline{B}(t) = \underline{F}(t) \quad (2)$$

where $\underline{\underline{M}}$ is a state independent matrix with constant elements. One technique for obtaining this $n \times n$ matrix is the method of statistical linearization applied earlier. Here we linearize both resonant and non-resonant interactions by projecting these interactions onto the linear macroscopic variables in the asymptotic steady state[178]. The projection operator P is defined to operate on a function $G(\underline{B})$ by

$$PG(\underline{B}) \equiv \sum_{\underline{K}} \frac{B_{\underline{K}}(t)}{\langle |B_{\underline{K}}|^2\rangle_{ss}} \int b_{\underline{K}}^* \, G(\underline{b})\, P_{ss}(\underline{b})\, d\Gamma(\underline{b}) \quad (3)$$

where $P_{ss}(\underline{b})$ is the steady state distribution discussed in lecture 22. The dynamic equation (2) therefore, describes the evolution of the gravity wave field in the neighborhood of the steady state and similar to (21.21) the elements of $\underline{\underline{M}}$ are given by

$$M_{\underline{K}\underline{K}'} = \delta_{\underline{K}-\underline{K}'}\ \gamma(\underline{K}) - \frac{\langle b^*_{\underline{K}}\, T_{\underline{K}'}(\underline{b})\rangle_{ss}}{\langle |b_{\underline{K}}|^2\rangle_{ss}} \quad . \tag{4}$$

The Fokker-Planck equation corresponding to the linear Langevin equation (2) is

$$\frac{\partial}{\partial t} P_M(\underline{b},t|\underline{b}_0) = \sum_{\underline{K}} \frac{\partial}{\partial b_{\underline{K}}}\left[\sum_{\underline{K}'} M_{\underline{K}\underline{K}'} b_{\underline{K}'} P_M(\underline{b},t|\underline{b}_0)\right] + cc$$
$$+ \sum_{\underline{K}} 2\tilde{D}_2(\underline{K})\ \frac{\partial^2}{\partial b_{\underline{K}}\partial b^*_{\underline{K}}} P_M(\underline{b},t|\underline{b}_0) \tag{5}$$

which for an initial delta function condition $P(\underline{b},t = 0|\underline{b}_0) = \delta(\underline{b}-\underline{b}_0)$ has the exact solution given by (1). The matrix $\underline{\underline{M}}$ describes the linear average evolution of the gravity wave field from its initial state $\underline{b}_0$. The subscript M is used on the distribution to denote its dependence on the linearization procedure used in going from (18.21) to (2).

The definition of the relative entropy (21.14), using (1) is now rewritten as

$$\Delta S(t) \cong -k_B \left\{ \langle \ln P_M;t\rangle_M - \langle \ln P_0;t\rangle_M \right\} \tag{6}$$

where the averages are here taken with respect to the distribution (1) as indicated by the M subscript on the brackets. If we continue to assume that the gravity wave field is spatially homogeneous near the steady state then $\underline{\underline{Q}}(t)$ remains diagonal and using the conservation of the number of states condition

$$\left\langle (\underline{b}-e^{-\underline{\underline{M}}t}\underline{b}_0)^{+}\, \underline{\underline{Q}}^{-1}(t)\, (\underline{b}-e^{-\underline{\underline{M}}t}\underline{b}_0);t\right\rangle_M = \left\langle \underline{b}^{+}\, \underline{\underline{Q}}^{-1}{}_{ss}\, \underline{b}\right\rangle_{ss} = n \tag{7}$$

Equation (6) simplifies to

$$\Delta S(t) = \tfrac{1}{2}k_B \sum_{\underline{K}} \left[\ln Q_{\underline{KK}}(t) - \ln Q_{\underline{KK}}(\infty)\right] \quad . \tag{8}$$

Introducing the mean square mode amplitude, often referred to as the average action density in mode $\underline{K}$,

$$J_{\underline{K}}(t) \equiv Q_{\underline{KK}}(t) = \left\langle |b_{\underline{K}}|^2;t\right\rangle_M \tag{9}$$

the time rate of change of the entropy excess in (8) becomes

$$\frac{d\Delta S(t)}{dt} = \tfrac{1}{2}k_B \sum_{\underline{K}} \frac{1}{J_{\underline{K}}(t)} \frac{\partial J_{\underline{K}}(t)}{\partial t} \tag{10}$$

yielding a time variation of $\Delta S(t)$ determined by the transport equation for the average action density of the gravity wave field.

Hasselmann[11] established, by a much different argument, an H-theorem for deep water gravity waves. In the above notation he postulated an expression for the entropy given by

$$S(t) = \tfrac{1}{2}k_B \iint_{-\infty}^{\infty} \ln J_{\underline{K}}(t)\, d^2K \tag{11}$$

which is the continuum limit of the first term in the sum in (8). However, as Hasselmann pointed out, the integral (11) does not converge. In order to prove that $\frac{dS(t)}{dt} \geq 0$ he restricted the accesible region of wavenumber space for the interaction and used the asymptotic properties of the interaction coefficients in his perturbation expansion for the transport equation $dJ_{\underline{K}}/dt$ to show that the integrand of the time derivative of (11) is a positive definite quantity. The steady state solution obtained by Hasselmann is exactly (22.16) with $\underline{\phi}_{\mu} = 0$, so that his invariants of the motion are the action, energy and momentum densities as we have obtained here in the *restricted* case.

An equation of the form (10) was also obtained in the context of Rosby waves propagating through a two-dimensional turbulent flow field by Carnevale[232] using an information theory argument, see also Refs. 221-225 for more general discussions of non-equilibrium probability densities based on information theory arguments. Carnevale demonstrates that in a quasi-geostrophic approximation to a planetary atmosphere, i.e., the so-called-β-plane model, that the definition of entropy leading to (10) yields a globally stable evolution of the fluid flow, i.e., $\frac{d\Delta S(t)}{dt} \geq 0$. To determine this condition he requires a closure hypothesis for the dynamic equation, just as did Hasselmann in the case of fluctuation gravity waves. The restriction to the linear response probability density P_M used here in the gravity wave field is related to the quasi-Guassian closure assumption employed by Hasselmann[11].

The time derivative of the action density may be obtained directly from the Fokker-Planck equation by multiplying (20.29) on the left by $|b_{\underline{K}}|^2$ and integrating by parts to obtain

$$\frac{\partial J_{\underline{K}}(t)}{\partial t} + 2\gamma_R(\underline{K})J_{\underline{K}}(t) = 2[|D_{\underline{K}}| + \tilde{D}_2(\underline{K}) + D_2(\underline{K})\; J_{\underline{K}}(t)] + 2\,\mathrm{Re}\left\langle b_{\underline{K}}^{*}\; T_{\underline{K}}^{R};t\right\rangle \tag{12}$$

Note that we have used the more general definition of the action density using the exact non-equilibrium probability density

$$J_{\underline{K}}(t) = \int |b_{\underline{K}}|^2\; P(\underline{b},t|\underline{b}_0)\; d\Gamma(\underline{b}) \tag{13}$$

in (12) and that we have been able to include the effect of the coherent pressure fluctuations; something we could not do in solving for the probability density directly. The transport equation for the action density (12) is *exact* within the limits that $\gamma_R(\underline{K})$ models the induced dissipation by the non-resonant interactions and $T_{\underline{K}}^R$ models the resonant interactions. With only the third order resonant interactions of (18.3) in $T^R_{\underline{K}}$ we have

$$\left\langle b^*_{\underline{K}} T^R_{\underline{K}}; t \right\rangle = \sum_{\underline{L},\underline{P},\underline{N}}^{R} i\Gamma^{\underline{K}\,\underline{N}}_{\underline{L}\,\underline{P}} \left\langle b_{\underline{L}} b_{\underline{P}} b^*_{\underline{N}} b^*_{\underline{K}}; t \right\rangle e^{i(\omega_K + \omega_N - \omega_L - \omega_P)t} \tag{14}$$

The coupling parameter Γ is real so that if the fourth order correlation function is also real, then *on-resonance* this quantity is completely imaginary and there is no contribution of the resonant nonlinear interactions to the transport of action in (12). This result explains the near Gaussian behavior observed for surface wave properties when the wave field is near its steady state. It is of interest for us here, however, to go beyond this approximation and determine the complex corrections to the fourth order averages in (14).

To calculate the fourth moment in (14) we examine the transport equation for this quantity with the correlation between the dynamic variables and the fluctuations neglected, i.e., we assume $\langle b^*_{\underline{K}} F_{\underline{K}}; t \rangle = 0$. The time derivative of the fourth order product of mode amplitudes is then

$$\begin{aligned} \frac{d}{dt} B_{\underline{L}} B_{\underline{P}} B^*_{\underline{N}} B^*_{\underline{K}} \cong\ & i\, B_{\underline{P}} B^*_{\underline{N}} B^*_{\underline{K}} \sum_{\underline{u},\underline{v},\underline{w}} \Gamma^{\underline{Lw}}_{\underline{uv}} B_{\underline{u}} B_{\underline{v}} B^*_{\underline{w}}\, e^{i\Delta\omega_L t} \\ & + i\, B_{\underline{L}} B^*_{\underline{N}} B^*_{\underline{K}} \sum_{\underline{u},\underline{v},\underline{w}} \Gamma^{\underline{Pw}}_{\underline{uv}} B_{\underline{u}} B_{\underline{v}} B^*_{\underline{w}}\, e^{i\Delta\omega_P t} \\ & - i\, B_{\underline{L}} B_{\underline{P}} B^*_{\underline{K}} \sum_{\underline{u},\underline{v},\underline{w}} \Gamma^{\underline{Nw}}_{\underline{uv}} B^*_{\underline{u}} B^*_{\underline{v}} B_{\underline{w}}\, e^{-i\Delta\omega_N t} \\ & - i\, B_{\underline{L}} B_{\underline{P}} B^*_{\underline{N}} \sum_{\underline{n},\underline{v},\underline{w}} \Gamma^{\underline{Kw}}_{\underline{uv}} B^*_{\underline{u}} B^*_{\underline{v}} B_{\underline{w}}\, e^{-i\Delta\omega_K t} \end{aligned} \tag{15}$$

with the primes denoting the proper wavevector restrictions on the sums and

$$\Delta\omega_j \equiv \omega_u+\omega_v-\omega_w-\omega_j \quad . \tag{16}$$

Equation (15) is obtained by taking the time derivative of each of the amplitudes in sequence and substituting from (18.3). We note that this technique has been used by Longuet-Higgins[233] to study the energy transfer near the peak of a narrow spectrum nonlinear wave. In his analysis he makes use of the random phase approximation and also sets the interaction strengths to be equal; assumptions we need not make here. The assumption that $B^*_{\underline{K}}$ and $F_{\underline{K}}$ in (8.21) are statistically independent uses the fact that the fluctuating flux is quadratic in the mode amplitudes and the statistical average is specified by the ensemble distribution (1) i.e., a Gaussain probability density, so that all odd-order moments vanish. A non-Gaussian probability density would of course bring in additional terms from this correlation as discussed in lecture 19, i.e., $\langle B^*_{\underline{K}}\, F_{\underline{K}};t\rangle \neq 0$ in general. Using the discrete action density $J_{\underline{K}}(t)$ in the ensemble average of (15) and keeping only those terms with a possible resonant frequency matching we obtain by averaging (15) with respect to the distribution (1),

$$\frac{d}{dt}\left\langle B_{\underline{L}}B_{\underline{P}}B^*_{\underline{N}}B^*_{\underline{K}}\right\rangle_M = 16i\left\{\Gamma^{\underline{LP}}_{\underline{NK}}\, J_{\underline{K}}\, J_{\underline{P}}\, J_{\underline{N}} + \Gamma^{\underline{PL}}_{\underline{NK}}\, J_{\underline{L}}\, J_{\underline{N}}\, J_{\underline{K}} - \Gamma^{\underline{NK}}_{\underline{LP}}\, J_{\underline{L}}\, J_{\underline{P}}\, J_{\underline{K}} - \Gamma^{\underline{KN}}_{\underline{PL}}\, J_{\underline{L}}\, J_{\underline{P}}\, J_{\underline{N}}\right\} e^{i(\omega_L+\omega_P-\omega_N-\omega_K)t} \quad . \tag{17}$$

Equation (17) can now be integrated from $-\infty$ to t and used to replace the nonlinear driving term in (12), i.e.,

$$\frac{\partial J_{\underline{K}}(t)}{\partial t} + 2[\gamma_R(\underline{K})-D_2(\underline{K})]\, J_{\underline{K}}(t) = 2\bar{D}(\underline{K}) + \mathrm{Real}\left\{32 \sum_{\underline{L},\underline{P},\underline{N}} \delta_{\underline{K}+\underline{N}-\underline{L}-\underline{P}}\, \Gamma^{\underline{KN}}_{\underline{LP}} \int_{-\infty}^{t} dt' \left[\Gamma^{\underline{KN}}_{\underline{PL}} J_{\underline{L}} J_{\underline{P}} J_{\underline{N}} + \Gamma^{\underline{NK}}_{\underline{LP}} J_{\underline{L}} J_{\underline{P}} J_{\underline{K}} - \Gamma^{\underline{PL}}_{\underline{NK}} J_{\underline{L}} J_{\underline{N}} J_{\underline{K}} - \Gamma^{\underline{LP}}_{\underline{NK}} J_{\underline{P}} J_{\underline{N}} J_{\underline{K}}\right] \times e^{i(\omega_L+\omega_P-\omega_N-\omega_K)(t-t')}\right\} . \qquad (18)$$

where

$$\bar{D}(\underline{K}) \equiv |D_2(\underline{K})| + \tilde{D}_2(\underline{K}) \quad . \qquad (19)$$

By assumption the mode amplitudes $B_{\underline{K}}(t)$ are slowly varying functions of time, therefore so too are their products. The integral (18) is then approximated by removing from the time integration the term in the square brackets which is of $O(\underline{B}^6)$. The integral over time yields a delta function in frequency plus the Cauchy principle value integral times i[234], i.e.,

$$\int_{-\infty}^{0} e^{i\omega t}\, dt = \pi\, \delta(\omega) - i\, P\left(\frac{1}{\omega}\right) \qquad (20)$$

where P (•) indicates the principle value integral. Equation (18) therefore reduces to

$$\frac{\partial J_{\underline{K}}(t)}{\partial t} + 2[\gamma_R(\underline{K})-D_2(\underline{K})]\, J_{\underline{K}}(t) = 2\overline{D}(\underline{K}) + 32\pi \sum_{\underline{L},\underline{P},\underline{N}} \delta_{\underline{K}+\underline{N}-\underline{L}-\underline{P}}\; \delta(\omega_L+\omega_P-\omega_N-\omega_K)\; \Gamma^{\underline{KN}}_{\underline{LP}}$$

$$\times \left\{ \left[\Gamma^{\underline{KN}}_{\underline{LP}} J_{\underline{N}} + \Gamma^{\underline{NK}}_{\underline{LP}} J_{\underline{K}} \right] J_{\underline{L}}\, J_{\underline{P}} - \left[\Gamma^{\underline{PL}}_{\underline{NK}} J_{\underline{L}} + \Gamma^{\underline{LP}}_{\underline{NK}} J_{\underline{P}} \right] J_{\underline{N}}\, J_{\underline{K}} \right\} . \qquad (21)$$

The coupling coefficients are symmetric under interchanges of $\underline{K}$ and $\underline{N}$ also of $\underline{L}$ and $\underline{P}$ and finally $\Gamma^{\underline{KN}}_{\underline{LP}} = \Gamma^{\underline{LP}}_{\underline{KN}}$ so that (21) in the continuum limit becomes, indicating the continuum action by $J(\underline{K})$ instead of $J_{\underline{K}}$,

$$\frac{\partial J_4}{\partial t} + 2[\gamma_R(\underline{K}_4)-D_2(\underline{K}_4)]\, J_4 = 2\overline{D}(\underline{K}_4) + 4\int d^2K_1 d^2K_2 d^2K_3\; D(1,2,3,4)$$

$$\times \left\{ J_1\, J_2\, [J_3 + J_4] - J_3\, J_4\, [J_1 + J_2] \right\} \qquad (22)$$

where we have introduced the notation $J_j \equiv J(\underline{K}_j)$; j = 1, 2, 2, 4 and defined the coefficient D(1,2,3,4) by

$$D(1,2,3,4) \equiv \delta(\underline{K}_4+\underline{K}_3-\underline{K}_2-\underline{K}_1)\; \delta(\omega_4+\omega_3-\omega_1-\omega_2)\; 8\pi \left(\Gamma^{\underline{K}_4\underline{K}_3}_{\underline{K}_1\underline{K}_2} \right)^2 . \qquad (23)$$

If we neglect the nonlinear terms in (22) then the steady state solution $\dot{J}_4 = 0$ requires that

$$\overline{D}(\underline{K}_4) = [\gamma_R(\underline{K}_4) - D_2(\underline{K}_4)]\; J_{ss}(\underline{K}_4) \tag{24}$$

where $J_{ss}(\underline{K}_4)$ is the steady state average action density. Inserting (24) into (22) yields

$$\frac{\partial}{\partial t} J_4(t) + 2[\gamma_R(\underline{K}) - D_2(\underline{K})][J_4(t) - J_{ss}] = 4\int d^2K_1 d^2K_2 d^2K_3 D(1,2,3,4)$$

$$\times\, [J_1J_2(J_3+J_4) - J_3J_4(J_1+J_2)] \tag{25}$$

so that as long as $[\gamma_R(\underline{K}) - D_2(\underline{K})] > 0$ the linear term induces a relaxation to the asypmptotic steady state. If for the sake of comparison we neglect the dissipative current in (25), the resulting expression is identical in *form* to the transport equation constructed by Hasselmann[11]. The scattering kernel in (25) differs from his however.

Assuming that $\gamma_R(\underline{K}) - D_2(\underline{K}) \simeq 0$ we obtain the integro-differential equation for the transport of action

$$\frac{1}{J_4}\frac{\partial J_4}{\partial t} = \int d^2K_1 d^2K_2 d^2K_3\; 4D(1,2,3,4)\; J_1J_2J_3\left[\frac{1}{J_4} + \frac{1}{J_3} - \frac{1}{J_1} - \frac{1}{J_2}\right]. \tag{26}$$

Taking the symmetric sum

$$\int \frac{d^2K_4}{J_4} \frac{\partial J_4}{\partial t} = \frac{1}{4} \sum_{j=1} \int \frac{d^2K_j}{J_j} \frac{\partial J_j}{\partial t} \tag{27}$$

and using the symmetry properties of (26) we obtain for the rate of increase of the entropy excess

$$\frac{d}{dt} \Delta S(t) = k_B \int dD(1,2,3,4)\; J_1 J_2 J_3 J_4 \left(\frac{1}{J_4} + \frac{1}{J_3} - \frac{1}{J_2} - \frac{1}{J_1}\right)^2 \quad , \tag{28}$$

where $dD(1,2,3,4) \equiv d^2K_1\, d^2K_2\, d^2K_3\, d^2K_4\, D(1,2,3,4)$. Since the coupling coefficient $\left(\Gamma^{\underline{K}_4 \underline{K}_3}_{\underline{K}_1 \underline{K}_2}\right)^2$ is positive definite, (28) is manifestly positive semi-definite. Thus, disorder increases for the system as a whole as the gravity wave field evolves, i.e., $\frac{d\Delta S}{dt} \geq 0$. Asymptotically the wave field is represented by the steady state solution to the Fokker-Planck equation.

The result of (28) for a long time was thought to stand in contradiction to the order observed in the neighborhood of the spectral peak of an evolving wind generated spectrum. This order was thought to be at variance with the idea of entropy maximization. The fallacy in interpreting the observations was in concentrating on one region of the wave system rather than considering the wave field as a whole. Webb[226] resolved the paradox by noting that the contributions to (28) can be separated into two classes of interaction. The first class is the *diffusion* of action in wave vector space and the second is a preferential transfer of action from the shorter waves to the long waves. The diffusive term is consistent with ones' notion of entropy and was specified by Webb to be

$$\frac{d}{dt}\Delta S(t)\bigg|_d \equiv k_B \int dD(1,2,3,4)\ J_1J_2J_3J_4 \left[\left(\frac{1}{J_1}-\frac{1}{J_3}\right)^2+\left(\frac{1}{J_2}-\frac{1}{J_4}\right)^2\right] \tag{29}$$

which is clearly positive semi-definite since $D(1,2,3,4) \geq 0$. The relation of (29) to the diffusion of action in wavenumber space had been previously established by West[235]. West showed that if we assume: 1) that $\underline{K}_1$ and $\underline{K}_4$ are small scale and $\underline{K}_2$ and $\underline{K}_3$ are large scale waves; 2) that the second order interactions of the small scale waves are negligible (weak interaction hypothesis) and 3) that using the $\underline{K}$ - delta function we can expand $J(\underline{K}_1)$ with $\underline{K}_1 = \underline{K}_2 + \underline{K}_4 - \underline{K}_3$ about $\underline{K}_4$ since $K_4 \gg |\underline{K}_2 - \underline{K}_3|$, we obtain from (24)

$$\begin{aligned}\frac{\partial J_4}{\partial t} &= \int d^2K_2d^2K_3\ D(\underline{K}_2+\underline{K}_4-\underline{K}_3,\underline{K}_2,\underline{K}_3,\underline{K}_4)\ J_2J_3\ [J(\underline{K}_4+\underline{K}_2-\underline{K}_3) - J_4] \\ &\cong \nabla_{\underline{K}_4}\cdot [\underline{\underline{D}}\cdot\nabla_{\underline{K}_4} J_4]\end{aligned} \tag{30}$$

where $\underline{\underline{D}}$ is a 2 x 2 matrix with elements $D_{\ell m}$ given by

$$D_{\ell m} \equiv \int d^2K_2d^2K_3\ D(\underline{K}_4,\underline{K}_2,\underline{K}_3,\underline{K}_4)\ J_2\ J_3\ (\underline{K}_2-\underline{K}_3)_\ell\ (\underline{K}_2-\underline{K}_3)_m \quad . \tag{31}$$

Thus the four wave resonant interaction has, for a two-scale process and weakly interacting waves, reduced to a linear diffusion equation in $\underline{K}$ - space for the high frequency gravity waves. The low frequency gravity waves determine the components of the diffusion matrix. An analogous expression has been obtained for the quadratic integro-differential equation describing the nonlinear interaction of undersea internal waves by McComas and Bretherton[236].

The second class of interactions Webb refers to as a pumping and is the nonlinear transfer of action from the high into the low wavenumber waves. These interactions are obtained by subtracting (29) from (28), i.e.,

$$\frac{d}{dt}\Delta S(t)\bigg|_p = 2k_B \int dD(1,2,3,4)\; J_1J_2J_3J_4\left(\frac{1}{J_1}-\frac{1}{J_3}\right)\left(\frac{1}{J_2}-\frac{1}{J_4}\right) \tag{32}$$

and can be of either sign. The diffusive term (30) therefore increases the disorder of the water wave system by diffusing action to all wave numbers, whereas (32) pumps action into the peak of the wind generated initial spectrum of water waves and can either increase or decrease the order in the system, but subject to the overall constraint

$$\frac{d\Delta S(t)}{dt}\bigg|_d + \frac{d\Delta S(t)}{dt}\bigg|_p \geq 0 \quad . \tag{33}$$

Webb[226] uses this separation of effects to explain the observation that the spectral peak in wind generated sea waves grows faster than expected from linear instability theory. The enhanced growth is due to the nonlinear transfer of energy from short to long waves. This mechanism is related to kinetic theory formulation of this effect in Lecture 18. The long waves receiving the energy become more ordered and the short waves supplying the energy become more disordered to compensate and maintain the monotonic increase in the total entropy excess with time. Webb also draws a number of conclusions about the transport dynamics of the wave field based on numerical calculations using (26) to explain the separation of effects into diffusion and pumping. Using the requirements that the energy and action densities are both conserved by the gravity wave interaction he argues that an anisotropy is found in the diffusive transfer because of the large value of the coupling coefficient for interactions involving two long and two short waves. This strong nonlinear coupling of the short to the long waves preferentially transfers action density from intermediate to low wavenumbers. It cannot transfer this action beyond the spectral peak, however, because the spectrum rapidly drops off beyond its peak value and the transfer is proportional to the energy level of the waves receiving the energy. The spectral peak thus tends to grow. This disorder created on the small scale therefore results in the generation of order on the large scale.

24. Closure of the Moment Transport Equations

In the preceeding lectures we derived the equation of evolution for the non-equilibrium water wave probability density. This is not a *directly measureable* quantity, however. In an experiment it is the set of fluctuating mode amplitudes $\{B_{\underline{K}}(t)\}$ which is measured as a function of time. An ensemble of such measurements then enables us to calculate the moments of $\underline{B}(t)$ as functions of time. The evolution of these moments can in turn be described by taking the appropriate averages of the dynamic equations (18.21). Alternatively we may define the non-equilibrium n^{th} moment of the mode $B_{\underline{K}}(t)$ using the probability density (20.29) as

$$\langle B_{\underline{K}}^n(t)\rangle \equiv \int b_{\underline{K}}^n \, P(\underline{b};t|\underline{b}_0)d\Gamma(\underline{b}) \tag{1}$$

The time derivative of (1) is

$$\frac{d}{dt}\langle B_{\underline{K}}^n \, (t)\rangle = \int b_{\underline{K}}^n \frac{\partial P}{\partial t}(\underline{b};t|\underline{b}_0) \, d\Gamma(\underline{b}) \tag{2}$$

from which it is clear that the time evolution of the nth moment of $B_{\underline{K}}(t)$ is determined by the time evolution of the probability density in phase space.

Replacing the time derivative in (2) by the phase space derivatives specified by the Fokker-Planck equation (20.31) yields the transport equation

$$\frac{\partial}{\partial t}\langle B_{\underline{K}}^n;t\rangle + n\gamma(\underline{K})\langle B_{\underline{K}}^n;t\rangle = 2n(n-1)\left[D_{\underline{K}}-\tilde{D}_2(\underline{K})\right]\langle B_{\underline{K}}^{n-2};t\rangle + n\langle B_{\underline{K}}^{n-1}\, T_{\underline{K}}^R;t\rangle \tag{3}$$

where it will be noted that the linear coefficient $\gamma_{\underline{K}}$ has been replaced by $\gamma(\underline{K})$ from (21.21) by envoking the arguments on the effects of the non-resonant interactions. In (3) we have adopted the general notation that the non-equlibrium average of a function $G(\underline{B})$ is written as

$$\langle G(\underline{B});t\rangle \equiv \int G(b)\; P(\underline{b},t|\underline{b}_0)\; d\Gamma(\underline{b}) \tag{4}$$

which we recall assumes that the time averages over the fluctuations in the original dynamic description of the the gravity wave evolution is replaced by an ensemble average. Again, only the resonant nonlinear interactions amoung gravity waves appear on the right hand side of (3).

In this study we restrict our attention to the lowest few moments since it is the first and second moments which are usually experimentally accessible. We determined in lecture 19 that the mean square surface displacement could be expressed as the wavevector integral over the second moment of the mode amplitudes and products of the second moment, i.e., the energy and action spectral densities. In the previous lecture we obtained a cubic integro-differential equation to describe the evolution of the action density of the gravity wave field near its asymptotic steady state. Here we are interested in studying some of the details flashed over in that discussion, e.g., the effect of the dissipative current on the evolution of the wave field.

Note the difference between the transport equation (3) for n=2, i.e.,

$$\frac{\partial}{\partial t}\left\langle B_{\underline{K}}^2;t\right\rangle + 2\gamma(\underline{K})\left\langle B_{\underline{K}}^2;t\right\rangle = 4\left[D_{\underline{K}}-\tilde{D}_2(\underline{K})\right] + 2\left\langle B_{\underline{K}}T_{\underline{K}}^R;t\right\rangle \tag{5}$$

and the transport equation for the action spectral density $\left\langle |B_{\underline{K}}|^2;t\right\rangle$, i.e..

$$\frac{\partial}{\partial t}\left\langle |B_{\underline{K}}|^2;t\right\rangle + 2\;\gamma_R(\underline{K})\left\langle |B_{\underline{K}}|^2;t\right\rangle = 2\left[|D_{\underline{K}}| + \tilde{D}_2(\underline{K})\right]$$
$$+\, 2D_2(\underline{K})\left\langle |B_{\underline{K}}|^2;t\right\rangle + 2\mathrm{Real}\left\langle B_{\underline{K}}^*\; T_{\underline{K}}^R;t\right\rangle . \tag{6}$$

The parameters $D_{\underline{K}}$, $D_2(\underline{K})$ and $\tilde{D}_2(\underline{K})$ determine the steady state spectral density, i.e. when $\frac{\partial}{\partial t}\left\langle |B_{\underline{K}}|^2\right\rangle_{ss} = 0$ we obtain,

$$|D_{\underline{K}}| + \tilde{D}_2(\underline{K}) = \left[\gamma_R(\underline{K}) - D_2(\underline{K})\right]\left\langle |B_{\underline{K}}|^2 \right\rangle_{ss} - \text{Real}\left\langle B^*_{\underline{K}}\, T^R_{\underline{K}} \right\rangle_{ss} \quad (7)$$

so that (6) can be written

$$\frac{\partial}{\partial t}\left\langle |\hat{B}_{\underline{K}}|^2;t \right\rangle + 2\left[\gamma_R(\underline{K}) - D_2(\underline{K})\right]\left\langle |\hat{B}_{\underline{K}}|^2;t \right\rangle = 2\ \text{Real}\left\langle \hat{B}^*_K\, T^R_K;t \right\rangle . \quad (8)$$

We have used the notation $\hat{B}_{\underline{K}} = B_{\underline{K}}(t) - \langle B_{\underline{K}} \rangle_{ss}$ so that (8) describes the evolution of the mean square variation in the mode amplitudes from the steady state. For $\gamma_R(\underline{K}) > D_2(\underline{K})$ this expression describes the relaxation of this perturbation back to the steady state and its coupling to other modes in the wave field. Whether the resonant nonlinear interactions dominate this process is determined in part by the relative size of the dissipation strength $\gamma_R(\underline{K})$ and the spectrum of coherent fluctuations. Looking more closely, the dissipation parameter spectrum $\gamma_R(\underline{K})$ is in its turn determined by the spectral strength of the incoherent pressure fluctuations, the energy influx from the high frequency waves and the energy transferred out of this spectral interval by the non-resonant interactions among the gravity waves. The existence of a steady state solution to (8) is still a matter of conjecture since we have *not* established that $\gamma_R(\underline{K}) > D_2(\underline{K})$ by direct calculation.

To facilitate discussion of the properties of the transport equation (6) we introduce a notation for the non-equilibrium *cumulants* of $\underline{B}(t)$. We recall that the cumulants or *semi-invariants* are simple algebraic functions of the moments. They were found in earlier lectures to simplify the discussion of the properties of the high order correlation functions of the mode amplitudes.[195] We restrict the discussion here to third order cumulants because we intend to develop a perturbation argument and truncate the analysis at third order in the perturbation parameter. The difficulties we encounter in solving the coupled set of transport equations give ample evidence of the need for further analysis in the study of such systems. The analytic difficulties are so severe in fact that many investigators have argued that the transport equation technique is a fruitless approach. This point of view is moot and whether or not true the technique is still widely used and therefore should be understood.

The first three semi-invariants for the water wave field can be written in terms of the deviation of the mode amplitude from its non-equilibrium average value, i.e. $\hat{B}_{\underline{K}}(t) \equiv B_{\underline{K}}(t) - \langle B_{\underline{K}};t\rangle$. The non-equilibrium average is defined by

$$S_{\underline{K}}(t) \equiv \langle B_{\underline{K}};t\rangle \quad , \tag{9a}$$

the two second cumulants by

$$S^{\underline{L}}_{\underline{K}}(t) \equiv \langle \hat{B}_{\underline{K}}\hat{B}^{*}_{\underline{L}};t\rangle = \langle B_{\underline{K}}B^{*}_{\underline{L}}\rangle - \langle B_{\underline{K}};t\rangle\langle B^{*}_{\underline{L}};t\rangle$$

$$S_{\underline{KL}}(t) \equiv \langle \hat{B}_{\underline{K}}\hat{B}_{\underline{L}};t\rangle = \langle B_{\underline{K}}B_{\underline{L}};t\rangle - \langle B_{\underline{K}};t\rangle\langle B_{\underline{L}};t\rangle \quad , \tag{9b}$$

and the third order cumulants by

$$S^{\underline{M}}_{\underline{KL}}(t) \equiv \langle \hat{B}_{\underline{K}}\hat{B}_{\underline{L}}\hat{B}_{\underline{M}};t\rangle$$

$$S_{\underline{KLM}}(t) \equiv \langle B_{\underline{K}}B_{\underline{L}}B_{\underline{M}};t\rangle \quad . \tag{9c}$$

Note that since the mode amplitudes are complex that $S^{\underline{KL}}(t) = S^{*}_{\underline{KL}}(t)$, etc., which is a bit more complex than the cumulants introduced earlier. We can invert the relations (9) to express the moments of the mode amplitudes in terms of cumulants, just as done in (19.18). The inverted relations are

$$\langle B_{\underline{K}};t\rangle = S_{\underline{K}}(t) \tag{10a}$$

$$\langle B_{\underline{K}}B^{*}_{\underline{L}};t\rangle = S^{\underline{L}}_{\underline{K}}(t) + S_{\underline{K}}(t)\, S^{\underline{L}}(t) \tag{10b}$$

$$\langle B_{\underline{K}} B_{\underline{L}} ; t \rangle = S_{\underline{KL}}(t) + S_{\underline{K}}(t)\, S_{\underline{L}}(t) \tag{10c}$$

$$\langle B_{\underline{K}} B_{\underline{L}} B^{*}_{\underline{M}} ; t \rangle = S^{\underline{M}}_{\underline{KL}}(t) + S_{\underline{KL}}(t)\, S^{\underline{M}}(t) + S^{\underline{M}}_{\underline{K}}(t)\, S_{\underline{L}}(t) + S_{\underline{K}}(t)\, S^{\underline{M}}_{\underline{L}}(t)$$

$$+ S_{\underline{K}}(t)\, S_{\underline{L}}(t)\, S^{\underline{M}}(t) \tag{10d}$$

$$\langle B_{\underline{K}} B_{\underline{L}} B_{\underline{M}} ; t \rangle = S_{\underline{KLM}}(t) + S_{\underline{KL}}(t)\, S_{\underline{M}}(t) + S_{\underline{KM}}(t)\, S_{\underline{L}}(t) + S_{\underline{K}}(t)\, S_{\underline{LM}}(t)$$

$$+ S_{\underline{K}}(t)\, S_{\underline{L}}(t)\, S_{\underline{M}}(t). \tag{10c}$$

Note how the moments decompose into a sum of distinct orderings of the cumulants, this is the linked cluster expansion property frequently observed in systems with many degrees of freedom[163,191] as mentioned earlier.

We now use the relationship between the moments and cumulants to rewrite the transport equation for the average mode amplitude, i.e., n=1 in (1) as

$$\frac{\partial S_{\underline{k}}(t)}{\partial t} + \gamma(\underline{k})\, S_{\underline{k}}(t) = \sum_{\underline{L},\underline{M},\underline{P}} i\, \Gamma^{\underline{KP}}_{\underline{LM}}\, \delta_{\underline{K}+\underline{P}-\underline{L}-\underline{M}} \Big\{ S^{\underline{P}}_{\underline{LM}}(t) + S_{\underline{LM}}(t)\, S^{\underline{P}}(t)$$

$$+ S^{\underline{P}}_{\underline{L}}(t)\, S_{\underline{M}}(t) + S^{\underline{P}}_{\underline{M}}(t)\, S_{\underline{L}}(t) + S_{\underline{L}}(t)\, S_{\underline{M}}(t)\, S^{\underline{P}}(t) \Big\} \tag{11}$$

where we have replaced the third moment using (10d). Here the second cumulants $S_{\underline{LM}}(t)$ and $S^{\underline{M}}_{\underline{L}}(t)$ measure the strength of the correlation between the first and second moments of the mode amplituds. The third cumulants $S^{\underline{P}}_{\underline{LM}}(t)$ measures the coupling of the third moments to the first two, etc. It may seem that we have used a round-about way of replacing the intractable set of mode rate equations (18.21) by an equally intractable set of transport equations (11); however, this is not entirely true. We can gain additional information about the system by examing the asymptotic behavior of the terms in (11) as we shall see.

In the approximations discussed in lectures 21 and 23 the steady state probability density was determined to have the form

$$P_{ss}(\underline{b}) = \frac{1}{(2\pi)^{n/2}\|\underline{\underline{Q}}_{ss}\|^{1/2}} \exp\left\{-\frac{1}{2}\, b^{+}\underline{\underline{Q}}_{ss}^{-1}\,\underline{b}\right\} \quad . \tag{12}$$

In this approximation the second order cumulants, for a zero centered steady state average mode amplitudes, are

$$\lim_{t\to\infty} S_{\underline{L}}^{\underline{M}}(t) = \left\langle |B_{\underline{L}}|^2\right\rangle_{ss} \delta_{\underline{L}-\underline{M}} \tag{13}$$

and

$$\lim_{t\to\infty} S_{\underline{LM}}(t) = 0 \ . \tag{14}$$

Note that in general (14) does not apply even for a Gaussian distribution. It is true here because of the particular form of the Fokker-Planck equation we assumed (21.23). Similarly the third order cumulants asymptotically become

$$\lim_{t\to\infty} S_{\underline{LMP}}(t) = 0 \tag{15}$$

as can be confirmed by direct calculation, i.e., by using the steady state probability density (12) to evaluate the terms in (9c). Near the statistical steady state we replace the cumulants higher than first in (11) by their steady state values and obtain

$$\frac{\partial}{\partial t}\, S_{\underline{K}}(t) + \gamma(\underline{K})\, S_{\underline{K}}(t) \cong 2i\sum_{\underline{L}} \Gamma\frac{\underline{KL}}{\underline{KL}}\, S_{\underline{L}}^{\underline{L}}\, S_{\underline{K}}(t) + \sum_{\underline{L},\underline{M},\underline{P}}^{R} i\Gamma\frac{\underline{KP}}{\underline{LM}}\, S_{\underline{L}}(t)\, S_{\underline{M}}(t)\, S^{\underline{P}}(t) \tag{16}$$

or introducing the frequency shift

$$-\delta\omega_{\underline{K}} \equiv 2\sum_{\underline{L}} \Gamma\frac{\underline{KL}}{\underline{KL}} \left\langle |B_{\underline{L}}|^2\right\rangle_{ss} \tag{17}$$

we obtain from (16)

$$\frac{\partial S_{\underline{K}}(t)}{\partial t} + (\gamma(\underline{K}) + i\delta\omega_{\underline{K}})\, S_{\underline{K}}(t) = \sum_{\underline{L},\underline{M},\underline{P}}^{R} i\Gamma\, \frac{\underline{KP}}{\underline{LM}}\, S_{\underline{L}}(t)\, S_{\underline{M}}(t)\, S_{\underline{P}}^{*}(t). \tag{18}$$

Note that (18) has the same *form* as the original hydrodynamic equations for the mode amplitudes $B_{\underline{K}}(t)$ in the absence of all the driving terms. The average effect of these drivers appears in the linear coefficient of (18); both in the change in growth properties and the shift in frequency due to the average nonlinear interactions. Thus, near the steady state the original fluid dynamic equations provide a good description of the *deterministic* evolution of the gravity wave field with an effective dissipation and frequency. There is still a question as to whether the average mode amplitudes are statistically stationary in time.

Consider another illustrative situation; one in which the turbulent wind has been gusting over a sufficiently long fetch that both the unresolved and resolved waves have come to their steady state levels. The mode amplitudes of this state are therefore given by $P_{ss}(\underline{b})$. At time t=0 we assume the wind field is switched to a new steady state level so the external supply of energy to support the water wave field is modified. The initial state of the wave field is therefore described by $P_{ss}(\underline{b}_o)$. We are interested in how the average mode amplitude $S_{\underline{K}}(t)$ evolves from this non-equilibrium initial state to its final steady state. The second cumulants at t=0 are thus,

$$S_{\underline{L}}^{\underline{M}}(0) = \langle |B_{\underline{L}}|^2 \rangle_{ss} \, \delta_{\underline{L}\underline{M}} \tag{19a}$$

$$S_{\underline{L}\underline{M}}(0) = 0 \tag{19b}$$

and the third cumulant is

$$S_{\underline{L}\underline{M}}^{P}(0) = 0 \tag{19c}$$

as are all the higher cumulants because of the Gaussian nature of the initial steady state distribution.

In the region of the prescribed steady state, i.e., near time $t = 0$, we insert the expression (19) into (11) and again obtain the transport equation (18) to describe the approximate evolution of the field. The correlation among pairs and triads of modes is, therefore, not important for the initial development of the average mode amplitude away from this particular initial state. These correlations were also found not to be important in the relaxation of the average mode amplitude in the neighborhood of its final steady state value (at least in this particular Fokker-Planck approximation). Given this dynamic behavior at the time extremes of the sea wave field evolution, one is tempted to speculate that (18) is approximately correct throughout the evolution of the system. However, it is shown by adopting an argument due to Zwanzig, Nordholm and Mitchell[237], that the initial fluctuations build up and then decay as the wave field evolves from its non-equlibrium initial state toward its final steady state value. It should be noted that although (11) is an "exact" description of the gravity wave field dynamics, that (18) is only valid near $t = 0$ and $t = \infty$ and the latter extreme only in the Fokker-Planck approximation, since it is in this approximation that the cumulants $S^{\underline{P}}_{\underline{L}\underline{M}}(t)$, $S_{\underline{L}\underline{M}}(t)$ and $S^{\underline{M}}_{\underline{L}}(t)$ have the nice properties (13)-(15).

We should also consider the fact that we neglected the coherent pressure fluctuations and set $\bar{D}_2(\underline{K}) \cong D_{\underline{K}}$ to obtain the Gaussian distribution (21.29). These assumptions are not consistent with the transport equations (3) and (6). There is, of course, no inconsistency for the $n = 1$ transport equations from the second moment onward however there is such an inconsistency.

25. Non-Homogeneous Transport Equation

In the Fokker-Planck description of the evolution of the gravity wave field the correlation time between fluctuations in the driving force is assumed to be vanishingly small, i.e., to be delta correlated in time. This implies that the process is Markoffian, that is, the evolution of the state variables are independent of earlier times. The Markoff approximation gives rise to the near steady state description of the evolution of the average mode amplitudes in the sea wave field (24.18). The mean mode amplitude is of interest in some contexts, but for $\gamma_R(\underline{K}) > 0$ it has an asymptotic state of zero except for the smallest $\underline{K}$, i.e., a drift current induced at the surface by the nonlinear interactions, the so-called Stokes' drift.[106,107] To describe the evolution of the wave field in greater detail we use a closure approximation and examine how this truncation of the moment hierarchy of equations affects the properties of the field.

In (24.3) we can see how the *nth* moment of the mode amplitude is coupled to (n - 2)-moment and the (n + 3)-moment; thus forming an infinite hierarchy of coupled moment equations. Truncating this hierarchy by *arbitrarily* expressing the (n + 3)-moments in terms of lower order moments is referred to as closure. The so-called quasi-Gaussian approximation used by a number of authors[11,103,109,64,159] to construct transport equations for the energy spectral density of the surface displacement is a second-order closure. This quantity is of interest since, as discussed in Lecture 14, the mean square surface displacement is given by

$$\langle \zeta^2(\underline{x},t) \rangle = \sum_{\underline{K}} F(\underline{K},t) = \sum_{\underline{K}} \frac{1}{V_K} \langle |B_{\underline{K}}(t)|^2 \rangle_{ss} \qquad (1)$$

In this approximation all cumulants of higher order than the second vanish identically. The general validity of this approximation has been argued for in the limit of vanishingly small nonlinearities by Benney and Saffman[165] and in the limit of very long times by Benney and Newell[166] among others.[162,167]

We will not be concerned further with the evoution of the mean mode amplitude of the gravity wave field here. This is because the resonant nonlinearities enter at third order in the dynamic equations for this field. For the gravity-capillary wave field the resonant nonlinearities enter at second order and introduce a more subtle effect. For this latter field a truncation of the transport equation for the average mode amplitude leads to an equation in which the dissipation parameter is replaced by a memory kernel. The argument establishing this result will not be presented here but it parallels the discussion given by Zwanzig, Nordholm and Mitchell[237] constructed for the transport equation for a turbulent velocity field. The Langevin description of the fluid flow has quadratic nonlinearities as does the gravity-capillary field. Thus although the details of the process differ, the result of truncating the hierarchy of transport equations at second order leads to a non-Markoffian description of the evolution of the average mode amplitude. Pursuing this discussion would lead us too far afield so we return to our investigation of the transport properties of the gravity wave field.

In Lecture 23 we briefly discussed the transport equation for the action density of the gravity waves assuming a statistically homogeneous wave field. Let us now consider the more general case of a field with non-homogeneous statistics. We construct the transport equation by multiplying the Fokker-Planck equation (20.32) on the left by the quantity $B_{\underline{L}} B^*_{\underline{M}}$, and integrate over phase space to obtain

$$\frac{\partial}{\partial t}\langle B_{\underline{L}} B^{*}_{\underline{M}};t\rangle + [\gamma(\underline{L}) + \gamma^{*}(\underline{M})]\langle B_{\underline{L}} B^{*}_{\underline{M}};t\rangle = \langle B_{\underline{L}} T^{*}_{\underline{M}};t\rangle + \langle B^{*}_{\underline{M}} T_{\underline{L}};t\rangle . \tag{2}$$

The transport equation (2) has a more direct physical interpretation if we introduce a spatially dependent energy spectral density by the expression

$$F(\underline{K},\underline{x},t) \equiv \sum_{\underline{\rho}} \frac{e^{i\underline{\rho}\cdot\underline{x}}}{[V_{\underline{K}+\underline{\rho}/2} V_{\underline{K}-\underline{\rho}/2}]^{1/2}} \langle B_{\underline{K}+\underline{\rho}/2} B^{*}_{\underline{M}-\underline{\rho}/2};t\rangle \tag{3}$$

where we have written $\underline{L} = \underline{K} + \underline{\rho}/2$ and $\underline{M} = \underline{K} - \underline{\rho}/2$.

The distribution (3) was used first in a quantum mechanical context by Wigner[239] to represent the overlap between two wave packets. Watson and West[103] and Willebrand[160] and later Alber[240] used (3) to represent a non-homogeneous spectrum of surface gravity waves. In each of these studies the gravity waves themselves were assumed to be in a statistical steady state so that the quasi-Gaussian approximation could be used to truncate the hierarchy of equations. The spatial dependence arose as a consequence of a modulation of this wave spectrum by an external non-uniform current. Here we do not assume the gravity wave field to be in a steady state, so that the spatial dependence is due to the fact that we are measuring the wave field at a finite fetch. Said differently, the wave field is in the process of evolving in both space and time and the energy in the spectral interval $(\underline{K}, \underline{K} + d\underline{K})$ has not achieved its steady state level at location $\underline{x}$ and time t. Note that we have obtained this transport equation by examining how the non-equilibrium probability density determined by the Fokker-Planck equation evolves.

We use the non-homogeneous spectral density (3) in (2) by introducing the centered wave numbers and multiplying the latter equation by $e^{i\underline{\rho}\cdot\underline{x}}$ and summing over $\underline{\rho}$ to obtain

$$\frac{\partial}{\partial t} F(\underline{K},\underline{x},t) + \sum_{\underline{\rho}} [\gamma(\underline{K} + \underline{\rho}/2) + \gamma^*(\underline{K} - \underline{\rho}/2)] \langle C_{\underline{K}+\underline{\rho}/2}\, C^*_{\underline{K}-\underline{\rho}/2};t\rangle e^{i\underline{\rho}\cdot\underline{x}}$$

$$= i \sum_{\underline{L}\,\underline{M}\,\underline{Q}\,\underline{\rho}}^{R} \left\{ \bar{\Gamma}^{\underline{K}+\underline{\rho}/2,\underline{Q}}_{\underline{L}\underline{M}} \langle C_{\underline{L}} C_{\underline{M}} C^*_{\underline{Q}} C_{\underline{K}-\underline{\rho}/2};t\rangle \right.$$

$$\left. - \bar{\Gamma}^{\underline{K}-\underline{\rho}/2,\underline{Q}}_{\underline{L}\underline{M}} \langle C^*_{\underline{L}} C^*_{\underline{M}} C_{\underline{Q}} C_{\underline{K}+\underline{\rho}/2};t\rangle \right\} e^{i\underline{\rho}\cdot\underline{x}} . \qquad (4)$$

For ease of writing we have introduced the normalized quantities into (4)

$$C_{\underline{L}} = B_{\underline{L}}/\sqrt{V_L} , \qquad (5)$$

$$\bar{\Gamma}^{\underline{K}\underline{Q}}_{\underline{L}\underline{M}} = \left(\frac{V_L V_M V_Q}{V_K}\right)^{1/2} \Gamma^{\underline{K}\underline{Q}}_{\underline{L}\underline{M}} . \qquad (6)$$

The linear term in (4) is the simplest to handle. We merely expand γ in a Taylor series about $\underline{K}$ making the simplifying assumption that the imaginary part of $\gamma(\underline{K})$ is given by the linear frequency ω_K. Thus

$$[\gamma_I(\underline{K} + \underline{\rho}/2) - \gamma_I(\underline{K} - \underline{\rho}/2.] = \omega_{|\underline{K}+\underline{\rho}/2|} - \omega_{|\underline{K}-\underline{\rho}/2|}$$

$$\cong \underline{\rho} \cdot \underline{V}_g(\underline{K}) + \frac{V_g(\underline{K})}{24K^2} \sum_{i\,j\,\ell}$$

$$\times \left\{ \frac{21}{4} \hat{\underline{K}}_i \cdot \hat{\underline{K}}_j\, \hat{\underline{K}}_\ell - \frac{3}{2} (\hat{K}_\ell \delta_{ij} + \hat{\underline{K}}_j\, \delta_{i\ell} + \hat{\underline{K}}_\ell\, \delta_{j\ell}) \right\}$$

$$\times \quad \rho_i \rho_j \rho_\ell \qquad (7)$$

where $\hat{\underline{K}}_\ell$ is a unit vector in the *ℓth* direction and ρ_ℓ is the *ℓth* component of $\underline{\rho}$. Using the relation between wavevectors in $\underline{K}$-space and operators in $\underline{x}$-space simply expressed for the function $F(\underline{x})$,

$$F(\underline{x}) = \int e^{i\underline{K}\cdot\underline{x}}\, \tilde{F}(\underline{K}) d^2K \tag{8}$$

by

$$-i\nabla_{\underline{x}} F(\underline{x}) = \int e^{i\underline{K}\cdot\underline{x}}\, \underline{K}\tilde{F}(\underline{K}) d^2K \tag{9}$$

we can replace the wavevector components ρ_j in (7) by $-i(\partial/\partial x_j)$ in configuration space. Thus we rewrite (4) as

$$\left[\frac{\partial}{\partial t} + \underline{V}_g(K)\cdot\nabla_{\underline{x}} + 2\,\gamma_R(\underline{K})\right] \quad F(\underline{K},\underline{x},t) = \frac{V_g(\underline{K})}{16K^2}\left[\frac{7}{2}\,(\underline{K}\cdot\nabla_{\underline{x}})^3 - 3\nabla_{\underline{x}}^2\underline{K}\cdot\nabla_{\underline{x}}\right] F(\underline{K},\underline{x},t)$$

$$+ \text{ nonlinear term (4)} \tag{10}$$

where we have kept only the first term from the expansion of the real part of $\gamma(\underline{K})$.

We now examine the nonlinear term from (4). This is the first time in this analysis that we need to decide which of the approximate distributions we are going to average over in evaluating the nonlinear terms. We choose the linearized distribution (23.1) so that we can factor the fourth order moments in (4). First we introduce the notation for the deviation of the mode amplitude from its steady state value

$$\hat{C}_{\underline{L}} \equiv C_{\underline{L}} - \langle C_{\underline{L}}\rangle_M = C_{\underline{L}} - [e^{-\underline{\underline{M}}t}\, \underline{C}(0)]_{\underline{L}} \tag{10}$$

so that the distribution (23.1) can be written

$$P_M(\underline{c},t|\underline{c}_0) = \frac{1}{(2\pi)^{n/2}\|\underline{\underline{Q}}'(t)\|^{1/2}} \exp\left\{-\frac{1}{2}\,\hat{\underline{c}}^+\underline{\underline{Q}}'^{-1}(t)\hat{\underline{c}}\right\} \tag{11}$$

where the prime on the correlation matrix indicates the change in normalization in going from the $\underline{b}$ to $\underline{c}$ variables. In terms of these difference variables we can write the fourth moments in (4) as

$$\langle c_{\underline{L}} c_{\underline{M}} c^*_{\underline{Q}} c^*_{\underline{K}} \rangle_M = \langle \hat{c}_{\underline{L}} \hat{c}_{\underline{M}} c^*_{\underline{Q}} c^*_{\underline{K}} \rangle_M + \langle \hat{c}_{\underline{L}} \hat{c}^*_{\underline{K}} \rangle_M \langle c_{\underline{M}} \rangle_M \langle c^*_{\underline{Q}} \rangle_M$$
$$+ \langle \hat{c}_{\underline{L}} \hat{c}^*_{\underline{Q}} \rangle_M \langle c_{\underline{M}} \rangle_M \langle c^*_{\underline{K}} \rangle_M + \langle \hat{c}_{\underline{M}} \hat{c}^*_{\underline{K}} \rangle_M \langle c_{\underline{L}} \rangle_M \langle c^*_{\underline{Q}} \rangle_M$$
$$+ \langle c_{\underline{L}} \rangle_M \langle c_{\underline{M}} \rangle_M \langle c^*_{\underline{Q}} \rangle_M \langle c^*_{\underline{K}} \rangle_M \tag{12}$$

so that using the zero-centered Gaussian distribution (11) we obtain

$$\langle c_{\underline{L}} c_{\underline{M}} c^*_{\underline{Q}} c^*_{\underline{K}} \rangle_M = \langle \hat{c}_{\underline{L}} \hat{c}^*_{\underline{Q}} \rangle_M \langle \hat{c}_{\underline{L}} \hat{c}^*_{\underline{K}} \rangle_M + \langle \hat{c}_{\underline{L}} \hat{c}^*_{\underline{K}} \rangle_M \langle \hat{c}_{\underline{M}} \hat{c}^*_{\underline{Q}} \rangle$$
$$+ \langle \hat{c}_{\underline{L}} \hat{c}^*_{\underline{K}} \rangle_M \langle c_{\underline{M}} \rangle_M \langle c^*_{\underline{Q}} \rangle_M + \langle \hat{c}_{\underline{L}} \hat{c}^*_{\underline{Q}} \rangle_M \langle c_{\underline{M}} \rangle_M \langle c^*_{\underline{K}} \rangle_M$$
$$+ \langle \hat{c}_{\underline{M}} \hat{c}^*_{\underline{K}} \rangle_M \langle c_{\underline{L}} \rangle_M \langle c^*_{\underline{Q}} \rangle_M + \langle c_{\underline{L}} \rangle_M \langle c_{\underline{M}} \rangle_M \langle c^*_{\underline{Q}} \rangle_M \langle c^*_{\underline{K}} \rangle_M . \tag{13}$$

Using the symmetry properties of the coupling coefficients and assuming that the average mode amplitude is homogeneous in space we can rewrite the nonlinear term in (4) as

$$i \sum_{\underline{L},\underline{M},\underline{Q},\underline{\rho}}^{R} \left\{ \bar{\Gamma}_{\underline{L}\underline{M}}^{\underline{K}+\underline{\rho}/2,\underline{Q}} \langle c_{\underline{L}} c_{\underline{M}} c^*_{\underline{Q}} c^*_{\underline{K}-\underline{\rho}/2} \rangle_M - \bar{\Gamma}_{\underline{L}\underline{M}}^{\underline{K}-\underline{\rho}/2,\underline{Q}} \langle c^*_{\underline{L}} c^*_{\underline{M}} c_{\underline{Q}} c_{\underline{K}+\underline{\rho}/2} \rangle_M \right\} e^{i\underline{\rho}\cdot\underline{x}}$$
$$\cong i \sum_{\underline{L}\,\underline{M}\,\underline{Q}\,\underline{\rho}} \left\{ \langle c_{\underline{L}} c^*_{\underline{Q}} \rangle_M \, 2\bar{\Gamma}_{\underline{L}\underline{M}}^{\underline{K}+\underline{\rho}/2,\underline{Q}} \langle \hat{c}^*_{\underline{M}} \hat{c}_{\underline{K}+\underline{\rho}/2} \rangle_M \right.$$
$$\left. - \langle c^*_{\underline{L}} c_{\underline{Q}} \rangle_M \, 2\bar{\Gamma}_{\underline{L}\underline{M}}^{\underline{K}-\underline{\rho}/2,\underline{Q}} \langle \hat{c}^*_{\underline{M}} \hat{c}_{\underline{K}+\underline{\rho}/2} \rangle_M \right\} e^{i\underline{\rho}\cdot\underline{x}} . \tag{14}$$

We introduce the normalization

$$\frac{1}{\Sigma_0}\sum_{\underline{\rho}'} e^{i\underline{\rho}'\cdot(\underline{x}-\underline{y})} = \delta(\underline{x}-\underline{y}) \tag{15}$$

where Σ_0 is the two-dimensional area of our gravity wave field, and write the inverse of (3) as

$$\langle C_{\underline{K}+\underline{\rho}'/2} C^*_{\underline{K}-\underline{\rho}'/2}\rangle_M = \frac{1}{\Sigma_0}\int d^2y\; e^{-i\underline{\rho}'\cdot\underline{y}}\, F(\underline{K},\underline{y},t) \tag{16}$$

so that (14) can be expressed as

$$(14) = \frac{2i}{\Sigma_0}\sum^{R}_{\underline{K}\,\underline{M}\,\underline{\rho}\,\underline{\rho}'}\int d^2y\; e^{-i\underline{\rho}'\cdot\underline{y}}\, F(\underline{K},\underline{y},t)\left\{\bar{\Gamma}^{\underline{K}+\underline{\rho}/2,\underline{Q}-\underline{\rho}'/2}_{\underline{Q}+\underline{\rho}'/2,\underline{K}}\,\langle\hat{C}_{\underline{M}}\hat{C}^*_{\underline{K}-\underline{\rho}/2}\rangle_M\right.$$

$$\left. - \bar{\Gamma}^{\underline{K}-\underline{\rho}/2,\underline{Q}+\underline{\rho}'/2}_{\underline{Q}-\underline{\rho}'/2,\underline{M}}\,\langle\hat{C}^*_{\underline{M}}\hat{C}_{\underline{K}+\underline{\rho}/2}\rangle_M\right\} \tag{17}$$

where the Kronecker delta for the resonant interactions fixes the value of $\underline{M}$ to be $\underline{K} + \underline{\rho}/2 \mp \underline{\rho}'$.

The final approximation we make here is to assume that the energy spectral density for $\hat{\underline{C}}$ does not differ significantly from that for $\underline{C}$ since they are identical in the steady state. Therefore using (16) again, we obtain

$$(14) = \frac{2i}{\Sigma_0^2}\sum_{\underline{Q}}\int d^2u\; F(\underline{Q},\underline{u})\int d^2v \sum_{\underline{\rho},\underline{\rho}'} e^{i\underline{\rho}\cdot(\underline{x}-\underline{u})}\, e^{i\underline{\rho}'\cdot(\underline{u}-\underline{v})}$$

$$\times\left\{\bar{\Gamma}^{\underline{K}+\underline{\rho}/2,\underline{Q}-\underline{\rho}'/2}_{\underline{Q}+\underline{\rho}'/2,\underline{K}+\underline{\rho}/2-\underline{\rho}'} F(\underline{K}-\underline{\rho}/2,\underline{v}) - \bar{\Gamma}^{\underline{K}-\underline{\rho}/2,\underline{Q}+\underline{\rho}'/2}_{\underline{Q}-\underline{\rho}'/2,\underline{K}-\underline{\rho}/2-\underline{\rho}'}\, F(\underline{K}+\underline{\rho}/2,\,\underline{v})\right\}. \tag{18}$$

We expect the wavevectors $\underline{\rho}$ and $\underline{\rho}'$ to have magnitudes of order $\Sigma_0^{-1/2}$ since the correlation between mode amplitudes vanish outside a narrow band of wave numbers, i.e., $\langle \hat{C}_{\underline{K}+\underline{\rho}/2} \hat{C}^*_{\underline{K}-\underline{\rho}/2} \rangle \cong 0$ for $|\underline{\rho}| >> \Sigma_0^{-1/2}$. We thus expect[103] that both $\underline{\rho}$ and $\underline{\rho}'$ are very small in magnitude compared with $\underline{K}$ and $\underline{Q}$. This implies that if one measures the energy of a given wave at a location in space, that in a few wavelengths the energy in that spectral interval will be completely uncorrelated from the initial measurement. We can, therefore, expand the coupling coefficients in (18) as well as the spectral densities in terms of $\underline{\rho}$ and $\underline{\rho}'$. We define the vectors $\underline{D}_1$ and $\underline{D}_2$ by the equations

$$\underline{D}_1 \equiv \nabla_{\underline{K}} \bar{\Gamma}^{KQ}_{\underline{KQ}} \tag{19a}$$

$$\underline{\rho}' \cdot \underline{D}_2 \equiv \lim_{\underline{\rho}'\to 0} \left\{ \bar{\Gamma}^{K,Q+\rho'/2}_{\underline{Q}-\underline{\rho}'/2,\underline{K}+\underline{\rho}'} - \Gamma^{K,Q-\rho'/2}_{\underline{Q}+\underline{\rho}'/2,\underline{K}-\underline{\rho}'} \right\} \tag{19b}$$

and keeping only the first order terms in the expansion of (18) we find that

$$(14) = -\sum_{\underline{Q}} \left\{ \bar{\Gamma}^{\underline{KQ}}_{\underline{KQ}} [\nabla_{\underline{x}} F(\underline{Q},\underline{x})]\cdot[\nabla_{\underline{K}} F(\underline{K},x)] \right.$$
$$\left. + F(\underline{K},\underline{x})[\underline{D}_2\cdot\nabla_x F(\underline{Q},\underline{x}) - \underline{D}_1\cdot\nabla_{\underline{x}} F(\underline{Q},\underline{x}) F(\underline{K},\underline{x})] \right. . \tag{20}$$

We can now express the time derivative of the spectral density in terms of itself in (10), i.e., close the equation.

We first express the discrete spectral density $F(\underline{K},\underline{x})$ in terms of the continuum spectral density $\Phi(\underline{K},\underline{x})$, i.e,

$$\Phi(\underline{K},\underline{x}) = \frac{(2\pi)^2}{\Sigma_0} F(\underline{K},\underline{x})$$

and introduce the distance parallel to $\hat{\underline{K}}$ as $s = \underline{K}\cdot\underline{x}$ to rewrite (10) in continuum form as

$$\left(\frac{\partial}{\partial t} + \frac{d\underline{x}}{dt}\cdot\nabla_{\underline{x}} + \frac{d\underline{K}}{dt}\cdot\nabla_{\underline{K}}\right)\Phi(\underline{K},\underline{x},t) + 2\gamma_R(\underline{K})\Phi(\underline{K},\underline{x},t)$$

$$= S(\underline{K},\underline{x})\Phi(\underline{K},\underline{x},t) - \frac{3}{16K^2}\frac{dx}{dt}\frac{\partial}{\partial s}\left(\nabla_\perp^2 - \frac{7}{6}\frac{\partial^2}{\partial s^2}\right)\Phi(\underline{K},\underline{x},t) \; . \qquad (21)$$

In (21) we have used the ray equations of Hamilton[48]

$$\frac{d\underline{x}}{dt} = \nabla_{\underline{K}}\Omega$$

$$\frac{d\underline{K}}{dt} = -\nabla_{\underline{x}}\Omega \qquad (22)$$

where the Hamiltonian has been replaced by the nonlinear dispersion relation

$$\Omega \equiv \omega_K - \int d^2Q \; \bar{\Gamma}^{KQ}_{\underline{KQ}} \; \Phi(\underline{Q},\underline{x},t) \qquad (23)$$

and the non-homogeneous coefficient by

$$S(\underline{K},\underline{x}) \equiv \int (\underline{D}_1 - \underline{D}_2)\cdot\nabla_{\underline{x}}\Phi(\underline{Q},\underline{x},t)\; d^2\underline{Q} \; . \qquad (24)$$

Equation (22) describes the paths along which energy propagates due to the change in the dispersion relation, i.e., they are the ray equations of geometrical optics. An equation similar to (22) has been previously used by Whitham[241], Kenyon[242] and Watson and West[103] to include coherent terms such as would be produced by a surface current $\underline{U}(\underline{x},t)$, i.e., $\Omega = \omega_K + \underline{K}\cdot\underline{U}(\underline{x},t)$. The refraction of the gravity wave spectrum by surface currents is included in the spectral density $\Phi(\underline{K},\underline{x},t)$ in (23) by means of a monochromatic wave of given strength under the integral, i.e.,

$$\Phi\ (\underline{Q},\underline{x},t) = \delta(\underline{Q}-\underline{K}_o)\ \underline{U}_o \cos\ (\underline{Q}\cdot\underline{x} - \omega_k t) \tag{25}$$

for the current produced by a swell of wavevector $\underline{K}_o$ and constant amplitude $\underline{U}_o$. The other terms in the integral which are incoherent, represent the influence of nonlinear wave interactions on the refraction and propagation of the gravity wave $\underline{Q}$.

The derivative terms obtained by expanding the linear frequency term represent the effects of diffraction. In general the derivative transverse to the direction of propagation is much greater than that parallel, so we neglect the $\frac{\partial^3}{\partial s^3}$ term relative to the $\frac{\partial}{\partial s}\ \nabla_\perp^2$ term in (21).

The first term in (21) can be put in a more familiar form by including coherent terms such as swell in the long wavelength spectrum. When this is done (24) can be replaced by

$$S(\underline{K},\underline{x}) = \nabla_{\underline{x}} \cdot \left\{-\underline{K}(\underline{K}\cdot\underline{U})/2K^2 + \int d^2Q\ (\underline{D}_1-\underline{D}_2)\ \Phi(\underline{Q},\underline{x},t)\right\} \tag{26}$$

where $\underline{U}$ is the external current. The coherent part of (26) represents the "radiation stress" introduced by Longuet-Higgins and Stewart.[243] This term may be transformed away if Φ is replaced by the action spectrum

$$J(\underline{K},\underline{x}) = \frac{2}{g}\ |\underline{V}_g(\underline{K})|\ \Phi\ (\underline{K},\underline{x}) \tag{27}$$

where $\underline{V}_g(\underline{K}) = \nabla_{\underline{K}}\omega_K$ is the linear group velocity. Substitution of (27) into (21) leads to the equation

$$\left(\frac{\partial}{\partial t} + \frac{d\underline{x}}{dt} \cdot \nabla_{\underline{x}} + \frac{d\underline{k}}{dt} \cdot \nabla_{\underline{K}}\right) J(\underline{K},\underline{x}) + \gamma_R(\underline{K}) J(\underline{K},\underline{x}) = \hat{S}(\underline{K},\underline{x}) J(\underline{K},\underline{x})$$

$$- \frac{1}{16K^2} V_g(\underline{K}) \frac{\partial}{\partial s} \nabla^2 J(\underline{K},\underline{x}) \qquad (28)$$

with

$$\hat{S}(\underline{K},\underline{x}) = \nabla_{\underline{x}} \cdot \int d^2Q \left[\underline{D}_1 - \underline{D}_2 - \frac{\underline{K}}{2K^2} C^{\underline{K},\underline{Q}}_{\underline{K},\underline{Q}} \right] \Phi(\underline{Q},\underline{x},t) \qquad (29)$$

There are a number of features of (28) that are worth stressing. First it is of lower order in the nonlinearity than the intro-differential equation constructed in lecture 23 and is probably of greater importance for describing the evolution of the wave field than the higher order terms in (23.25). Secondly, there is a cross spectral coupling in the wave frequency leading to both a refraction and diffraction of the energy spectral density. Finally, there is a modification of the radiation stress term due to the incoherent interactions with the ambient spectrum of waves.

26. Example Calculations

Here we present some simple test calculations to provide insight into the complicated non-homogeneous transport equation (25.28). In particular we examine the effects of refraction and diffraction in this expression. The group velocity of the spectral density in a wavevector interval ($\underline{k}, \underline{k} + d\underline{k}$) is obtained from (25.22) and (25.23) to be

$$\frac{d\underline{x}}{dt} = \underline{V}_g(K) + \underline{U}_s(\underline{x},t) - \int d^2K \; \underline{D}_1 \; \Phi(\underline{K},\underline{x},t) \tag{1}$$

where we have included a swell generated coherent current. The first term in (1) is the group velocity of a linear wave $\nabla_{\underline{k}} \, \omega_k$; the second is the current produced by a swell passing under the wave of interest and the third term is the influence of the nonlinear interactions of the wave of interest with the fluctuating ambient field of waves. The ambient wave spectrum will in general be asymmetrical owing to the non-constant influence of the wind and the occurrence of obstructions along the path of evolution. Thus, the total group velocity can have a component induced by the nonlinear interactions which is not parallel to $\underline{k}$.

An alternate derivation of the nonlinear term in (1) may prove to be useful. Consider a 'test wave' interacting with a spectrum of ocean waves in a statistically uniform ocean in the absence of swell. This model has its origin in kinetic theory, but has been shown to be of value in an oceanographic context by a number of investigators[169,244]. We imagine the test wave to be mechanically generated with identical characteristics for each of a sequence of observations. We then write

$$B_{\underline{\ell}} = B_{\underline{\ell}}^{(o)} + \delta B_{\underline{k}}\ \delta_{\underline{\ell}-\underline{k}} \tag{2}$$

where $B_{\underline{\ell}}^{(o)}$ is a random variable describing the ambient sea and $\delta B_{\underline{k}}$ represents a small amplitude test wave. We substitute (2) into the mode rate equation (18.21) and obtain

$$\dot{B}_{\underline{k}}^{(o)} + \delta\dot{B}_{\underline{k}} + i\omega_K\left[B_{\underline{k}}^{(o)} + \delta B_{\underline{k}}\right] = i\sum_{\underline{\ell}\,p\,\underline{n}}^{R} \Gamma_{\underline{\ell p}}^{\underline{kn}} \Big\{ B_{\underline{\ell}}^{(o)} B_{\underline{p}}^{(o)} B_{\underline{n}}^{(o)^*}$$

$$+ \delta B_{\underline{k}}^{*}\, B_{\underline{\ell}}^{(o)} B_{\underline{p}}^{(o)} \delta_{\underline{n}-\underline{k}} + \delta B_{\underline{k}} B_{\underline{p}}^{(o)} B_{\underline{n}}^{(o)} \delta_{\underline{\ell}-\underline{k}}$$

$$+ \delta B_{\underline{k}}\, B_{\underline{\ell}}^{(o)} B_{\underline{n}}^{(o)} \delta_{\underline{p}-\underline{k}} + O(\delta B^2) \Big\} \tag{3}$$

where we have turned off the wind and allowed sufficient time to elapse that the high frequency waves have damped to zero. Removing the equation for the $B_{\underline{k}}^{(o)}$ terms we have the linear expression for the test wave

$$\delta\dot{B}_{\underline{k}} + i\omega_k\, \delta B_{\underline{k}} = i \sum_{\underline{\ell}\,\underline{p}\,\underline{n}}^{R} \Gamma_{\underline{\ell p}}^{\underline{kn}} \Big\{ \delta B_{\underline{k}}^{*}\, B_{\underline{\ell}}^{(o)} B_{\underline{p}}^{(o)} \delta_{\underline{n}-\underline{k}} + \delta B_{\underline{k}} B_{\underline{\ell}}^{(o)} B_{\underline{n}}^{(o)} \delta_{\underline{p}-\underline{k}}$$

$$+ \delta B_{\underline{k}}\, B_{\underline{p}}^{(o)} B_{\underline{n}}^{(o)^*} \delta_{\underline{\ell}-\underline{n}} \Big\} \quad . \tag{4}$$

We now average the $\underline{B}^{(o)}$ term over a steady state ensemble of fluctuations, i.e., we assume the ambient sea to have reached a steady state with the wind field at time t = 0 when we shut off the wind and generate the test wave. For simplicity, we assume that the test wave and the ambient field are statistically independent, so that using $P_{ss}(\underline{b})$ given by (21.29), we obtain

$$\frac{d}{dt}\,\delta B_{\underline{k}} + i\left\{\omega_k - \sum_{\underline{K}} 2\Gamma_{\underline{kK}}^{\underline{kK}} \left\langle |B_{\underline{K}}^{(o)}|^2 \right\rangle_{ss}\right\} \delta B_{\underline{k}} = 0 \tag{5}$$

In the continuum limit the frequency in (5) is given by

$$\Omega_{\underline{k}} = \omega_k - \int d^2K\, \overline{\Gamma}_{\underline{kK}}^{\underline{kK}}\, \Phi_{ss}(\underline{K}) \tag{6}$$

where $\Phi_{ss}(\underline{K})$ is the steady state continuum ambient wave spectrum. It is clear that $\delta B_{\underline{k}}$ varies harmonically in time with frequency $\Omega_{\underline{k}}$.

To illustrate the implications of (6) we consider the modification in the frequency of waves very much shorter than the wind generated spectral peak ($\underline{K}_0$), i.e., $k \gg K_0$. The dominant contribution to the integral in this case comes from the spectral peak so we use the Phillips equilibrium spectrum

$$\begin{aligned}\Phi(\underline{K}) &= \frac{.004}{2\pi K^4}\,(1 + \hat{\underline{K}}\cdot\hat{\underline{W}})\;; \qquad K_0 \le K \le k_\gamma \\ &= 0 \qquad K < K_0\;,\quad K > k_\gamma\end{aligned} \tag{7}$$

to evaluate (6). In (7), K_o and k_γ are the long and short wavelength cutoffs of the spectrum, respectively, and $\hat{\underline{W}}$ is a unit vector in the direction the wind is blowing. The coupling coefficient $\Gamma_{\underline{k},\underline{K}}^{\underline{k},\underline{K}}$ for k >> K is found to be

$$\Gamma_{\underline{kK}}^{\underline{kK}} \cong -\frac{7}{4}\,\underline{k}\cdot\underline{K}\,\omega_k \quad . \tag{8}$$

Substituting (8) into (6) yields

$$\Omega_{\underline{k}} = \omega_k + \frac{.007}{2\pi}\,k\,\sqrt{g}\int_{K_o}^{k_\gamma}\int_{0}^{2\pi}\frac{dK}{K^{3/2}}\,d\theta_K\,\cos(\theta_K-\theta_k)\left[1+\cos(\theta_K-\theta_W)\right] \tag{9}$$

which integrates to

$$\Omega_{\underline{k}} = \omega_k\left[1 + .007\left(\frac{k}{K_o}\right)^{\frac{1}{2}}\cos(\theta_k-\theta_W)\right] . \tag{10}$$

The group velocity for the wave is obtained from (10) to be

$$\nabla_{\underline{k}}\Omega_{\underline{k}} = \underline{V}_g(k) + .014\,\cos(\theta_k-\theta_W)\;V_g(\underline{K}_o) \tag{11}$$

where $\underline{V}_g(K_o)$ is the linear group velocity of the spectral peak and is parallel to the wind direction $\hat{\underline{W}}$. Thus the group velocity of the $\underline{k}$-wave changes by $\pm$ .014 $V_g(K_o)$ from traveling with the wind to traveling against the wind.

As another example we study the induced correlation properties of a short gravity wave by a spectrum of long wavelength gravity waves.[245] We use a simple model to evaluate the auto-correlation function $C_{\underline{k}}(\underline{\rho},\tau)$ which is defined as

$$C_{\underline{k}}(\underline{\rho},\tau) \quad \frac{\left\langle B_{\underline{k}+\underline{\rho}/2}(t+\frac{\tau}{2})\, B^*_{\underline{k}-\underline{\rho}/2}(t-\frac{\tau}{2})\right\rangle}{\left\langle |B_{\underline{k}}(t)|^2\right\rangle} \tag{12}$$

The function $C_{\underline{k}}(\underline{\rho},\tau)$ vanishes for times very much greater than the correlation time τ_c and distances very much greater than the correlation distance $d_c \sim 1/\rho_{max}$.

The principle wave-wave interaction we consider here is that of gravity waves of relatively high wavenumber $\underline{k}$ with a set of gravity waves $\underline{K}_1, \underline{K}_2, \ldots$ near the spectral cut-off. We may assume that τ is small compared with the period of the dominant wave and also that $2\pi/\rho$ is small compared with the corresponding wavelength. Thus, we shall be able to express $B_{\underline{k}\pm\underline{\rho}/2}(t \pm \tau/2)$ as functions of $B_{\underline{k}}(t)$ and $B_{\underline{K}_1}(t) \ldots B_{\underline{K}_n}(t)$. To perform the statistical averages indicated in (12) we shall assume that the ambient wave field is in the statistical steady state described in (21.29).

The mode rate equation for the short wavelength test wave $B_{\underline{k}}(t)$ is given by the second order coupling term in (5.16)

$$\dot{B}_{\underline{k}}(t) + i\,\omega_k\,B_{\underline{k}}(t) = \sum_{\underline{\ell},\underline{p}} \delta_{\underline{k}-\underline{\ell}-\underline{p}} \left\{ \Gamma_{\underline{\ell}\,\underline{p}}^{\underline{k}}\,B_{\underline{p}}(t) + \Gamma_{\underline{\ell}}^{\underline{k},-\underline{p}}\,B_{-\underline{p}}^{*}(t) \right\} B_{\underline{\ell}}(t)\ . \tag{13}$$

We now construct a wave packet using

$$B_{\underline{k}}(\underline{x},t) \equiv \sum_{\underline{\rho}(\underline{K}_0)} B_{\underline{k}+\underline{\rho}}(t)\, e^{i\underline{\rho}\cdot\underline{x}} \tag{14}$$

where $\underline{K}_0$ is the peak of the wind generated ambient spectrum and the sum over $\underline{\rho}$ extends over a wavevector interval somewhat greater than $\underline{K}_0$, but very small compared with $\underline{k}$, i.e., $k \gg K_0$. Thus $B_{\underline{k}}(\underline{x},t)$ represents the Fourier amplitude of a wave of wavevector $\underline{k}$ localized on the face of the dominant wave. The $\underline{x}$ - dependence locates the amplitude along the wave. We now multiply the terms in (13) by $e^{i\underline{\rho}\cdot\underline{x}}$ and sum $\underline{\rho}$ over this restricted interval to obtain

$$\dot{B}_{\underline{k}}(\underline{x},t) + i\,\omega_k\,B_{\underline{k}}(\underline{x},t) = \sum_{\underline{K}} \left[2\Gamma_{\underline{k}\,\underline{K}}^{\underline{k}+\underline{K}}\,B_{\underline{K}}(t)e^{i\underline{K}\cdot\underline{x}} + \Gamma_{\underline{k}}^{\underline{k}-\underline{K},\underline{K}}\,B_{\underline{K}}^{*}(t)e^{-i\underline{K}\cdot\underline{x}} \right]$$

$$\times\ B_{\underline{k}}(\underline{x},t) \tag{15}$$

where $\underline{\ell} \approx \underline{k}$ in (13) and $k \gg K$ for those terms of interest in (15). The asymptotic forms of the coupling coefficients are given by

$$\Gamma_{\underline{k}\,\underline{K}}^{\underline{k}+\underline{K}} = -\,\frac{1}{2}\,\Gamma_{\underline{k}}^{\underline{k}-\underline{K},\underline{K}} \cong \frac{1}{4}\,\underline{k}\cdot\underline{K}\;V_K \tag{16}$$

so we may write (15) as

$$\dot{B}_{\underline{k}}(\underline{x},t) + i\, W_{\underline{k}}(\underline{x},t)\, B_{\underline{k}}(\underline{x},t) = 0 \tag{17}$$

where the non-homogeneous frequency is given by

$$W_{\underline{k}}(\underline{x},t) = \omega_k + \frac{i}{2} \sum_{\underline{K}} \underline{k}\cdot\underline{K}\; V_K \left[B_{\underline{K}}(t)\, e^{i\underline{K}\cdot\underline{x}} - B^*_{\underline{K}}(t)\, e^{-i\underline{K}\cdot\underline{x}}\right] \tag{18}$$

The approximate integral of (17) is given by

$$B_{\underline{k}}(\underline{x},t\pm\tau/2) = B_{\underline{k}}(\underline{x},t)\, e^{\mp i W_{\underline{k}}(\underline{x},t)\tau/2} \tag{19}$$

where τ is a time interval very much shorter than a period of the dominant wave, i.e., more on the order of $2\pi/\omega_k$. Inverting the Fourier expansion for the wave packet we have for the mode amplitude of interest

$$B_{\underline{k}}(t \pm\tau/2) = \sum_{\underline{\ell}} \frac{1}{\Sigma_o} \int d^2x\; e^{i(\underline{\ell}-\underline{k})\cdot\underline{x}}\; e^{\mp iW_{\underline{\ell}}(\underline{x},t)\tau/2}\, B_{\underline{\ell}}(\underline{x},t) \tag{20}$$

so that the numerator in the auto-corrleation function (12)with ρ=0

$$\left\langle B_{\underline{k}}(t+\tau/2)B^*_{\underline{k}}(t-\tau/2)\right\rangle = \sum_{\underline{\ell},\underline{\ell}'(\underline{K}_o)} \frac{1}{\Sigma_o^2} \int d^2x d^2x'\; e^{i[(\underline{\ell}-\underline{k})\cdot\underline{x} - (\underline{\ell}'-\underline{k})\cdot\underline{x}']}$$

$$\times \left\langle e^{-i[W_{\underline{\ell}}(\underline{x},t) + W_{\underline{\ell}'}(\underline{x}',t)]\tau/2}\, B_{\underline{\ell}}(\underline{x},t)\, B^*_{\underline{\ell}'}(\underline{x}',t)\right\rangle. \tag{21}$$

To simplify (21) we make a number of assummptions: i) the fluctuations in the short waves are statistically independent of the fluctuations in the waves near the spectral peak so that the average in (21) factors, i.e.,

$$\left\langle e^{-i[W_{\underline{\ell}}(\underline{x},t) + W_{\underline{\ell}'}(\underline{x}',t)]\tau/2} B_{\underline{\ell}}(\underline{x},t)B^{*}_{\underline{\ell}'}(\underline{x}',t)\right\rangle = \left\langle e^{-i[W_{\underline{\ell}}(\underline{x},t) + W_{\underline{\ell}'}(\underline{x}',t)]\tau/2}\right\rangle \times \left\langle B_{\underline{\ell}}(\underline{x},t)B^{*}_{\underline{\ell}'}(\underline{x}',t)\right\rangle$$

ii) the range of values in the $\underline{\ell}$ and $\underline{\ell}'$ sums is very restricted so that we may approximate the mean square value of the wave packet amplitude by the mean square value of the central mode, i.e.,

$$\left\langle B_{\underline{\ell}}(\underline{x},t)B^{*}_{\underline{\ell}'}(\underline{x}'t)\right\rangle \cong \left\langle B_{\underline{k}}(\underline{x},t)B^{*}_{\underline{k}}(\underline{x}',t)\right\rangle \delta_{\underline{\ell}-\underline{\ell}'}$$

and finally, iii) again because of the narrowness of the spectral sum we can replace the shift in the actual frequency by the shift in the test wave frequency

$$W_{\underline{\ell}}(\underline{x},t) \cong W_{\underline{k}}(\underline{x},t) \quad .$$

Assumptions (i) through (iii) enable us to obtain a spatial delta function

$\delta(\underline{x}-\underline{x}')$ in (21) so that the expression reduces to

$$\left\langle B_{\underline{k}}(t+\tau/2)B^{*}_{\underline{k}}(t-\tau/2)\right\rangle = \frac{1}{\Sigma_0}\int d^2x \left\langle e^{-iW_{\underline{k}}(\underline{x},t)\tau}\right\rangle\left\langle|B_{\underline{k}}(\underline{x},t)|^2\right\rangle \tag{22}$$

Then since the spatial fluctuations in $B_{\underline{k}}(\underline{x},t)$ are homogeneousthe auto-correlation function (12) when $\rho=0$ reduces to

$$C_{\underline{k}}(\tau) = \frac{1}{\Sigma_0}\int d^2x \left\langle e^{-iW_{\underline{k}}(\underline{x},t)\tau}\right\rangle . \tag{23}$$

We have assumed the ambient wave field to be in a statistical steady state so that the average in (23) is independent of $\underline{x}$ and t and we can write

$$C_{\underline{k}}(\tau) = \left\langle e^{-iW_{\underline{k}}(0,0)\tau}\right\rangle_{ss} . \tag{24}$$

Using the probability density (21.18) we can evaluate (24) to be

$$C_{\underline{k}}(\tau) = e^{-i\omega_{\underline{k}}\tau}\int P_{ss}(\underline{b}) \exp\left\{-i\,\tau \sum_{\underline{K}} \underline{k}\cdot\underline{K}\; V_{\underline{K}}\left[b_{\underline{K}} - b^{*}_{\underline{K}}\right]\right\} d\Gamma(\underline{b}) \tag{25}$$

which immediately integrates to

$$C_{\underline{k}}(\tau) = \exp\left\{-i\omega_{\underline{k}}\tau - \frac{\tau^2}{4}\sum_{\underline{K}}(\underline{k}\cdot\underline{K})^2\, V_{\underline{K}}^2 \left\langle|b_{\underline{K}}|^2\right\rangle_{ss}\right\} . \tag{26}$$

Introducing the quantity $\sigma_{\underline{K}}^{2}$ by

$$\sigma_{\underline{k}}^{2} \equiv \sum_{\underline{K}}(\underline{k}\cdot\underline{K})^2\, V_{\underline{K}}^{2}\;\frac{1}{2}\left\langle|b_{\underline{K}}|^2\right\rangle_{ss} = \sum_{\underline{K}}(\underline{k}\cdot\underline{K})^2 V_{\underline{K}}F_{ss}(\underline{K}) \tag{27}$$

we can rewrite (26) as

$$C_{\underline{k}}(\tau) = e^{-i\omega_k\tau}\, e^{-\sigma_{\underline{k}}^2\tau^2/2} \quad . \tag{28}$$

Here $1/\sigma_{\underline{k}}$ is the spectral width of the test wave as observed by taking the Fourier transform in time of the auto-correlation function

$$\tilde{C}_{\underline{k}}(\omega) = \int_{-\infty}^{\infty} e^{i\omega\tau}\, C_{\underline{k}}(\tau)d\tau \tag{29}$$

to obtain

$$\tilde{C}_{\underline{k}}(\omega) = \exp\left[-(\omega-\omega_k)^2/2\sigma_{\underline{k}}^2\right] \tag{30}$$

Based on an analogy with quantum mechanics one can interpret $1/\sigma_{\underline{k}}$ as the decorrelation time of the short gravity wave induced by the spectrum of long gravity waves.

The above result can be used to rewrite the auto-correlation function (12) in the factored form

$$C_{\underline{k}}(\underline{\rho},\tau) = C_{\underline{k}}(\tau)\; C_{\underline{k}}(\underline{\rho}) \tag{31}$$

$$\text{where } C_{\underline{k}}(\underline{\rho}) \equiv \frac{\langle B_{\underline{k}+\underline{\rho}/2}(t)B^*_{\underline{k}-\underline{\rho}/2}(t)\rangle}{\langle|B_{\underline{k}}(t)|^2\rangle} \quad . \tag{32}$$

Thus, by taking the Fourier transform of (31) in wave vector and time we find the wavenumber-frequency spectrum $F(\underline{k},\underline{x},\Omega,t)$, i.e.

$$F(\underline{k},\underline{x},\Omega,t) = \sum_{\underline{\rho}} e^{i\underline{\rho}\cdot\underline{x}} \int \frac{d\tau}{4\pi} e^{-i\Omega\tau} C_{\underline{k}}(\tau) C_{\underline{k}}(\underline{\rho}) <|B_{\underline{k}}(t)|^2>_{ss} \qquad (33)$$

Substituting from (30) and (25.3) we obtain

$$F(\underline{k},\underline{x},\Omega,t) = F(\underline{k},\underline{x},t)\; e^{-(\Omega-\omega_k)^2/2\sigma_{\underline{k}}^2} \qquad (34)$$

from which it is clear that the initially narrow spectral line at $\Omega=\omega_k$ has been broadened by interacting with the long wavelength gravity waves.

To obtain the above simple result it was necessary to assume that τ was a short time interval so that (19) was a reasonable approximate solution to (17). If we lift this restriction we can write the exact solution to (17) as

$$B_{\underline{k}}(\underline{x},t) = e^{-i\int_{t_0}^{t} W_{\underline{k}}(x,t')dt'}\; B_{\underline{k}}(\underline{x},t_0) \quad . \qquad (35)$$

We use the exact solution (35) to study the auto-correlation of the wave height at two separate space-time points on the ocean surface. For this purpose we introduce the complex envelope function

$$Z_{\underline{k}}(\underline{x},t) = \sum_{\underline{q}(\underline{k})} e^{i\underline{q}\cdot\underline{x}}\; B_{\underline{q}}(\underline{x},t) \qquad (36)$$

where $\underline{q}$ is restricted to a wave vector region near $\underline{k}$. If this restriction is relaxed and $\underline{q}$ is extended over all $\underline{k}$ - space then $Z_{\underline{k}}(\underline{x},t) \rightarrow Z(\underline{x},t)$ the complex surface displacement. In terms of the exact solution to (17) the complex wave packet amplitude is given by

$$Z_{\underline{k}}(\underline{x},t) = \sum_{\underline{q}(\underline{k})} \exp \left| i \left[\underline{q}\cdot\underline{x} - \int_{t_0}^{t} W_{\underline{q}}(\underline{x},t'|dt'\right] \right| B_{\underline{q}}(\underline{x},t_0) \quad . \tag{37}$$

Thus if we continue to assume that the modulated phase $W_{\underline{q}}(\underline{x},t)$ and wave packet amplitudes $B_{\underline{q}}(\underline{x},t)$ are statistically independent, then employing the same factoring used in (21) we obtain

$$<Z_{\underline{k}}(\underline{x}+\underline{r}/2,t+\tau/2)\, Z^{*}_{\underline{k}}(\underline{k}-\underline{r}/2,t-\tau/2)>_{ss} = \sum_{\underline{q}(\underline{k})} C_{\underline{q}}(\underline{r},\tau)<|B_{\underline{q}}(\underline{x},t_0)|^2>_{ss} e^{i\underline{q}\cdot\underline{r}} \tag{38}$$

where the space-time correlation function is

$$C_{\underline{q}}(r,\tau) \equiv \left\langle e^{i\left[\int_{t_0}^{t-\tau/2} W_{\underline{q}}(\underline{x}-r/2)t')dt' - \int_{t_0}^{t+\tau/2} W_{\underline{q}}(\underline{x}+\underline{r}/2,t')dt'\right]}\right\rangle_{ss} \quad .$$

We evaluate the average indicated in (39) using the steady state distribution function (21.29). Letting the initial time recede to the remote past with an "adiabatic switch-on" boundary condition, i.e. the long wave length gravity wave current is turned on slowly enough in the distant past that transients are not generated, we then have

$$\int_{t_0}^{t} [W_{\underline{q}}(\underline{x},t')-\omega q]dt' = -\sum_{\underline{k}} \underline{q}\cdot\hat{\underline{k}}\, [X(\underline{k}) \cos (\underline{k}\cdot\underline{x}-\omega_k t)$$

$$- Y(\underline{k}) \sin (\underline{k}\cdot\underline{x}-\omega_k t)]$$

where $B_{\underline{k}}(t)=[X(\underline{k})+iY(\underline{k})]e^{-i\omega_k t}$. The steady-state probability density $P_{ss}(b)$ factors in terms of the real and imaginary parts of $b_{\underline{k}}(=x_{\underline{k}}+iy_{\underline{k}})$, so that after some algebraic manipulation (39) reduced to

$$C_{\underline{q}}(\underline{r},\tau) = \prod_{\underline{k}} \int P_{ss}(x_{\underline{k}})P_{ss}(y_{\underline{k}})dx_{\underline{k}}dy_{\underline{k}} \exp\left| 2i\underline{q}\cdot\hat{\underline{k}}\, X_{\underline{k}} \sin(\underline{k}\cdot\underline{x}-\omega_k t)\right.$$

$$\times \sin 1/2\,(\underline{k}\cdot\underline{r}-\omega_k t) - 2i\underline{q}\cdot\hat{\underline{k}}\, y_{\underline{k}} \cos(\underline{k}\cdot\underline{x}-\omega_k t)\sin\tfrac{1}{2}(\underline{k}\cdot\underline{r}-\omega_k t)$$

$$\left. - i\omega_q\tau\right| . \tag{41}$$

Performing the indicated averages in (41) and introducing the steady state spectral density $F_{ss}(\underline{k})$ we obtain

$$C_{\underline{q}}(\underline{r},\tau) = \exp\left|-i\omega_q t - 2\sum_{\underline{k}} (\underline{q}\cdot\hat{\underline{k}})^2 F_{ss}(\underline{k}) \sin^2\tfrac{1}{2}(\underline{k}\cdot\underline{r}-\omega_k t)\right| . \tag{42}$$

If we set $\underline{r}=0$ in (42) and assume that the period of the dominant wave is much longer than the integration time, i.e., $\omega_k\tau<<2$, then $C_{\underline{q}}(o,\tau)\rightarrow C_{\underline{q}}(\tau)$ as given by (28).

If we set $\tau=0$ and consider distances short compared to the wavelength of the dominant wave, i.e., $\underline{k}\cdot\underline{r}<<1$, (42) reduces to

$$C_{\underline{q}}(\underline{r},o) \cong \exp\left|-i\omega_q t - \tfrac{1}{2}\sum_{\underline{k}} (\underline{q}\cdot\underline{k})^2 F_{ss}(\underline{k})(\hat{\underline{k}}\cdot\underline{r})^2\right| \tag{43}$$

We can also write (43) in the form

$$C_{\underline{q}}(\underline{r},0) \cong \exp\left| -i\omega_q\tau - r^2\Lambda_{\underline{q}}^2/2 \right| \tag{44}$$

where $\Lambda_{\underline{q}}^{-1}$ is interpreted as the decorrelation distance defined by

$$\Lambda_{\underline{q}}^2 \equiv \sum_{\underline{k}} (\underline{q}\cdot\underline{k})^2\, F_{ss}(\underline{k})\, \cos^2(\theta_r-\theta_k) \tag{45}$$

where $\cos(\theta_r-\theta_k)$ is the angle between $\underline{r}$ and $\underline{k}$. Thus for short times $\omega_k\tau \ll 1$ and short distances $\underline{k}\cdot\underline{r} \ll 1$, we can write the general auto-correlation function of the wave packet as

$$C_{\underline{q}}(\underline{r},\tau) = \exp\left| -i\omega_q\tau - \sigma_{\underline{q}}^2\tau^2/2 - \Lambda_{\underline{q}}^2\, r^2/2 \right| . \tag{46}$$

We can therefore write (38) as

$$\left\langle Z_{\underline{k}}(\underline{x}+\underline{r}/2, t+\tau/2) Z_{\underline{k}}^*(\underline{x}-\underline{r}/2, t-\tau/2) \right\rangle_{ss} = \sum_{\underline{q}(\underline{k})} e^{-(\underline{q}\cdot\underline{r}-\omega_q\tau)}$$

$$\times\ e^{-\sigma_{\underline{q}}^2\tau^2/2-\Lambda_{\underline{q}}^2 r^2/2}$$

$$\times \left\langle |B_{\underline{q}}(\underline{x},t_0)|^2 \right\rangle_{ss} \tag{47}$$

and Fourier transforming (47) in $\underline{r}$ and τ yields

$$\int e^{-i(\underline{\rho}\cdot\underline{r}-\Omega\tau)}d^2r\,d\tau\,\frac{1}{2}\left\langle Z_{\underline{k}}(\underline{x}+\underline{r}/2,t+\tau/2)Z^*_{\underline{k}}(\underline{x}-\underline{r}/2,t-\tau/2)\right\rangle_{ss}$$

$$\cong \sum_{\underline{q}(\underline{k})} e^{-(\Omega-\omega_q)^2/2\sigma_{\underline{q}}^2}\; e^{-(\underline{k}-\underline{q})^2/2\Lambda_{\underline{q}}^2}\, F(\underline{q},\underline{x},t) \qquad (48)$$

so that the spectrum of the initial test wave packet is broadened in both frequency and wave number due to interactions with the ambient steady-state spectrum of long wavelength gravity waves.

27. Areas for Further Study

At the outset of these lectures I promised to review the limitations encountered in our studies of the physics of the generation, evolution and decay of deep water gravity waves. In this final lecture I will attempt to make good on that promise. In the main we have been concerned with understanding the dominant interactions among deep water gravity waves in the context of weak interaction theory. We have demonstrated that both the spectral evolution properties and the statistics of the gravity wave field are determined by the nonlinear interactions and must be determined in a self-consistent manner. But this is anticipating the following summary of the properties of a deep water gravity wave field.

A gravity wave field in isolation constitutes a Hamiltonian system (cf. (3.18)) and the dynamics of the waves are determined by Hamilton's equations (cf. (3.13)). The canonical field variables are the free surface displacement $\zeta(x,t)$ and the velocity potential at the free surface $\phi_s(x,t)$. Viscosity and surface tension can be ignored in constructing this Hamiltonian, but may be important in determining the generation properties by the wind field as we mentioned. The time derivative of the vertical displacement of the ocean surface $\partial\zeta(\underline{x},t)/\partial t$ that enters into the Hamiltonian (3.18) must satisfy the kinematic boundary condition at the free surface i.e., the vertical surface motion must match the vertical water motion. This condition leads to an infinite series when expressed in terms of the canonical field variables (cf. (3.33)) and results in a Hamiltonian that can be ordered in powers of the nonlinear interactions between the field variables. The convergenence properties of this series have not been established so that the very foundation of weak interaction theory is taken as an article of faith.

The purist will find this state of affairs unsatisfactory. The pragmatist however, will observe that for a specified initial condition one can use the Hamiltonian series to calculate the evolution of the wave field forward in time. A comparison of the predicted properties of this field with laboratory experiments have been made and a large number of the calculated properties have been observed. Therefore one can adopt the point of view that because the results of many water wave experiments can be described using weak interaction theory that the theory has been vindicated throughout much of the physical range of interest. Establishing the convergence and stability properties of the weak interaction series is of more than academic interest however, if one is ever to obtain a theory of strong interactions, eg. to describe wave breaking. Recalling the many surprises concerning the stability of finite amplitude water waves we discussed that were a direct consequence of scientists' failure to distinguish between Levi-Civita's proof of the convergence of the Stokes expansion and the stability properties of that expansion (which had not been examined). We therefore present as the first of the many problems remaining that one needs to determine the analytic properties of the weak interaction series.*

We here adopt the pragmatist's viewpoint and accept the weak interaction series (cf. (3.34)) as a valid representation of the gravity wave field. (At least to some order of approximation). To calculate the evolution of the wave field we expand the field variables ($\phi_s(\underline{x},t)$, $\zeta(\underline{x},t)$) in discrete Fourier series normalized over a large horizontal area and impose periodic boundary conditions. Then by applying Parsavel's theorem we obtain an eigenmode re-

---

*Note that there are a number of alternative expansions in the literature whose equivalence has not been established. Presumably if these expansions can be shown to converge then they must be equivalent.

presentation of Hamiltons' equations (cf. (4.14)). The nonlinear interactions among the field variables appear as product of mode amplitudes in this representation, being interpreted as the physical scattering of three or four waves at a time. One automatically obtains a wave vector matching condition by using Fourier modes and this is interpreted as a conservation of momentum condition during each of the individual scatterings. Most of these interactions result in a periodic interchange of energy among the modes. However when there is a corresponding matching of the frequencies during an interaction a net energy exchange can occur. This is the phenomenon of nonlinear resonance as described by generic arguments in analytic mechanics and has been observed in water tank experiments. These resonant interactions are assumed to dominate the evolution of water wave fields.

The resonant interactions in the gravity wave field involves four modes at a time and have been shown to conserve energy, action and momentum. The proofs have been based on discrete wave fields, however, in which the number of waves is finite. The case for the corresponding continuum situation, with its infinite number of degrees of freedom, has not been made. The resonant interaction among gravity wave modes leads to the secular growth of new modes and also provides the eventual quenching of this resonant growth by establishing a feedback mechanism. Although these interactions provide many of the interesting phenomena observed in water tanks, such as the existence of envelope solitons, their dominance and indeed their existence in the wind generated wave fields of the ocean has not been established. In fact evidence is accumulating that seems to indicate that non-resonant interactions, i.e., those involving forced waves that are not on the linear dispersion curve (cf. Figure (6.1)), are equal in importance to the resonantly interacting free waves, i.e., those that are on the dispersion curve. The determination of the relative contribution of these two classes of waves to the evolution of wind generated gravity waves is an open question.

The physics of the water wave generation process resembles a cascade of energy from the high frequency water waves, which supports the stress of wind at the surface, to the longer wavelength gravity waves. The effects of surface tension and viscosity are important for the high frequency waves and in part determines the efficiency of the coupling between the air-flow and the sea interface. The pressure fluctuations at the water surface, due to turbulent eddies in the wind field, change Hamilton's deterministic equations for the wave field into a set of nonlinear *stochastic mode rate equations*. The coupling mechanism between the wind field and the surface wave field is not understood. One usually assumes a linear coupling between the ocean surface and the ambient environment. This environment consists of the air-flow in the immediate neighborhood of the surface and the sea water just below the surface. Various models of the growth rates of wind generated waves have been proposed. That of Miles and Phillips describes the growth of gravity-capillary waves very well, but underpredicts the growth rate of longer wavelength gravity waves. The model of West and Seshadri, which phenomenologically treats the air-sea coupling parameter as a stochastic quantity, gives a growth rate for gravity waves in closer agreement with field data. There is as yet no fundamental theory to justify the fluctuations used in the air-sea coupling parameter, however.

The mechanisms referred to above induce an instability in the surface wave field leading to an exponential growth of wave energy. Presumably this exponentiation is quenched by viscous dissipation and the nonlinear interactions among the waves. To date there is no dynamic calculation of the nonlinear mode rate equations which establishes the existence of an asymptotic steady state for the high frequency waves. Recall that we are talking about the solution of a nonlinear stochastic rate equation. The author has applied a linearization technique to the stochastic equations and obtained a self-consistent expression for the asymptotic steady state of the energy spectral density for the gravity-capillary waves and the renormalized dissipation rate. Both these quantities depend on the average nonlinear interaction. This result is encouraging but should not be taken as a replacement of the actual solution to the nonlinear problem.

The combined efforts of pressure fluctuations and the coupling between the rapidly varying gravity-capillary waves in their steady state and the evolving gravity waves leads to the complicated system of stochastic, nonlinear mode rate equations (18.21). The coupling between wavelength scales employed in this model is not rigorous since it is unclear how to fill-in the assumed gap in the spectrum [cf. (18.2)]. For a model problem in which the long waves are generated by a source distinct from that which generates the high frequency waves the model is probably a good one. It is not adequate for describing a continuously evolving wave field in which there are no spectral gaps, or rather its adequacy has as yet to be demonstrated. A second difficulty with this model is that the high frequency waves, although modulated by the gravity waves, are always in a steady state. This is reasonable if the relaxation rates of these waves is much faster than the characteristic rates of the gravity waves. Again the relaxation rates of the gravity-capillary waves have only been *calculated* using a linearization of the dynamic equations. A more direct determination of the relaxation times remains to be done.

It is quite clear that because the mode rate equations are driven by a stochastic flux of action (energy) that the mode amplitudes are themselves stochastic quantities. If the flux is assumed to be determined by a zero-centered

Gaussian process then for a linear rate equation the mode amplitude would be Gaussian, as is the case for a linear Langevin equation. For a nonlinear rate equation the statistics of the mode amplitudes are non-Gaussian, regardless of the assumed statistics of the driving flux. It has been known for a long time that in the steady state the statistics of the gravity wave field are non-Gaussian. However, the measured deviation from Gaussianity, [cf. Figure 19.1] was felt to be negligibly small by most theorists and so a fourth order cummulant discard (quasi-Gaussian) approximation was invariably employed in considering the evolution of the energy spectral density of surface gravity waves. In these lectures we have attempted to demonstrate that the statistical properties of the steady state cannot and should not be used to determine the statistics of the gravity wave field far from the steady state.

We have demonstrated that the solution to even the simplest equation of evolution for the probability density, i.e. the Fokker-Planck equation, only yields the standard Gaussian distribution under a very restricted set of assumptions. These assumptions, which essentially require that the steady state probability density be independent of the nonlinear interactions, are shown to lead to the steady state distribution function derived by Hasselmann using a different technique. In a sense the probability density has been linearized in a very crude fashion. A more systematic linearization procedure (statistical linearization) leads to a Gaussian distribution which is *not* the one usually employed. This distribution depends parametrically on the average nonlinear interaction among the gravity waves and provides a renormalized dissipation rate for these waves. The resulting transport equation evolves irreversibly towards an asymptotic steady state at a rate determined by the average nonlinear interaction, [cf. (23.25)] .

Thus, although many of the questions regarding the generation and evolution of gravity waves on the ocean surface have not been answered in the present set of lectures, I think it is quite proper to say that weak interaction theory does provide a context for the further examination of these questions in a systematic way. I look forward to participating in the future resolution of the remaining problems.

ACKNOWLEDGEMENTS

I would like to thank J.A.L. Thomson and K. Watson for contributing to my understanding of physical oceanography and their mathematical description of wave fields. I also thank K. Lindenberg and V. Seshadri for the opportunity to work with them and acknowledge the pleasure it has been to explore the statistical properties of physical systems with them. I also thank H. Yuen for supplying the photographs used in Section 9.

REFERENCES

1. Sir H. Lamb; Hydrodynamics, 6th ed. Dover (1945).
2. J. J. Stoker; Water Waves, Interscience, N.Y. (1957).
3. L. D. Landau and E. M. Lifshitz; Fluid Mechanics, Pergamon Press, London (1959).
4. B. Kinsman; Wind Waves, Prentice Hall, N.J. (1965).
5. O. M. Phillips; The Dynamics of the Upper Ocean, 2nd ed., Cambridge Univ. Press, London-New York (1977).
6. P. H. LeBlond and L. A. Mysak; Waves in the Ocean. Elsevier Scientific Pub. Co., Amsterdam-Oxford-N.Y. (1978).
7. S. A. Kitiagorodskii; Physics of Air-Sea Interaction, Engl. Transl., Jerusalem, Israel Prog. Sci. Transl. (1973).
8. V. P. Krasitiskii and M. M. Zaslavskii; Boundary Layer Met., 14, 199 (1978).
9. T. P. Barnett and K. E. Kenyon; Rep. Prog. Phys. 38, 667 (1975).
10. P. H. LeBlond and L. A. Mysak; SIAM Review 21, 289 (1979).
11. K. Hasselmann; J. Fluid Mech. 12, 481 (1962); ibid 15, 273 (1963); ibid. 15, 385 (1963).
12. A. Favre and K. Hasselmann; Turbulent Fluxes Through the Sea Surface, Wave Dynamics, and Prediction, Plenum Press, N.Y. (1977).
13. W. H. Munk, G. R. Miller, F. E. Snodgrass and N. F. Barber; Phil. Trans. A, 255, 505 (1963).
14. G. G. Stokes; Scientific Papers 1, 227 (1880), J. H. Mitchell; Phil. Mag. 36, 430 (1893).
15. M. S. Longuet-Higgins; J. Fluid Mech. 16, 138 (1963).
16. E. C. Lafond; in The Sea vol. I, Interscience Publ. (1962).

17. M. S. Longuet-Higgins and R. W. Stewart; J. Fluid Mech. 10, 529 (1961).

18. F. K. Ball; J. Fluid Mech. 19, 465 (1964).

19. J. A. Thomson and B. J. West; J. Phys. Ocean. 5, 736 (1975)

20. D. H. Peregrine; Adv. Appl. Mech. 16, 9 (1976).

21. W. H. Munk; J. Mar Res. 14, 302 (1955).

22. H. Jeffreys; Proc. Roy. Soc. A 107, 189 (1924); Proc. Roy. Soc. A 110, 341 (1925).

23. O. H. Shemdin; College of Engineering, University of Florida, Tech. Rep. No. 4.

24. M. L. Banner and O. M. Phillips; J. Fluid Mech. 65, 647 (1974)

25. J. Wu; J. Fluid Mech. 68, 49 (1975)

26. O. M. Phillips; J. Fluid Mech. 9, 163 (1960)

27. D.J. Benney; J. Fluid Mech. 14, 577 (1962).

28. M.S. Longuet-Higgins; J. Fluid Mech. 12, 321 (1962).

29. K. Hasselmann; in Basic Developments in Fluid Mechanics, ed. by M. Holt, vol. 2, 117 (1968) Academic Press.

30. L.J.F. Broer; Appl. Sci. Res. 30, 430 (1974).

31. J. W. Miles; J. Fluid Mech. 83, 153 (1977).

32. M. Milder; J. Fluid Mech. 83, 159 (1977).

33. J. Moser; Stable and Random Motions in Dynamic Systems, Princeton Univ. Press, Princeton, N.J. (1973).

34. B. Chirkov; Physics Rep. 52, 264 (1979).

35. See eg. the review papers in Topics in Nonlinear Dynamics, ed. S. Jorna AIP Conf. Proc. 46, AIP, N.Y. (1979)

36. V. Seshadri, B.J. West and K. Lindenberg; to appear in Physics A.

37. M. Tabor; to appear in Advances in Chemical Physics, (1981).

38. M. S. Longuet-Higgins and N. D. Smith; J. Fluid Mech. 25, 417 (1966).

39. L. F. McGoldrick, O. M. Phillips, N. Huang and T. Hodgson; J. Fluid Mech. 25, 437 (1966).

40. T. B. Benjamin and J. E. Feir; J. Fluid Mech. 27, 417 (1967)

41. T. B. Benjamin; Proc. Roy. Soc. Lond: A 299, 59 (1967).

42. D. J. Benney; Mathematical Problems in the Geophysical Sciences, publ. by AMS (1970).

43. V. H. Chu and C. C. Mei; J. Fluid Mech. 41 873 (1970); ibid. 47, 337 (1971).

44. A. Davey; J. Fluid Mech. 53, 769 (1972).

45. B.J. West, K. M. Watson and J. A. Thomson; Phys. Fluids 17, 1059 (1974).

46. B.M. Lake, H.C. Yuen, H. Rungaldier and W.E. Ferguson; J. Fluid Mech. 83, 49 (1977).

47. E. Fermi, J. Pasta and S. Ulan; Los Alamos Scientific Lab Rep. LA-1940 (1955).

48. W.R. Hamilton; Third Supplement to an Essay on the Theory of Systems of Rays; reprinted in Mathematical Papers vol. 1, 164 (1931) Cambridge Univ. Press.

49. J. L. Lagrange; Miscellanea Tourinensia (1760) [Oeuvres, Paris 1867-90]

50. L. D. Landau and E. M Lifshitz; Statistical Physics, Pergamon Press, London (1954).

51. G. Green; in Mathematical Papers. ed. N.M. Ferrers, Chelsea Publ. Co., N.Y. (1955).

52. T. Keyes and I. Oppenheim; Phys. Rev. A 7, 1384 (1973); ibid. 8, 937 (1973).

53. R. Fox; Physics Rep. 48, 181 (1978).

54. B. H. Lavenda; Thermodynamics of Irreversible Processes, Halsted Press, John Wiley (1978).

55. S. R. deGroot and P. Mazur; Non-Equilibrium Thermodynamics, North-Holland, Amersterdam-London (1962).

56. L. Euler; Hist de l' Acad de Berlin (1755).

57. Navier; Mem' de l'Acad. des Sciences, 389 (1822), G.G. Stokes; Camb. Trans. 287 (1854)

58. P. Langevin; C. R. Acad. Sci. Paris, 530 (1908).

59. D.T. Mashizama and H. Mori; J. Stat. Phys. 18, 385 (1978).

60. M.V. Berry and K. E. Mount; Rep. Prog. Phys. 35, 315 (1972).

61. C. Eckart; Rev. Mod. Phys. 20, 399 (1948).

62. J. H. Van Vleck and D. L. Huber; Rev. Mod. Phys. 49, 939 (1977).

63. R. Pierels; Ann. Phys. 3, 1055 (1929).

64. D.R. Crawford, P. G. Saffman and H.C. Yuen; preprint (1980).

65. M. J. Lighthill; Proc. Roy. Soc. A, 299, 28 (1967).

66. G.B. Whitham; Comm. Pure and App. Math. XIV, 675 (1961); J. Fluid Mech 9, 347 (1960); ibid. 22, 273 (1965).

67. G.B. Whitham; Linear and Nonlinear Waves, John Wiley & Sons, N.Y. (1974).

68. H. Yuen and B. M. Lake; Phys. Fluids 18, 956 (1975).

69. E.J. Plate; Proceed. 9th Symp. on Naval Hydro., Paris (1972); see also ref. (12).

70. N.N. Bogoliubov and Y.A. Mitropolsky; Asymptotic Methods in the Theory of Nonlinear Oscillators, Gordon and Breach (1961)

71. A.H. Nayfeh and D.T. Mook; Nonlinear Oscillations, John Wiley (1979).

72. K.M. Watson, B.J. West and B.I. Cohen; J. Fluid Mech. 27, 185 (1976).

73. S.A. Thorpe; Ph.D. Dissertation, Univ. Cambridge (1965).

74. D.J. Benney; Studies in Appl. Math. 55, 93 (1976); ibid. 60, 27 (1979).

75. A. Newell; SIAM J. Appl. Math. 35, 650 (1978).

76. A. Messiah; Quantum Mechanics, John Wiley, N.Y. (1962).

77. E. Fermi; Rev. Mod. Phys. 4, 87 (1932).

78. E. H. Lieb and D. C. Mathis; editors, Mathematical Physics in One Dimension, Academic Press (1966).

79. A.S. Bakai; Sov. Phys. JETP 28, 140 (1969).

80. P. Debye; Ann. Phys. 39, 789 (1912).

81. A.I. Khinchin; Mathematical Foundations of Statistical Mechanics, trans. by G. Gamow, Dover, N.Y. (1949).

82. R.M. Muira; SIAM Review 18, 412 (1976).

83. C.S. Gardner, J.M. Greene, M.D. Kurskal and R.M. Muira; Phys. Rev. Lett. 19, 1095 (1967).

84. B.B. Kadomtsev and V.I. Karpman; Sov. Phys. USPEKHI, 14, 40 (1971).

85. O.M. Phillips; J. Fluid Mech. 2, 417 (1957).

86. J. W. Miles; J. Fluid Mech. 3, 185 (1957).

87. J. W. Miles; J. Fluid Mech. 7, 469 (1960).

88. F. W. Dobson; J. Fluid Mech. 48, 91 (1971).

89. B. J. West and V. Seshadri; to be published in JGR.

90. F. W. Dobson; J. Fluid Mech. 48, 91 (1971).

91. M. Lax; Rev. Mod. Phys. 32, 25 (1960).

92. M. Lax; Rev. Mod. Phys. 38, 541 (1966).

93. V. Seshadri and B. J. West; in progress.

94. N. G. van Kampen; Physics Rep. 24C, 173 (1976).

95. E. Noether; Goett. Nachr., 235 (1918).

96. C. Lanczos; The Variational Principles of Mechanics, 4th Edition University of Toronto Press, Toronto-Buffalo (1974).

97. A. S. Monin and A. M. Yaglom; Statistical Fluid Mechanics; Mechanics of Turbulence, MIT Press, Mass. (1965).

98. G. K. Batchelor; The Theory of Homogeneous Turbulence, Cambridge Univ. Press, London (1959).

99. H. Tennekes and J. L. Lumley; A First Course in Turbulence, MIT Press, Mass. (1972).

100. S.A. Thorpe; J. Geophys. Res. 80, 328 (1975).

101. V. Ye. Zakharov; Prikl. Mekh. i Tekhn. Fiz. No. 2, 86 (1968).

102. A. S. Monin, V. M. Kamenkovich and V. G. Kort; Variability of the Ocean, trans. J. L. Lumley, Wiley-Interscience, N.Y. (1974).

103. K.M. Watson and B.J. West; J. Fluid Mech 70, 815 (1975).

104. G.G. Stokes; Trans. Cam. Phil. Soc. 8, 441 (1847).

105. T. Levi-Civita; Math. Ann. 93, 264 (1925).

106. B.L. Weber and D.E. Barrick; J. Phys. Ocean. 7, 3 (1977).

107. D.E. Barrick and B.L. Weber; J. Phys. Ocean, 7, 11 (1977).

108. J.C. Luke; J. Fluid Mech. 27, 395 (1967).

109. D. Holliday; J. Fluid Mech. 83, 737, (1977).

110. J. Ford and J. Waters; J. Math. Phys. 4, 1293 (1963).

111. H.C. Corben and P. Stehle; Classical Mechanics, 2nd Ed., John Wiley, N.Y. (1960); M. Berry in ref. (35).

112. K. Lindenberg, V. Seshadri and B.J. West; Phys. Rev. A, 22, 2171 (1980).

113. H. Goldstein; Classical Mechanics, Addison-Wesley Press, Cambridge (1950).

114. L. Brillouin; Scientific Uncertainty and Information Theory, Pt. 2, Academic Press (1964).

115. A.N. Kolmogorov; Dokl. Akad. Nauk. SSSR 98, 527 (1954).

116. V.I. Arnold; Russ. Math. Sur. 18, 9 (1963); ibid. 18, 85 (1963).

117. J. Moser; Nachr. Akad. Wiss. Grottingen, II, Math. Phys. Kl. 1 (1962).

118. O.M. Phillips; Proc. Roy. Soc. A299, 104 (1967).

119. L. J. Tick; J. Math. Mech. 18, 643 (1959).

120. N. E. Huang and C. C. Tung; J. Fluid Mech. 75, 337 (1976).

121. A. Masuda, Y.Y. Kuo and H. Mitsuyasu; J. Fluid Mech. 92, 717 (1979).

122. A. Ramamonjiarisoa and M. Coantic; C. R. Acad. Sci. 282B, 111 (1976).

123. A. Ramamonjairisoa, A. Baddy and I. Choi; in ref. (10).

124. B. M. Lake and H. C. Yuen; J. Fluid Mech. 88, 33 (1978).

125. G. J. Komen; J. Phys. Oceang. (1980).

126. M. S. Longuet-Higgins; in A Voyage of Discovery, 393, Pergamon Oxford (1977).

127. G. J. Komen; J. Geophys. Res. 85, 3311 (1980).

128. F. P. Bretherton; J. Fluid Mech. 20, 457 (1964).

129. H.C.Yuen and W.E. Ferguson,Jr., Phys. Fluids 21, 1275 (1978); ibid,2116 (1978). P.Bryant, preprint

130. B. I. Cohen, K. M. Watson and B. J. West; Phys. Fluids 19, 345 (1976).

131. A. Davey and K. Stewartson; Proc. Roy. Soc. London A 338, 172 (1974).

132. K. Lonngren and A. Scott; Solitons in Action, Academic Press, N.Y. (1978).

133. A. C. Scott, F. Y. F. Chu and D. W. McLaughlin; Proc. IEEE 61, 144 (1973).

134. N. J. Zabusky and M. D. Kruskal; Phys. Rev. Lett. 15, 240 (1965).

135. D. J. Korteweg and G. de Vries; Philos. Mag. Ser. 5, 39, 422 (1895).

136. G. B. Whitham; J. Fluid Mech. 27, 399 (1967).

137. M. J. Lighthill; J. Inst. Maths. Appl. 1, 1 (1965).

138. J. E. Feir; Proc. Roy Soc. London A 283, 54 (1967).

139. A. Hasagawa and F. Tappert; Appl. Phys. Lett. 23, 171 (1973).

140. M. Froissart; ed. Hyperbolic Equations and Waves, Springer-Verlag, Berlin (1970).

141. V. E. Zakharov and A. B. Shabat; Zh. Eksp. Theor. Fiz. 61, 118 (1971), [Sov. Phys.-JETP 34, 62 (1972)].

142. P. Bryant; preprint

143. V. E. Zakharov, J. Appl. Mech. and Tech. Phys. 2, 190 (1968).

144. W. J. Plant and J. W. Wright; J. Fluid Mech. 82, 767 (1977).

145. G. R. Valenzuela and J. W. Wright; J. Geophys. Res. 31, 5795 (1976).

146. M. S. Longuet-Higgins; J. Fluid Mech. (1969).

147. A. N. Kolmogorov: Dokl. Akod. Nauk. SSR 98, 527 (1957); Proc. Intern. Congr. of Math., Amsterdam 1, 315 (1958).

148. H. Poincáre; Les Méthodes nouvelles de le mécánique céleste, Paris, (1892). reprinted by Dover Publ., N.Y. (1957).

149. J. Meiss and K. M. Watson; in ref. (34).

150. V. Seshadri and B.J. West; in preparation

151. W.G. van Dorn; J. Mar. Res. 12, 249 (1953).

152. M.S. Longuet-Higgins; Physics Fluids 12, 737 (1969).

153. K. Hasselmann; J. Fluid Mech. 50, 189 (1971).

154. B.J. West; to be published

155. K. Hasselman, T.P. Barnett, E. Bouws, H. Carlson, D.E. Cartwright, K. Enke, J. A. Ewing, H. Gienapp, D. E. Hasselmann, P. Kruseman, A. Meerburg, P. Muller, D.J. Olbers, K. Richter, W. Sell and H. Wolden, 1973; Measurements of Wind-Wave Growth and Swell Decay During the Joint North Sea Wave Project (JONSWAP), Deutsche Hydrogr. Z., Suppl. A(8°), No. 12.

156. G.J.R. Garrett and J. Smith; J. Phys. Oceanog. 6, 925 (1976).

157. M.J.H. Fox; Proc. Roy. Soc. London A, 349, 467 (1976).

158. Lord Rayleigh; Proc. Lond. Math. Soc. 1, 357 (1873).

159. G. Valenzuela and Laing; J. Fluid Mech. 54, 507 (1972).

160. J. Willebrand; J. Fluid Mech. 70, 113 (1975).

161. C.S. Cox and W. H. Munk; J. Mar. Res. 13, 198 (1954); J. Optical Soc. Amer. 44, 838 (1954).

162. K. Hasselmann; Proc. Roy. Soc. London A299, 77 (1967).

163. J.G. Kirkwood, F.P. Buff and M. S. Green; J. Chem. Phys. 17, 988 (1949).

164. D.M. McQuarrie; Statistical Mechanics, Harper and Row, N.Y. (1976).

165. D.J. Benney and P.G. Saffman; Proc. Roy. Soc. A, 289, 301 (1966).

166. D.J. Benney and A.C. Newell; J. Math. and Phys. 46, 363 (1967).

167. A.C. Newell; Rev. of Geophysics 6, 1 (1968).

168. Boltzmann

169. J.D. Meiss, N. Pomphrey and K. M. Watson; Proc. Natl. Acad. Sci. USA, 76, 2109 (1979); N. Pomphrey, J.D. Meiss and K. M. Watson; J. Geophys. Res. 85, 1085 (1980).

170. C. Garrett and W. Munk; J. Geophys. Res. 80, 291 (1975).

171. A.J. Sutherland; J. Fluid Mech. 33, 545 (1968).

172. J.B. Bole and E.Y. Hsu; J. Fluid Mech. 35, 657 (1969).

173. R. Brown; Phil. Mag. 6, 161 (1829) and Edinburgh J. of Sci. 1, 314 (1829).

174. J. Ingenhousz; Dictionary of Scientific Biology (ed. C.C. Gillespie), Scribners, 11 (19

175. A. Einstein; Ann. d. Physik 17, 549 (1905); ibid. 19, 371 (1906).

176. J. Perrin; Brownian Movement and Molecular Reality, Taylor and Francis, London (1910).

177. Th. Svedverg; The Existence of the Molecules, Leipzig (1912).

178. B.J. West, K. Lindenberg and K. Shuler; J. Stat. Phys. 18, No. 2, 217 (1978).

179. A. Budgor and B.J. West; Phys. Rev. A, 17, 370 (1978).

180. A. Budgor, K. Lindenberg and K. Shuler; J. Stat. Phys. 15, 375 (1976).

181. V.I. Arnold; Mathematical Methods of Classical Mechanics, Springer-Verlag, N.Y. (1978).

182. J. Ford; in Lectures in Statistical Physics, ed. W.C. Schieve, and J. S. Turner, Springer-Verlag, Berlin (1974).

183. M.C. Wang and G.E. Uhlenbeck; Rev. Mod. Phys. 17, 323 (1945).

184. K. Pierson; Nature 77, 294 (1905).

185. J.C. Kluyver; Konink. Acad. Wetenshop. te Amsterdam 14, 325 (1905).

186. Lord Rayleigh; Nature 72, 318 (1905).

187. G.N. Watson; Quar. J. Math. Oxford 10, 266 (1939).

188. E.W. Montroll and B.J. West; in Fluctuation Phenomena, ed. J.L. Lebowitz and E.W. Montroll, North Holland (1979).

189. Lord Rayleigh; Phil. Mag. Series 5, 10 (60), 73 (1880).

190. M.S. Longuet-Higgins; J. Mar. Res. 11, 245 (1952).

191. M. Lax; Rev. Mod. Phys. 38(2), 359 (1966).

192. R. W. Stewart; in Boundry Layers and Turbulence, Phys. of Fluids Supp. AIP, N.Y. (1967).

193. B.J. West, K. Lindenberg and V. Seshadri; Physica 102A, 470 (1980).

194. F.J. Dyson; Phys. Rev. 75, 486 (1949).

195. I. Oppenheim, K. Shuler and G. Weiss; Stochastic Processes in Chemical Physics, MIT Press (1978).

196. B. V. Gnedenko and A. N. Kolmogorov; Limit Distributions for Sums of Independent Random Variables; translated from Russian, annotated and revised by K. L. Chung, Addison-Wesley, Mass. (1954).

197. R. Kubo; J. Math. Phys. 4, 174 (1963).

198. K. Lindenberg, V. Seshadri, K. Shuler and B. J. West; J. Stat. Physics, 23, 755 (1980).

199. K. Lindenberg, V. Seshadri and B. J. West; Phys. Rev. A 22, 2171 (1980).

200. S. A. Yermakov and YE.N. Pelinovkiy; Izvestiya, Atmos. and Ocean Phys. 13, No. 5, 373 (1977).

201. W. C. Keller and J. W. Wright; Radio Sci. 10, 139 (1975).

202. M. S. Longuet-Higgins; J. Fluid Mech. 17, 459 (1963); Proc. Trans. Roy. Soc. London A249, 322 (1957); Phil. Trans. Roy. Soc. London A250, 158 (1957).

203. M. S. Longuet-Higgins; Radio Sci. 68D, 1049 (1963).

204. O. M. Phillips; J. Fluid Mech. 11, 143 (1961).

205. B. Kinsman; Chesapeake Bay Inst. Tech. Rep. 19, (1960).

206. R. Kubo; in Fluctuations, Dissipation and Resonances in Magnetic Systems, ed. D. ter Haar, Plenum Press Inc., N.Y. (1962).

207. R. Kubo; J. Phys. Soc. Japan 17, 1100 (1967).

208. U. Frisch; in Probabilistic Methods in Applied Mathematics; ed. A. T. Bharucha-Reid, Academic Press, N.Y. (1968).

209. B. J. West, A. Bulsara, K. Lindenberg, V. Seshadri and K. Shuler; Physica A97, 211 (1979); ibid. 97, 234 (1979).

210. K. Lindenberg, K. Shuler, V. Seshadri and B. J. West; to appear in A. T. Bharucha-Reid series.

211. D. C. Leslie; Developments in the Theory of Turbulence, Charendon Press, Oxford (1973).

212. S. Tolman; The Principles of Statistical Mechanics, Oxford Univ. Press (1938).

213. S. Mukamel, I. Oppenheim and J. Ross; Phys. Rev. A17, 1988 (1978).

214. R. H. Terweil; Physics 74, 248 (1974).

215. P. Glansdorff and I. Prigogine; Thermodynamic Theory of Structure, Stability and Fluctuations, Wiley-Interscience, N.Y. (1971).

216. J. W. Gibbs; Collected Works, Scribner, N.Y. (1902).

217. A. Einstein; Ann. Physik 33, 1275 (1910).

218. R. F. Green and H. B. Callen; Phys. Rev. 83, 1231 (1951).

219. G. Nicolis and A. Bablozantz; J. Chem. Phys. 51, 2632 (1969).

220. I. Prigogine; Introduction to the Thermodynamics of Irreversible Processes, Interscience, N.Y. (1967).

221. Y. Alhassid and R. D. Levine; Phys. Rev. A18, 89 (1978).

222. M. B. Faist, R. D. Levine and R. B. Bernstein; J. Chem. Phys. 66, 511 (1979).

223. R. D. Levine, R. B. Bernstein, P. Kahana, I. Procaccia and E. T. Upchurch; J. Chem. Phys. 64, 796 (1976).

224. M. Moreau; J. Math. Phys. 19, 2494 (1978).

225. E. T. Jaynes; Phys. Rev. 106, 620 (1957); ibid. 108, 171 (1957).

226. D. J. Webb; in ref. (10); also Deep Sea Research 25, 279 (1977).

227. L. Onsager; Nuovo Cimento 6, Supp., 279 (1949).

228. R. Kraichnan; J. Fluid Mech. 67, 155 (1975).

229. R. Piccirelli, Phys. Rev. 175, 77 (1968).

230. L. Onsager; Phys. Rev. 37, 405 (1931), ibid. 38, 2265 (1931)

231. J. H. Weare and I. Oppenheim; Physica 72, 1 (1974).

232. G. F. Carnevale; Statistical Dynamics of Non-Equilibrium Fluid Systems; Ph.D. Thesis, Harvard (1979).

233. M. S. Longuet-Higgins; Proc. Roy. Soc. Lond. A341, 311 (1976).

234. M. L. Goldberger and K. M. Watson; Collision Theory, John Wiley, N.Y. (1964).

235. B. J. West; Phys. Fluids, 21, 1448 (1978).

236. C. H. McComas and F. P. Bretherton; J. Geophys. Res. 82, 1397 (1977).

237. R. Zwanzig, K.S.J. Nordholm, W. C. Mitchell; Phys. Rev. A, 5, 2680 (1972).

238. E. Wigner; Phys. Rev. 40, 749 (1932).

239. I. E. Alber; Proc. Roy. Soc. Lond. A363, 525 (1978).

240. G. B. Withman; Comm. Pure. Appl. Math. 14, 675 (1961).

241. K. E. Kenyon; Deep-Sea Res. 18, 1023 (1971).

242. M. S. Longuet-Higgins and R. W. Stewart; J. Fluid Mech. 10, 529 (1961).

243. P. Müller and D. J. Olbers; J. Geophys. Res. 80, 3848 (1975).

244. This model was constructed by K. M. Watson and B. J. West; Physical Dynamics Report RADC-TR-74-267.

M. Holt

Numerical Methods in Fluid Dynamics

1977. 107 figures, 2 tables. VIII, 253 pages
(Springer Series in Computational Physics)
ISBN 3-540-07907-6

Contents: General Introduction. – The Godunov Schemes. – The BVLR Method. – The Method of Characteristics for Three-Dimensional Problems in Gas Dynamics. – The Method of Integral Relations. – Telenin's Method and the Method of Lines.

The first part of this monograph is concerned with numerical problems in gas dynamics. The discussion of finite difference methods is concentrated on hyperbolic systems. The author describes the present status of two approaches developed in the USSR, both based on the method of characteristics: the method of Godunov and the BVLR method due to Rusanov and coworkers. Other techniques treated in this volume are due to Butler and Sauer. In later chapters the author describes the methods of integral relations introduced by Dorodnitsyn, Telenin's method and the method of Lines – techniques based on polynomial or series representation to the unknowns – all applied to problems in fluid dynamics. The presentation is made for graduate students in mechanical engineering and applied mathematics with basic knowledge of fluid mechanics. Many applications and samples of numerical solutions of model problems are presented.

Springer-Verlag
Berlin
Heidelberg
New York

Vol. 114: Stellar Turbulence. Proceedings, 1979. Edited by D. F. Gray and J. L. Linsky. IX, 308 pages. 1980.

Vol. 115: Modern Trends in the Theory of Condensed Matter. Proceedings, 1979. Edited by A. Pekalski and J. A. Przystawa. IX, 597 pages. 1980.

Vol. 116: Mathematical Problems in Theoretical Physics. Proceedings, 1979. Edited by K. Osterwalder. VIII, 412 pages. 1980.

Vol. 117: Deep-Inelastic and Fusion Reactions with Heavy Ions. Proceedings, 1979. Edited by W. von Oertzen. XIII, 394 pages. 1980.

Vol. 118: Quantum Chromodynamics. Proceedings, 1979. Edited by J. L. Alonso and R. Tarrach. IX, 424 pages. 1980.

Vol. 119: Nuclear Spectroscopy. Proceedings, 1979. Edited by G. F. Bertsch and D. Kurath. VII, 250 pages. 1980.

Vol. 120: Nonlinear Evolution Equations and Dynamical Systems. Proceedings, 1979. Edited by M. Boiti, F. Pempinelli and G. Soliani. VI, 368 pages. 1980.

Vol. 121: F. W. Wiegel, Fluid Flow Through Porous Macromolecular Systems. V, 102 pages. 1980.

Vol. 122: New Developments in Semiconductor Physics. Proceedings, 1979. Edited by F. Beleznay et al. V, 276 pages. 1980.

Vol. 123: D. H. Mayer, The Ruelle-Araki Transfer Operator in Classical Statistical Mechanics. VIII, 154 pages. 1980.

Vol. 124: Gravitational Radiation, Collapsed Objects and Exact Solutions. Proceedings, 1979. Edited by C. Edwards. VI, 487 pages. 1980.

Vol. 125: Nonradial and Nonlinear Stellar Pulsation. Proceedings, 1980. Edited by H. A. Hill and W. A. Dziembowski. VIII, 497 pages. 1980.

Vol. 126: Complex Analysis, Microlocal Calculus and Relativistic Quantum Theory. Proceedings, 1979. Edited by D. Iagolnitzer. VIII, 502 pages. 1980.

Vol. 127: E. Sanchez-Palencia, Non-Homogeneous Media and Vibration Theory. IX, 398 pages. 1980.

Vol. 128: Neutron Spin Echo. Proceedings, 1979. Edited by F. Mezei. VI, 253 pages. 1980.

Vol. 129: Geometrical and Topological Methods in Gauge Theories. Proceedings, 1979. Edited by J. Harnad and S. Shnider. VIII, 155 pages. 1980.

Vol. 130: Mathematical Methods and Applications of Scattering Theory. Proceedings, 1979. Edited by J. A. DeSanto, A. W. Sáenz and W. W. Zachary. XIII, 331 pages. 1980.

Vol. 131: H. C. Fogedby, Theoretical Aspects of Mainly Low Dimensional Magnetic Systems. XI, 163 pages. 1980.

Vol. 132: Systems Far from Equilibrium. Proceedings, 1980. Edited by L. Garrido. XV, 403 pages. 1980.

Vol. 133: Narrow Gap Semiconductors Physics and Applications. Proceedings, 1979. Edited by W. Zawadzki. X, 572 pages. 1980.

Vol. 134: $\gamma\gamma$ Collisions. Proceedings, 1980. Edited by G. Cochard and P. Kessler. XIII, 400 pages. 1980.

Vol. 135: Group Theoretical Methods in Physics. Proceedings, 1980. Edited by K. B. Wolf. XXVI, 629 pages. 1980.

Vol. 136: The Role of Coherent Structures in Modelling Turbulence and Mixing. Proceedings 1980. Edited by J. Jimenez. XIII, 393 pages. 1981.

Vol. 137: From Collective States to Quarks in Nuclei. Edited by H. Arenhövel and A. M. Saruis. VII, 414 pages. 1981.

Vol. 138: The Many-Body Problem. Proceedings 1980. Edited by R. Guardiola and J. Ros. V, 374 pages. 1981.

Vol. 139: H. D. Doebner, Differential Geometric Methods in Mathematical Physics. Proceedings 1981. VII, 329 pages. 1981.

Vol. 140: P. Kramer, M. Saraceno, Geometry of the Time-Dependent Variational Principle in Quantum Mechanics. IV, 98 pages. 1981.

Vol. 141: Seventh International Conference on Numerical Methods in Fluid Dynamics. Proceedings. Edited by W. C. Reynolds and R. W. MacCormack. VIII, 485 pages. 1981.

Vol. 142: Recent Progress in Many-Body Theories. Proceedings. Edited by J. G. Zabolitzky, M. de Llano, M. Fortes and J. W. Clark. VIII, 479 pages. 1981.

Vol. 143: Present Status and Aims of Quantum Electrodynamics. Proceedings, 1980. Edited by G. Gräff, E. Klempt and G. Werth. VI, 302 pages. 1981.

Vol. 144: Topics in Nuclear Physics I. A Comprehensive Review of Recent Developments. Edited by T.T.S. Kuo and S.S.M. Wong. XX, 567 pages. 1981.

Vol. 145: Topics in Nuclear Physics II. A Comprehensive Review of Recent Developments. Proceedings 1980/81. Edited by T. T. S. Kuo and S. S. M. Wong. VIII, 571-1.082 pages. 1981.

Vol. 146: B. J. West, On the Simpler Aspects of Nonlinear Fluctuating. Deep Gravity Waves. VI, 341 pages. 1981.

Selected Issues from

Lecture Notes in Mathematics

Vol. 728: Non-Commutative Harmonic Analysis. Proceedings, 1978. Edited by J. Carmona and M. Vergne. V, 244 pages. 1979.

Vol. 729: Ergodic Theory. Proceedings 1978. Edited by M. Denker and K. Jacobs. XII, 209 pages. 1979.

Vol. 730: Functional Differential Equations and Approximation of Fixed Points. Proceedings, 1978. Edited by H.-O. Peitgen and H.-O. Walther. XV, 503 pages. 1979.

Vol. 731: Y. Nakagami and M. Takesaki, Duality for Crossed Products of von Neumann Algebras. IX, 139 pages. 1979.

Vol. 733: F. Bloom, Modern Differential Geometric Techniques in the Theory of Continuous Distributions of Dislocations. XII, 206 pages. 1979.

Vol. 735: B. Aupetit, Propriétés Spectrales des Algèbres de Banach. XII, 192 pages. 1979.

Vol. 738: P. E. Conner, Differentiable Periodic Maps. 2nd edition, IV, 181 pages. 1979.

Vol. 742: K. Clancey, Seminormal Operators. VII, 125 pages. 1979.

Vol. 755: Global Analysis. Proceedings, 1978. Edited by M. Grmela and J. E. Marsden. VII, 377 pages. 1979.

Vol. 756: H. O. Cordes, Elliptic Pseudo-Differential Operators – An Abstract Theory. IX, 331 pages. 1979.

Vol. 760: H.-O. Georgii, Canonical Gibbs Measures. VIII, 190 pages. 1979.

Vol. 762: D. H. Sattinger, Group Theoretic Methods in Bifurcation Theory. V, 241 pages. 1979.

Vol. 765: Padé Approximation and its Applications. Proceedings, 1979. Edited by L. Wuytack. VI, 392 pages. 1979.

Vol. 766: T. tom Dieck, Transformation Groups and Representation Theory. VIII, 309 pages. 1979.

Vol. 771: Approximation Methods for Navier-Stokes Problems. Proceedings, 1979. Edited by R. Rautmann. XVI, 581 pages. 1980.

Vol. 773: Numerical Analysis. Proceedings, 1979. Edited by G. A. Watson. X, 184 pages. 1980.

Vol. 775: Geometric Methods in Mathematical Physics. Proceedings, 1979. Edited by G. Kaiser and J. E. Marsden. VII, 257 pages. 1980.

Vol. 779: Euclidean Harmonic Analysis. Proceedings, 1979. Edited by J. J. Benedetto. III, 177 pages. 1980.

Vol. 780: L. Schwartz, Semi-Martingales sur des Variétés, et Martingales Conformes sur des Variétés Analytiques Complexes. XV, 132 pages. 1980.

Vol. 782: Bifurcation and Nonlinear Eigenvalue Problems. Proceedings, 1978. Edited by C. Bardos, J. M. Lasry and M. Schatzman. VIII, 296 pages. 1980.

Vol. 783: A. Dinghas, Wertverteilung meromorpher Funktionen in ein- und mehrfach zusammenhängenden Gebieten. Edited by R. Nevanlinna and C. Andreian Cazacu. XIII, 145 pages. 1980.

Vol. 786: I. J. Maddox, Infinite Matrices of Operators. V, 122 pages. 1980.

Vol. 787: Potential Theory, Copenhagen 1979. Proceedings, 1979. Edited by C. Berg, G. Forst and B. Fuglede. VIII, 319 pages. 1980.

Vol. 791: K. W. Bauer and S. Ruscheweyh, Differential Operators for Partial Differential Equations and Function Theoretic Applications. V, 258 pages. 1980.

Vol. 792: Geometry and Differential Geometry. Proceedings, 1979. Edited by R. Artzy and I. Vaisman. VI, 443 pages. 1980.

Vol. 793: J. Renault, A Groupoid Approach to C*-Algebras. III, 160 pages. 1980.

Vol. 798: Analytic Functions, Kozubnik 1979. Proceedings. Edited by J. Ławrynowicz. X, 476 pages. 1980.

Vol. 799: Functional Differential Equations and Bifurcation. Proceedings 1979. Edited by A. F. Izé. XXII, 409 pages. 1980.

Vol. 801: K. Floret, Weakly Compact Sets. VII, 123 pages. 1980.

Vol. 802: J. Bair, R. Fourneau, Etude Géometrique des Espaces Vectoriels II. VII, 283 pages. 1980.

Vol. 804: M. Matsuda, First Order Algebraic Differential Equations. VII, 111 pages. 1980.

Vol. 805: O. Kowalski, Generalized Symmetric Spaces. XII, 187 pages. 1980.

Vol. 807: Fonctions de Plusieurs Variables Complexes IV. Proceedings, 1979. Edited by F. Norguet. IX, 198 pages. 1980.

Vol. 809: I. Gumowski and Ch. Mira, Recurrences and Discrete Dynamic Systems. VI, 272 pages. 1980.

Vol. 810: Geometrical Approaches to Differential Equations. Proceedings 1979. Edited by R. Martini. VII, 339 pages. 1980.

Vol. 816: L. Stoica, Local Operators and Markov Processes. VIII, 104 pages. 1980.

Vol. 819: Global Theory of Dynamical Systems. Proceedings, 1979. Edited by Z. Nitecki and C. Robinson. IX, 499 pages. 1980.

Vol. 821: Statistique non Paramétrique Asymptotique. Proceedings, 1979. Edited by J.-P. Raoult. VII, 175 pages. 1980.

Vol. 826: Ph. G. Ciarlet et P. Rabier, Les Equations de von Kármán. VI, 181 pages. 1980.

Vol. 827: Ordinary and Partial Differential Equations. Proceedings, 1978. Edited by W. N. Everitt. XVI, 271 pages. 1980.

Vol. 831: Representation Theory I. Proceedings, 1979. Edited by V. Dlab and P. Gabriel. XIV, 373 pages. 1980.

Vol. 832: Representation Theory II. Proceedings, 1979. Edited by V. Dlab and P. Gabriel. XIV, 673 pages. 1980.

Vol. 836: Differential Geometrical Methods in Mathematical Physics. Proceedings, 1979. Edited by P. L. García, A. Pérez-Rendón, and J. M. Souriau. XII, 538 pages. 1980.

Vol. 837: J. Meixner, F. W. Schäfke and G. Wolf, Mathieu Functions and Spheroidal Functions and their Mathematical Foundations Further Studies. VII, 126 pages. 1980.

Vol. 838: Global Differential Geometry and Global Analysis. Proceedings 1979. Edited by D. Ferus et al. XI, 299 pages. 1981.

Vol. 840: D. Henry, Geometric Theory of Semilinear Parabolic Equations. IV, 348 pages. 1981.

Vol. 841: A. Haraux, Nonlinear Evolution Equations- Global Behaviour of Solutions. XII, 313 pages. 1981.

Vol. 842: Séminaire Bourbaki vol. 1979/80. Exposés 543–560. IV, 317 pages. 1981.

Vol. 843: Functional Analysis, Holomorphy, and Approximation Theory. Proceedings. Edited by S. Machado. VI, 636 pages. 1981.

Vol. 845: A. Tannenbaum, Invariance and System Theory: Algebraic and Geometric Aspects. X, 161 pages. 1981.

Vol. 849: P. Major, Multiple Wiener-Itô Integrals. VII, 127 pages. 1981.